“This is the most complete and accurate critique of the fossils of the so-called ‘ape-men.’ The myth of human evolution is completely demolished. The last chapter repudiating the compromising views of many modern evangelicals on this subject is illuminating. Lubenow has made an invaluable contribution to scientific biblical creationism.”

Henry M. Morris, president, Institute for Creation Research

“On the question of biological, especially human, origins, Lubenow is not content to merely quote biblical theory (if I may use that word). Like a true scholar he researches in depth the literature in the scientific journals, sifting the evidence, searching out the areas open to interpretation. . . . He does his homework so thoroughly that he makes someone like me who would carry on a dialogue with him (as we did on creationism vs. Darwinism) also do his homework. . . . He is a pleasure to fence with intellectually.”

Michael Charney, emeritus professor of anthropology,
affiliate professor of zoology, Colorado State University

“Ever since my ninth-grade science teacher showed our class fossils of the four ‘missing links,’ I have wished for a book like this. *Bones of Contention* is factual, scientific, and accurate and exposes the current frauds the evolutionists are using to deceive yet another generation of young people that human evolution is a fact. This book is well written, timely, and interesting to read. Lubenow’s revised edition is even better than the first.”

Tim LaHaye, coauthor, the Left Behind series

“There’s no better creationist researcher I know who has studied and critiqued man’s supposed evolutionary ancestry more thoroughly than Professor Lubenow. After reading this masterful book, only a hardened heart would still accept human evolution and reject divine creation.”

Ken Ham, president, Answers in Genesis

Marvin L. Lubenow is professor of Bible, theology, and apologetics at Southern California Bible College and Seminary in El Cajon, California. He has spent more than thirty-five years researching the human fossil issue and frequently speaks and writes to defend the creationist position.

BONES OF CONTENTION

A Creationist Assessment of Human Fossils

Revised and Updated

MARVIN L. LUBENOW

BakerBooks
Grand Rapids, Michigan

Published by Baker Books
a division of Baker Publishing Group
P.O. Box 6287, Grand Rapids, MI 49516-6287
www.bakerbooks.com

Printed in the United States of America

Library of Congress Cataloging-in-Publication Data

Lubenow, Marvin L.
Bones of contention : a creationist assessment of human fossils / Marvin L. Lubenow.—Rev. and updated.
p. cm.
Includes bibliographical references and index.
ISBN 0-8010-6523-2 (pbk.)
1. Fossil hominids. 2. Creationism. I. Title.
GN282.5.L82 2004
599.93′8—dc22 2004010942

To Enid Arlene,
who deserves far more in life than a
husband intrigued with human fossils,

and

to Nettie Wilhelmena Gaulke Lubenow,
who dedicated me to Christ before I was born.
I was not aware of this until I made my own commitment to Christ.

Contents

Definitions, Illustrations, and Charts

DEFINITIONS

ILLUSTRATIONS

CHARTS

PREFACE

I AM BOTH HUMBLED by and grateful for the reception of the original edition of *Bones of Contention.* The first edition has had eleven printings and has been translated into Spanish and Japanese.

The central theme of the first edition was that human evolution is false. The scientific and fossil evidence presented in that edition is as valid today as it was when it was first published. That edition also showed that the famous "parade" of small apelike forms gradually evolving into modern humans was a myth. Although evolutionists still insist on the truth of human evolution, they are finally beginning to admit that that famous parade was fiction. They are very quiet about that admission, however, and the general public is unaware of it.

Largely rewritten, this second edition carries on the tradition of the first. But it goes much further. It includes the latest developments in the field of human evolution and the latest fossil discoveries. It includes stunning developments by creationist scientists that reveal that the whole of evolution is dead.

This second edition differs significantly from the original edition in three areas. The first difference is the emphasis on human evolution as being intrinsically racist. In evolution's early days, "scientific racism," which claimed white Europeans were superior to other races, was heralded as scientific truth. Now that we are aware of the evils of racism, the racism of evolution is being camouflaged. The model of choice in human evolution, the Out of Africa or African Eve Model, lacks solid fossil evidence. However, it is preferred because it better hides the racism present in all evolutionary models.

The second major difference in this revised edition is the presence of much archaeological and cultural evidence to support the true humanity of the Neandertals and *Homo erectus.* This inclusion was necessary because both of these groups of true humans are accused of being culture thin and hence not fully evolved humans. When evolutionists divide ancient humans into separate groups to establish a false evolutionary sequence, they demonstrate that evolution is actually a divider of humans.

Third, in this edition I have not hesitated to quote creationist scientists wherever appropriate. In the original edition, my plan was to demonstrate that human evolution was false using only evolutionist sources. While proving human evolution false was not difficult to do, the implication was that the book would have more "authority" if I used only evolutionist sources. The idea that evolutionists are scholars and creationists are just religious bigots could not be more misguided. It is becoming increasingly evident that the so-called "Bible-Science War" is not a science war but a philosophical war. Creation scientists are coming into their own and are proving to be far more objective and fair than their evolutionist counterparts. I honor my brothers and sisters in the Creation Research Society, the Institute for Creation Research, Answers in Genesis, and other creationist organizations for the outstanding job they are doing in bringing science out of the Dark Ages of paganism and back onto a solid foundation.

The first two sections of this edition, similar to the first chapters of the original edition but updated, are intended to give readers a background in the field of the human fossils (paleoanthropology) in order to better understand the rest of the book. Using the educational principle of "going from the known to the unknown," the sections center on Java Man, perhaps the only fossil that most people know.

Although the fossil dates used in this book and on the charts speak in terms of millions of years, I do not accept those dates. I march to the tune of a different drummer, based on the data set forth in the early chapters of Genesis. The purpose of this book is to demonstrate that even when the human fossils are placed on time charts according to evolutionists' dates for these fossils, the results do not support human evolution but conflict with it.

I want to thank Dr. Kenneth Cumming, dean of the graduate school, Institute for Creation Research, for reading chapter 23 on Neandertal mtDNA and for offering valuable suggestions. I also want to thank one of my sweethearts, my granddaughter, Liz Laribee, for her drawings of Acheulean hand axes in chapter 12.

Marvin L. Lubenow
Southern California Bible College and Seminary
El Cajon, California
Fall 2004

SECTION I

A Really Weird Science

I don't need to tell you how things are, Miss Franklin. Non-scientists think of science as universal. Celestial, even. But science is terrestrial. Territorial. Political.

—William Nicholson
Nature 422 (20 March 2003): 259

INTRODUCTION TO SECTION I

THE ASSIGNMENT

A FEW WEEKS AGO, my wife and I were eating lunch at a sidewalk café in downtown Orange, California. The café is rather nondescript, but it features one of my favorite dishes, Cuban orange chicken.

Three men sat down at the table next to us. Their conversation was animated, but we paid no attention until I heard the word *creationist*. We listened—it was impossible not to. They were denigrating their supervisor. At first I wasn't sure if they were denigrating him because he was their supervisor, or because he was a creationist, or both. It immediately became clear when one of the men, in exasperation, bellowed, "And do you know what's scary? There are fifty million of those crazy creationists in the country!"

Roaring with laughter, I pointed my finger at him and said, "You want to know what's *really* scary?"

Surprised, he said simply, "What?"

"You're sitting next to one! But I hope you have better luck tomorrow."

He looked sheepish but said nothing. My wife mentioned that he didn't have enough class to apologize for making a scene in public. But I will never forget the look on his face.

Although I wanted to, I did not feel it wise to engage in further conversation. Neither he nor the other men showed even the slightest desire to do so. Had we done so, however, I suspect that he would have known very little about the creationism he so despised.

I think this incident was hilarious. But when I tell it, people say, "Weren't you upset that the man was insulting your belief as a creationist?"

I tell them, "No! Although I have no idea who he was, I have been praying that he will come to know Jesus Christ as his savior."

The incident illustrates two things: (1) Creationism today is very much at the center of public interest, and (2) Creationism is a very emotional issue. And because of the incredible breakthroughs creationists have made in demonstrating that the radiometric dating methods are invalid (see Section VI), creationism will become even more popular and even more of a public issue in the future.

But creationism itself is not what causes emotions to flare. The term *creationism* literally screams "God!" That's why it is so emotional. And that's why many people who do not personally know God embrace evolution. People tend to believe what they want to believe. Evolutionists think their reason is "scientific." The real reason is deeper. Hating creationism is just one of the ways that many twenty-first-century humans have of showing their hatred for God. Believe it or not, I think that's a good thing. Let me explain.

Some years ago we lived in Wayne, Michigan. Between us and downtown Detroit was the city of Dearborn, well-known because it is the headquarters of the Ford Motor Company. The mayor of Dearborn at that time was Orville Hubbard. Hubbard was known to have political aspirations. This was the 1960s, when race riots in Detroit spilled over into Dearborn. To bring order, Hubbard called out the National Guard. News reporters, interviewing Hubbard about his decision, asked, "Aren't you afraid that this will bring you bad publicity?" Hubbard replied, "There's no such thing as bad publicity."

Hubbard was saying that there is something worse than *bad* publicity, and that is *no* publicity. There may be exceptions, but I believe this concept is also true for creationism. I believe that there is something worse than people bad-mouthing creationism. It would be far worse if they simply didn't care about it. That type of person is the hardest to reach.

The moral of the story is simple. When someone speaks disparagingly about any aspect of your Christian faith, put that person on your prayer list. Because he or she is thinking about the issue and has feelings about it, you have been introduced by the Holy Spirit to a "live" prospect!

There are some people who have no idea what the Bible teaches on creationism. They believe that it is just ancient superstition or myth. Darwin was incredibly ignorant of what the Bible teaches on it, and most of his disciples follow in his train. In the Bible, God reveals himself as being all-powerful and all-loving. That's why he created the universe and us. My experience has been that when people understand the powerful logic and rationality of biblical creationism, it at least makes sense to them—whether they accept it or not.

The theory of human evolution is all-pervasive in our society. This book was written to show that it is a false theory, both philosophically and scientifically. The problem is that human evolution and the human fossils can be unbelievably confusing. My job as a teacher is to make it understandable. My approach in this book is the same as in the college course I designed on the subject. Let me describe it.

My task was to communicate the true nature of the human fossil record to young college students. I easily could have said, "The human fossil record is completely compatible with special creation but is contrary to human evolution. It actually demonstrates that humans did not evolve." Based upon more than forty years of study and research in this field, I know this is true. But it is difficult to impress that fact upon young students when scientific superstars like Stephen Jay Gould and Carl Sagan have proclaimed that evolution is a fact that no rational person doubts.

As professor of Bible and theology/apologetics at a Christian college, I was naturally expected by the students to uphold the biblical view of human origins. However, students commonly think that a Christian has only two options: (1) take the Bible's teachings on faith by going against the "evidence," or (2) accept the findings of "science" (read "evolution") and question the Bible—at least the early chapters of Genesis. In contrast to those two options, I had in my possession hard evidence that the biblical account of human origins was valid and that human evolution was false. I needed to communicate that evidence.

The problem was threefold. First, young college students have no background in paleoanthropology (the technical term for the study of human fossils). They have been told repeatedly that the fossils support human evolution, and they are seldom given reasons to doubt it. Second, the study of human fossils demands a knowledge of human anatomy, a lot of Latin terms, and reams of often-confusing scientific literature. Thus, the field is a no-man's-land for most people. Third, in the face of the incredible propaganda assault by the evolutionary establishment, it is hard to believe the contrary voice of an unknown person.

Since teaching students to do research was part of our collegiate program, I decided to risk seeing just what the students could do. I assigned each of the thirty students five human or australopithecine (the name of our alleged evolutionary ancestors) fossils to research as their semester project. Only fossils that were fully accepted as legitimate by the scientific community were included.

The rules were as follows: (1) The students were to spend a minimum of eight hours of research on each of the five fossils. (2) They were to consult at least five evolutionist sources on each fossil. They were forbidden to use creationist material. (3) They were to determine the date the evolutionists have assigned to each fossil. (4) They were to determine the category assigned to each fossil by evolutionists (such as anatomically modern *Homo sapiens,* Neandertal, archaic or early *Homo sapiens, Homo erectus, Homo habilis,* or some form of australopithecine). (5) They were to write a one-page paper on each of their fossils, outlining these findings, and make copies for distribution to the class. (6) Each paper must contain at least five documented sources.

Although the literature speaks of the human fossils as being hundreds of thousands or even millions of years old, as a young-earth creationist, I do not accept those dates. In fact, I will demonstrate that they are nothing but science fiction. However, it is impossible to deal with these fossils without using those evolutionist dates. I trust that you, the reader, will keep that in mind.

Knowing the human evolution literature as I do, the results of my assignment were quite predictable. After about the third week, a number of students came to me complaining that they could not find much agreement among evolutionists regarding either the date or the category of some of their fossils. Many of the fossil discoveries made in the first half of the twentieth century have very questionable dates. In spite of that, many of these fossils have become the backbone of the concept of human evolution. I could have told the students this in a class lecture, but it wouldn't have had the impact upon them that their own research provided.

Further, many important fossils are the subject of intense controversy among evolutionists regarding the date, the category, or both. The two matters are sometimes related. The category to which a fossil is assigned sometimes determines the date assigned to it. Or the date of the fossil sometimes determines the category to which it is assigned. In our college, that would be called cheating. Certainly it is not an unbiased approach in interpreting human fossils. I suggested to the students that if they could find at least two evolutionists who agreed on the matter, they could use that date or that category assignment of their fossil in their paper.

When I made the fossil assignments, I loaded the dice a bit. I purposely assigned the more obscure fossils to students whom I knew, from previous classes, were good at research. Weeks later this group started coming to me and complaining that they could not find any information on their fossils. Some of them had made quite an extensive search, going to San Diego's Museum of Man and to the libraries or the anthropology departments at the University of California, San Diego, and San Diego State University. A few students even implied that a particular fossil I had assigned to them did not exist. Laughingly, I was accused of sending them on a snipe hunt. I had expected that reaction. I wanted them to discover firsthand that there are many legitimate fossil discoveries about which it is very difficult to obtain information.

When I was satisfied that a student had spent at least eight hours searching for information on a particular fossil without success, I gave him or her copies of the materials in my own files on that fossil, materials taken from scientific journals, to write the assigned report. It would have been impossible to duplicate my forty years of research on the human fossils in the time allotted.

I do not wish to imply that the difficulty in finding material on many of the human fossils represents some kind of evolutionist plot. Only the most sensational fossil discoveries receive much publicity. Most discoveries are reported in professional journals and are quite unknown to all but some in the anthropological community. The fossils mentioned in popular presentations of human evolution represent just a small portion of the total fossil material that has been discovered.

Throughout the semester, as we studied the matter of human creation versus human evolution, questions would arise. "Why are the footprints found by Mary Leakey at Laetoli, Tanzania, assigned to Lucy-like, apish fossils when the evolutionists themselves state that they are virtually identical to modern human footprints?" Or, "Why are many recent Australian fossils assigned to *Homo sapiens* when the evolutionists state that they are almost identical to the older *Homo erectus* fossils from Java?" Eventually the students began to answer their own questions. In the evolutionist mind, the fossils exist to serve evolution—not objective science. It didn't take long for some of the more curious and resourceful students to figure that out.

Then came "Fossil Day." The students brought their reports to class and placed their fossils on the master chart according to the evolutionists' dating and category. As the process unfolded, it became increasingly obvious to the students that the fossils did not show human evolution over time. If human evolution were true, the fossils should have fallen roughly in a line going from the australopithecines, through some form of *Homo habilis,* on up to *Homo erectus,* then through some form of early *Homo sapiens,* and finally on to anatomically modern *Homo sapiens* (that's big, beautiful you and me). Instead, the fossils were all over the place without any definite evolutionary progression. Even using the evolutionists' own dates and assignments, it became obvious to the students that the fossil evidence actually *falsified* the concept of human evolution. No lecture or series of lectures by me could have been as convincing as the research done by the students themselves. Nothing I could have said would have impacted the students as much as the stark reality of the human fossil record.

When I tell people about this project, some people are horrified. "Aren't you taking quite a chance? What if it doesn't work out the way you expect? Wouldn't that embarrass you as a professor?" In spite of the ambiguities of some of the fossils, there are so many fossil discoveries involved that the total testimony is always clear. It always works out that way! The ambiguities have little effect on the outcome. The key is to attempt to study all of the human fossils that have been discovered rather than just those that evolutionists choose to show us in trying to prove human evolution. That is why you seldom, if ever, find *complete* fossil charts in books on human evolution.

Future fossil discoveries will not substantially change the picture. Future discoveries cannot nullify the objective evidence already unearthed. This message is not what we hear from a hundred different voices coming at us from a dozen different directions. But the human fossils themselves tell the real story. This is the story I hope to share with you.

Welcome to a really weird science.

CHAPTER 1

"Show Me Your Fossils; I'll Show You Mine"

PEOPLE STARE. As they approach the table lined with human skulls, the mood is one of silence and incredible wonder. When someone finally dares to break the silence, I know instinctively what the question will be. I have heard it hundreds of times. "Are they real?"

When I inform the questioners that the skulls are plaster casts of original fossils, the mood changes to relief that they are not in the presence of death. However, even my assurance that the skulls are accurate and expensive casts of the original fossil material doesn't restore the mystique that was obvious before the question was asked.

The very thought that a professor at a Christian college would possess thirty *original* human fossils reveals the magnitude of the misconception that exists in the mind of the public regarding these fossils. It represents the first myth about human evolution that I want to discuss.

Although I have visited most of the major natural history museums in the United States and some overseas, I have never seen an original human fossil. Neither have most of the anthropologists who teach human evolution in our universities. Neither have you. In fact, you may not have even seen a picture of an original fossil. What you thought were pictures of original fossils may have been pictures of reproductions.

No prisoner on death row is under greater security than those ancient relics called human fossils. Most of the original fossils are sequestered inside vaults of concrete or stone and accessible only through massive steel doors, the type you would expect to see at the First National Bank. Few can even see them—let alone study them.

This process of seclusion was true with the original 1856 Feldhofer Cave Neandertal. "The skull and the bones were Fuhlrott's private property, and he did not show them to many. Only very few scholars in Britain and on the Continent had seen the skull or obtained a cast." Even Rudolf Virchow, the greatest medical man of his time, "could only study the remains in Fuhlrott's house after gaining access from his wife when Fuhlrott was away."[1] William King never saw the original fossils, although he is the one who, in naming them *Homo neanderthalensis* in 1864, declared them to represent a different species from modern humans. Darwin never saw these or any other fossil humans, although he published an entire book on human evolution in 1871. Thomas Huxley, Darwin's bulldog, never saw the original fossils either, although he described them in his famous 1863 work, *Man's Place in Nature.*[2] That should dispense with the concept that human evolution was based upon fossil evidence.

HOMINID

The word is used by the evolutionist community to mean "humans and their evolutionary ancestors." It includes the genus *Homo*, the genus *Australopithecus*, and all creatures in the family *Hominidae*. As an evolutionist term it is meaningless in a creationist worldview. The creationist counterpart would include the terms humans and non-human primates. I use the term *human* in this book to refer to those who are descendants of the biblical Adam.

Germany built a two-story museum to celebrate the fossil skull known as Steinheim Man, discovered in 1933. Visitors, however, see only plastic replicas. The fossil itself is kept in a small safe several miles away. This safe is set into the thick stone wall of a 250-year-old military arsenal outside Stuttgart. The fossil's former home was a bank vault. The story is told that when scientists came to study the fossil, they were blindfolded, driven to the bank, and unmasked only when safely inside so that they would not even know the location of the bank. "While it was never described in great detail, this fossil played a central role in various evolutionary models."[3]

The director of paleontology, National Museums of Kenya in Nairobi, is Dr. Meave G. Leakey. Many of the fossils housed there were found by her and her husband, Richard, and their teams of national workers. The fossils are kept in the Hominid Room, which has reinforced concrete walls and is designed to withstand conventional bomb blasts. Leakey and one other trusted museum staff member are the only ones who have keys to the room. Inside the room are locked boxes with hinged lids containing the fossils, which rest on form-fitted blocks of foam rubber.

Most of the South African fossils reside at the Transvaal Museum in Pretoria. They are kept in a strong room known as the Red Cave because of the three-foot-thick walls that are painted red. This vault was originally designed to house valuable documents. The fossils rest on red velvet placed over foam-rubber lining.

The *Homo erectus* fossils from Java, some of the most important fossils in the world in determining models of human evolution, suffer from a similar problem at the hands of their curator, Teuku Jacob (Gadjah Mada University).

> These fossils, the prized objects of Jacob's collection, are rarely seen, even by professionals in the fossil-hunting business. Scholars with serious research programs have to apply to Jacob for permission even to see them, let alone touch them, for scientific study. And even those few who succeed in obtaining official permission have to wait for Jacob's final OK, for he alone is permitted to remove the fossils from the safes.[4]

Jacob has an assistant, Angus, who is a trained anatomist. Yet even he is not allowed access to the vault where the fossils are kept. Nor is there any possibility that Angus would be allowed to study the fossils on his own or write a paper on them. Jacob maintains that there is a "committee" that determines access to the fossils. Veteran fossil hunters understand that Jacob himself is the committee.

Milford Wolpoff (University of Michigan) tells the story of the most complete, and one of the most important, *Homo erectus* skulls yet discovered, Sangiran 17. A long-standing feud between two Javanese investigators (he did not give names, but one of them almost certainly is Jacob), famous for guarding their fossils like jealous lovers, had kept the skull largely hidden from the paleontological community until Wolpoff discovered it in a laboratory in Java and assembled it.[5]

Why this incredible secrecy and security? Besides the element of raw power in controlling access, these fossils (certainly the human ones) are the remains of our ancestors. They are priceless treasures of human history. Their discovery has been the result of hard work, great expense, and often incredible luck. They are irreplaceable. How would one replace a fossil that has been lost or damaged beyond repair? Where would one go to find another just like it? Since in paleoanthropology and archaeology "quantity makes for quality" in the study of human variation, finding a similar fossil does not make up for the loss of the first one.

Furthermore, many of the fossils are extremely delicate. Sometimes their teeth will shatter at the slightest impact. Chunks of bone may flake off at the scratch of a fingernail. Some of the fossils are not completely fossilized, meaning that the organic material has not been completely replaced by inorganic minerals. Even the air in fossil rooms is maintained at a constant temperature and humidity to minimize contraction and expansion that could crack the fossils.

Unfortunately, some fossils have been lost, such as the original Peking Man fossils, lost in 1941 at the outbreak of World War II. Although we have plaster casts of them made by Franz Weidenreich, their loss is still keenly felt. Many other *Homo erectus* fossils (the present classification of Peking Man) have been subsequently discovered, but these new ones have not made up for the lost information on human variation in early populations that the original Peking Man fossils would have provided.

Because of their incalculable value and fragile nature, the original human fossils are so protected that the total number of people who have access to them is actually fewer than the total number of heads of state in the world today. However, there was one brief, glorious moment when this condition did not exist.

In 1984, the American Museum of Natural History in New York sponsored its famous Ancestors exhibit, in which more than forty of these original fossils were brought together for the first time ever for the public to view and for scholars to study. Obviously, security had top priority. Each fossil was accompanied by the curator of its home museum. Special agents met them at Kennedy International Airport and whisked them through a special section of customs without even opening the containers housing the fossils. Black Cadillac limousines with police escorts rushed them to the American Museum. When the fossils were put on public display, they were placed behind one-inch laminated acrylic panels in batter-proof, bulletproof, electronically monitored exhibit cases. Even work on the subway line under the museum was halted until after the exhibit to protect the fossils from vibration.

Although many nations, such as China, Australia, Tanzania, Kenya, and Ethiopia (where Lucy is kept), refused to send their fossils and expose them to risk, the exhibit was considered a resounding success. For the very first time, scholars from all over the world were able to study the originals side by side.

Half of a million people were able to view them. To everyone's relief, nothing was broken. But, because of the high risk involved, most authorities predict that such a "family gathering" will never take place again.

If the risk to the fossils was so great, why was this "family gathering" held even once? The idea of having the world's leading paleoanthropologists study these fossils was just an afterthought to the main purpose of allowing the public to view the original fossils. What situation could loom so large as to pry these fossils loose from the security of their shelters and expose them to public view? The answer: the rising threat of creationism![6]

Eric Delson, John Van Couvering, and Ian Tattersall, American Museum scientists who were largely responsible for the Ancestors exhibit, admit that the creationist assault on evolutionary biology was a matter of "great and growing concern" at the museum. They go on to say that the primary purpose of the exhibit was to show people, lay and professional, the evidence for evolution. They refrained from making any kind of political statement regarding human evolution lest they "dignify" the challenge of "creation science."[7]

Bernard Wood (George Washington University), writing in *Nature*, one of the most prestigious science journals in the world, states that the Ancestors exhibit was the response of the American Museum to creationist attempts to influence both public opinion and legislators in their "attack on the foundations of all scientific endeavor—namely reason and evidence."[8]

It is obvious that, in spite of the decibels, communication is not taking place. The problem is not with the fossils. It is with the interpretation of the fossils. Delson makes this naive comment: "How can you be anti-evolution when you see so much tangible evidence of our own roots?"[9]Evolutionists apparently believe that all one has to do is look at the fossils to experience a "born-again" conversion to evolution. They seem oblivious to the fact that the human fossils can be arranged another way, a better way. To show that way is the purpose of this book.

PALEOANTHROPOLOGY

Anthropology is the Greek word for "the study of man." *Paleo* means "old." Paleoanthropology is the study of fossil humans. The term replaces the older term *human paleontology.*

Except for that one glorious moment in the summer of 1984, the original hominid fossils are not generally available for study, even by paleoanthropologists. In fact, Milford Wolpoff is said to have seen more of the original hominid fossil material than any other paleoanthropologist, although even he has not seen all of it. On the other hand, Ian Tattersall and Niles Eldredge (American Museum of Natural History), who have written extensively on the human fossil record, confess that up to the time of the exhibit they had seen only a fraction of the available material. They go on to say that it is not comforting to realize that many of the statements by others regarding human evolution "are similarly removed from the original data."[10] Even the Ancestors exhibit displayed only a tiny fraction of the total material that has been recovered.

One would assume that those who have the proper academic credentials and are able to travel to where the original fossils are housed would have access to them. However, this is not always the case. Science writer Roger Lewin quotes Donald Johanson, the discoverer of Lucy, as agreeing that sometimes "only those in the inner circle get to see the fossils; only those who agree with the particular interpretation of a particular investigator are allowed to see the fossils."[11]

In the light of Johanson's behavior, his complaint is a bit humorous. Paleoanthropologist Adrienne Zihlman (University of California, Santa Cruz) tells of writing to Johanson when he was at the Cleveland Natural History Museum and asking permission to see the fossils (including Lucy) that he had discovered in Ethiopia. He replied that he would grant permission only if he were allowed to review any article she might write before she sent it to a journal. She interpreted this as his insisting that he must approve it. Since she felt that this was a form of censorship, she declined. The result was that she didn't get to study those fossils before they were sent back to Ethiopia, where they are permanently housed.[12] Zihlman also suspects that there were times when she was denied access to fossil collections because she is a female worker in a male-dominated field.[13]

In spite of some obvious cases of injustice, I do not wish to imply that this lack of general access to the original fossil material is some sort of evolutionist plot. The problem rests with the basic nature of the material itself: its fragility and its irreplaceability. Most of the fossil material, especially the newer material, is housed in the particular country in which it was found. As ancestor remains, these fossils are national treasures of incredible value. In some countries, the protection of these fossils seems to be far more important than the study of them.

This lack of access, however, has important implications for the study of human origins. It means that paleoanthropology is in the strange situation of being a science in which most of its workers do not have access to the material upon which their science is based. They are at least one step removed from the objects of their study.

What, then, do they work with? They use reproductions made of plaster or some other material. This means that the authority of the statements paleontologists make regarding fossils depends on the quality of the casts with which they work. Obviously, one cannot make a universal statement about the general quality of these casts. That quality depends on the accuracy of the molds used, the type of material used, the care taken in making the casts, and other factors.

It is possible to have fossil reproductions that are of excellent quality. The Peking Man casts are said to be of such quality. C. Loring Brace (University of Michigan) tells the story of a tiny piece of new Peking Man cranial material that was found many years after the other originals were lost. This new piece fit perfectly into the space on the cast of the original from which the new piece had come.

The classic illustration that casts can be far from ideal is the account of the fraudulent Piltdown Man fossils. Piltdown Man was a combination of a late-model human cranium and a piece of the lower jaw of an orangutan. The teeth of the orangutan mandible had been filed down to make them look human and to match those in the upper jaw of the cranium. Louis Leakey, in his book *Adam's Ancestors*, tells of several attempts to make a detailed study of the original Piltdown fossils. On each occasion when he visited the British Museum to do so, he was given the original fossils for just a few moments. They were then taken away, and he was given casts to work on. The file marks on the orangutan teeth were visible on the originals, but they were not visible on the casts.[14]

Given the unavailability of the originals, casts are the next best medium of study. Yet it is common knowledge that casts or reproductions, while giving a general impression of the original, often lack the detail of the original. Becky A. Sigmon (University of Toronto) says there is a general feeling among paleoanthropologists "that casts should not be used as resource material for a scientific paper."[15] However, there is another problem with the use of casts. Casts of only a small percentage of the total fossil material and less than half of the most important fossil material are available for study. This in itself is a serious situation. It would seem to place a degree of contingency on all conclusions reached in the study of human origins.

Descriptions of fossils in the scientific literature, although a poor substitute for casts, are probably the most common tools used in the study of the human fossil material. Since only the original fossils should be used in the writing of such papers, this would seem to place serious limitations on their preparation. Unfortunately, seldom do authors of such papers indicate what their sources were: the literature, casts, or the original fossil material. Milford Wolpoff, commenting on the value of the 1984 Ancestors exhibit, which allowed him and others to compare points of difference between fossils by seeing them side by side, says, "You can't do that properly through the literature."[16]

Perhaps the best example of the problem facing paleoanthropology is that many of the scholars who felt that casting technology was now able to provide copies as good as the originals, after studying the originals in the American Museum exhibit, admitted "that technology still has a long way to go."[17] The crowning blow came at the beginning of the public display. The precision mounts for the original fossils were carefully prepared based on casts supplied in advance. When the original fossils were placed in those mounts, most of them did not fit. No better illustration could be found showing that "casts are no substitute for originals."[18]

The problem of very limited access to the fossil originals does not apply just to the fossils that have been a part of the evolutionist arsenal for many years. It applies even more to newly discovered fossils. This problem involves the time between the original discovery of a fossil or fossil assemblage and the time when the discoverer has made his full report to the scientific community about his determination of their age and classification. Let me illustrate with a specific example.

In December 1993, seventeen dental, cranial, and postcranial (body bones below the skull) fossils were discovered at Aramis, Middle Awash, Ethiopia, by a team lead by Tim White (University of California, Berkeley) and Berhane Asfaw (Ethiopian Ministry of Culture). The fossils were believed to represent a new australopithecine species. They were named *Australopithecus ramidus* and were described in the 22 September 1994 issue of *Nature.*[19] However, in the 4 May 1995 issue of *Nature*, a notice appeared stating that this fossil material was considered different enough to be assigned to a new genus: *Ardipithecus ramidus.*[20]

The *Ardipithecus ramidus* fossils are housed in the National Museum, Addis Ababa, Ethiopia, and have been under study by Berhane Asfaw and Tim White ever since their discovery. There is an unwritten law in paleoanthropology that those who discover fossils have broad control over their access until they are fully studied and described. Many nations, including Ethiopia, have also passed specific laws to that effect.

In 2000, Ian Tattersall and Jeffrey Schwartz (University of Pittsburgh) requested permission to study and photograph the *Ardipithecus ramidus* fossils for an atlas of early human relatives they were preparing, similar to their earlier work, *Extinct Humans.* Their request was denied. In early 2002, they were finally granted email permission from an Ethiopian official to come to Addis Ababa and study and photograph the fossils. When Berhane Asfaw came to the Museum to work on the fossils and learned that Tattersall and Schwartz had been given permission to study them, he reminded the museum director of the Ethiopian law allowing discoverers of fossils to deny access to them.

Tattersall, sitting in the fossil room outside the locked safe, was livid. "What are you trying to hide?" Asfaw replied, "You don't know how we suffered in the field to get these fossils. You have to give us a chance to study them first."[21]

Attempts the next day were equally unsuccessful. Tattersall and Schwartz returned to the United States without even seeing the *Ardipithecus ramidus* fossils, let alone studying them. This was nine years after the fossils had first been discovered and eight years after they had been initially described in *Nature.*

As strange as this sounds, there are legitimate reasons for the law. It took three years of continuous excavation just to get all of the *Ardipithecus ramidus* fossils out of the ground because of their delicate nature. Fossil hunting involves laboring in some of the most unforgiving areas of the world. Hardships include disease, wild animals, military coups, heat, and long months spent away from family. Rewards include fame and (sometimes) fortune.

There are also very few fossil hunters compared to the many people who want to study fossils and publish about them. A longtime member of Richard and Meave Leakey's teams, Alan Walker (Pennsylvania State University), tells about showing some people a brand-new hominid fossil he had just found. They asked him if they could write it up. "Why would you do all this to get robbed?" he asks.[22]

However, there is a way of getting access to new and important fossils: Find a new and important fossil yourself. Michel Brunet (University of Poitiers, France) is apparently the only person who has seen all of the earliest hominids, or at least seen casts of them. His passport is a very old fossil that he found in Chad. Since others want to see his fossil for comparison, they let him see theirs. It's becoming a buddy system: "I'll show you my fossil if you show me yours."

John Fleagle (State University of New York, Stony Brook) has summed up the problem. "The big awkwardness right now is when someone announces they have found a specimen that overturns everything we know, but almost no one has seen it."[23]

Since the original fossils are virtually beyond access even to most who teach and write in the field of paleoanthropology, and only a few of the fossils are available as reproductions, and reproductions are not recommended in the preparation of scientific papers, and those scientific papers themselves cannot adequately convey differences between fossils, the "science" of paleoanthropology seems to have a problem.

The myth in the minds of the public is that the human fossil material is readily available and is thoroughly studied by all who teach and write on the subject. The truth is that paleoanthropology is in the awkward position of being a science that is several steps removed from the very evidence upon which it claims to base its findings.

TAXONOMY

Taxonomy is the science of the classification of living things. The common classification system used today involves classification according to structure. Humans are classified as follows:

Phylum	Humans are chordates
Class	Humans are mammals
Order	Humans are primates
Family	Humans are in the family Hominidae
Genus	Humans are in the genus *Homo*
Species	Humans are in the species *sapiens*
Subspecies	Humans are in the subspecies *sapiens*

Modern humans are classified as *Homo sapiens sapiens*. Only the genus term is capitalized. Since these terms are Latin, all are italicized.

Homo. Humans are the only living forms in the genus *Homo.* Biblically, there are no creatures past or present who would qualify for the genus *Homo*, or "true humans," other than descendants of Adam.

sapiens. The first *sapiens* in the classification refers to the species level. The Latin term means "wise." The species level is the level of reproduction and of reality. The higher levels are constructs of the human mind to bring order out of the complex world in which we live. The scientific term *species* is a very involved concept that has yet to be defined with finality. The biblical word *kind* is not a synonym of *species* and should not be confused with it.

sapiens. The second *sapiens* in the classification refers to the subspecies level. The racial distinctions of humans are so slight that they are well below the subspecies level. All humans belong to the same subspecies, the same species, and the same genus. This amazing unity of the human family is in itself strong evidence for creation and against evolution.

CHAPTER 2

A FAIRY TALE FOR GROWN-UPS

It may be more useful to regard the study of evolution as a game rather than as a science.

—SHERWOOD L. WASHBURN
(UNIVERSITY OF CALIFORNIA, BERKELEY)[1]

BERNARD WOOD, one of the foremost authorities on the human fossils, writes, "We are certain as anyone can ever be in biology that modern humans are more closely related to chimpanzees than any other living animal."[2] Bernard Wood may be a bit too optimistic about the certainty of humans evolving from chimpanzees.

Science writer Carl Zimmer, former senior editor of *Discover*, writes in the September 2003 issue about a seldom-discussed subject: "Great Mysteries of Human Evolution."[3] He calls them the big questions, the things that are unknown about human evolution. He lists eight questions for which evolutionists have no certain answers.

1. Who was the first hominid?
2. Why do we walk upright?
3. Why are our brains so big?

4. When did we first use tools?
5. How did we get modern minds?
6. Why did we outlive our relatives?
7. What genes make us human?
8. Have we stopped evolving?

Without definitive answers to these questions, there can be no certainty that humans evolved from chimpanzees. There are many theories regarding each of these questions, but Zimmer is right when he states that evolutionists do not have a definitive answer for even one of them. What, then, are we to make of Bernard Wood's statement, "We are certain as anyone can ever be in biology that modern humans are more closely related to chimpanzees than any other living animal"? His statement is arrogant. It is fiction. It is a lie. It shows that evolutionists abuse and ignore the scientific methods. It proves that in the field of evolution, "Philosophy Rules!"

Speaking at an annual convention of the American Association for the Advancement of Science (AAAS) in San Francisco, the late Carl Sagan explained how science works. "The most fundamental axioms and conclusions may be challenged," and the prevailing hypothesis "must survive confrontation with observation." "Appeals to authority," he said, "are impermissible," and "experiments must be reproducible."[4] This, of course, is the concept of science that the general public has.

It is hard to believe that Sagan delivered his lecture with a straight face. He believed, like all others in the scientific establishment believe, that evolution is "science." Yet no one in that establishment is allowed to challenge evolution. Evolution is nonobservable. Evolutionists often appeal to the authority of the scientific community regarding the fact of evolution. And there are no repeatable experiments that are able to confirm evolution.

If anyone abused the rules by which scientists are supposed to work, it was Sagan himself. He gave lip service to the accepted methodology of science. However, when presenting his views on the evolution of everything, he gave the public a freewheeling fantasy in which one could not separate science from science fiction. The result is that all of it is accepted as science by the undiscerning public. This sort of thing is permitted because scientists are considered to be the high priests of our society, paragons of objectivity who have no philosophical axes to grind. Hence, the public is often fed a rich diet of philosophy under the guise of science.

In the same lecture, Sagan then made a comment that is both true and profound: "Not all scientific statements have equal weight." He cited Newtonian dynamics, the first and second laws of thermodynamics, and the law of angular momentum as being on extremely sound footing because of the millions of experiments and observations that have been performed on their reliability. Sagan's remark about scientific statements having various weights based on the data backing them is obvious. But few people put it into practice.

Human evolution allegedly took place in the past over vast periods of time. Evolutionists readily admit that evolutionary processes work so slowly that they are not observable over the lifetime of one individual or even over the successive lifetimes of hundreds of generations. In other words, there are no *direct* observations or experiments that can confirm the process of human evolution.

On a scale from zero to ten, it is then possible to assign relative values to various scientific statements based on the number of direct experiments and observations involved. If, based on Sagan's statement, we assign a value of "ten" to Newtonian dynamics and the laws of thermodynamics because of the millions of confirming experiments and observations, what value can we assign to statements regarding human evolution when there are no direct observations to back them up? The only value to assign to those statements is "zero."

One of the lines of evidence promoting the concept of human evolution involves studies on living chimpanzees—their behavior, genetic makeup, and anatomy. All of these studies are fundamentally flawed. The flaw is known in logic as begging the question. In begging the question, you assume to be true the very thing you are trying to prove. Let me illustrate.

A man was observed walking down a street in Chicago, snapping his fingers. Finally, someone was driven by curiosity to ask him why he repeatedly snapped his fingers. "It keeps the elephants away," the man replied.

"Why, man, there aren't any elephants within ten thousand miles of this place!" responded his questioner.

"Pretty effective, isn't it?" exclaimed the man. He first assumed that snapping his fingers kept elephants away. He then used the absence of elephants to prove that his snapping worked. To presuppose the truth of what you are trying to prove is the illogical practice of begging the question.

Evolutionists first assume that humans and chimpanzees evolved from a common ancestor. They then use superficial similarities between humans and chimpanzees to prove their assumption. In a better world, evolutionists would be required to take a course in logic. Studies on chimpanzees could cast light on human nature only if evolution were first proven to be true. If evolution is not true, chimpanzee studies, although valuable in their own right, are worthless in shedding light on human origins.

Another logical fallacy of evolutionists is their seeming failure to understand the difference between scientific and historical evidence. Again, let me illustrate. It is believed that during the American Revolution, George Washington and his men crossed the Delaware River to attack the city of Trenton. How would you go about proving that event? If you used the scientific method, you would do research on boats, measure the width and flow of the river, do studies on the rowing of boats, and perhaps even row across the Delaware River yourself. Would all of this data prove that Washington crossed the Delaware? No. Scientific evidence is not what is needed. Historical evidence, such as records of eyewitnesses or of persons closely associated with those involved, is what is needed. All the scientific method could prove is the *possibility* that Washington crossed the Delaware, not that he actually did so.

G. A. Kerkut (University of Southampton) discusses this problem of historical versus scientific evidence as it pertains to evolution. The problem concerns all phases of evolution, and if evolution in general did not take place, humans did not evolve either. Kerkut first lists seven basic assumptions in evolutionary theory, which, he claims, are seldom mentioned in discussions on the subject. He then states:

> The first point that I should like to make is that these seven assumptions by their nature are not capable of experimental verification. They assume that a certain series of events has occurred in the past. Thus though it may be possible to mimic some of these events under present-day conditions, this does not mean that these events must therefore have taken place in the past. All that it shows is that it is *possible* for such a change to take place. Thus to change a present-day reptile into a mammal, though of great interest, would not show the way in which the mammals did arise. Unfortunately we cannot bring about even this change; instead we have to depend upon limited circumstantial evidence for our assumptions.[5]

All experiments performed with present-day animals, plants, or biological molecules are logically flawed. They cannot prove or even support the alleged evolutionary processes of the past. The extensive use of present-day experiments to try to demonstrate evolution reveals that evolutionists do not understand the difference between scientific and historical evidence.

Another major line of evidence used to support the concept of human evolution is the fossil record. We have all seen pictures of the impressive sequence allegedly leading to modern humans—those small, primitive, stooped creatures gradually evolving into big, beautiful you and me. What is not generally known is that this sequence, impressive as it seems, is a very artificial and arbitrary arrangement because (1) Some fossils are selectively excluded if they do not fit well into the evolutionary scheme; (2) Some human fossils are arbitrarily downgraded to make them appear to be evolutionary ancestors when they are in fact true humans; and (3) Some nonhuman fossils are upgraded to make them appear to be human ancestors.

A major section of this book will consider the human fossil evidence. At this point I merely want to emphasize a phenomenon that seems almost universally unrecognized: *Any series of objects created by humans (or God) can be arranged in such a way as to make them look as if they had evolved when in fact they were created independently by an intelligent being.* The fact that objects can be arranged in an "evolutionary" sequence does not prove that they have an evolutionary relationship or that any of them evolved from any of the others.

In a certain graduate course I took in paleontology at a state university, the professor attempted to teach us the concepts of taxonomy and the construction of those familiar evolutionary family trees. He handed each student a packet of about 150 metal objects such as nails, tacks, and paper clips. Utilizing the various rules of evolutionary taxonomy, such as small to large, simple to complex, and generalized to specialized, we were each expected to arrange these objects in evolutionary order. Starting with generalized nails, we went on to nails gradually increasing in size and then branching off into various specialized types of nails and tacks. Naturally, no two students in the class arranged their objects in exactly the same way, although there was an overall similarity. When the project was finished, we all had created a beautiful series of phylogenetic trees showing the "evolution" of nails, tacks, and paper clips.

What I found fascinating about the project was that, as we played with our object lesson, no one sensed that the illustration was totally invalid; it had no relationship to reality. Each of the objects that we had arranged in such a convincing evolutionary sequence had in fact been individually created for a specific purpose by humans. There was no actual evolutionary relationship between them. We were able to arrange them in an "evolutionary" sequence even though none of them had evolved. That fact did not seem to dawn on anyone in the class, including the professor.

Because of the problems I have discussed in these chapters, creationists feel that paleoanthropology does not deserve the same status in science that is accorded fields like chemistry or physics. When we creationists make such statements, we are often accused of being antiscience. We are accused of degrading science and yet using its benefits in our private lives. But we are merely saying what Carl Sagan said: "Not all scientific statements have equal weight."

It is refreshing to know that some within the scientific community have been open-minded enough to acknowledge the quality-of-information problem in human origins. This field is the scene of much prejudice, subjectivity, and emotionalism in the interpretation of the human fossils and in the construction of phylogenetic trees. The professionalism and objectivity found in other areas of science have been conspicuously absent in this area.

Sir Peter Medawar, who won a Nobel Prize in medicine, spoke of paleoanthropology as "a comparatively humble and unexacting kind of science."

Andrew Hill (Yale University), following up on Medawar's comment, added, "It has certainly been possible to get away with being an unexacting practitioner." Hill observes that the casual student could easily discern the rules of paleoanthropology. They are: "Every new fossil hominid specimen is the most important ever found and solves all known phylogenetic problems; every new hominid specimen is completely different from all previous ones, no matter how similar; every new hominid specimen is a new species, and probably new genus, and therefore deserves a new name."[6]

Along this same theme of calling each new fossil discovery a new species or genus, Jeffrey H. Schwartz observes, "Often, type specimens [the first significant fossil of a new type] lack morphological detail and taxonomic judgements are based on nonbiological criteria, e.g., time and/or geography."[7]

Elsewhere, Andrew Hill has written, "Compared to other sciences, the mythic element is greatest in paleoanthropology."[8] Ian Tattersall, commenting on the remarks of a writer who took liberal swipes at many of the views of paleoanthropologists, admits, "As she has noticed, palaeoanthropologists are fond of telling each other 'Just-So' stories; and once in a while a little needling of this kind does no harm at all."[9]

Commenting on the lack of accessibility to newly discovered fossils by independent observers (the condition we described in chapter 1), Milford Wolpoff charges, "When the only people who can comment are the discoverers or friends of the discoverers, there is no sense of independent observer. We're not practicing science. We're practicing opera."[10]

Many books on primate and human evolution are written by nonspecialists. Jonathan Marks (Yale University) has noticed this also. Reviewing a book on primate origins and evolution, he writes:

> Biological anthropology is a field distinctive for its curious attitude towards dilettantism [amateurs in anthropology]. In most scientific disciplines, dilettantism is flatly abjured. Yet in biological anthropology the proffered opinions of ornithologists [bird specialists] or fruitfly geneticists . . . are sometimes taken as authoritative. There are even university textbooks written solely by journalists. As far as I know, this is not the case in organic chemistry or endocrinology.[11]

Paleoanthropologists and the Laws of Evidence

I have noticed in evolutionist writers both a lack of logic and an inability to weigh evidence properly. Legal experts have also recognized this fact. Some years ago, Harvard-trained lawyer Norman Macbeth wrote a book, *Darwin*

Retried.[12] After studying evolution for many years, Macbeth, who is not a creationist, concluded that there were serious gaps in the evidence for evolution and errors in the reasoning of evolutionists. He claimed that evolution itself had become a religion. The alleged evidence for evolution, he charged, was not of the quality that would stand up in a court of law.

A similar conclusion was reached in a recent book, *Darwin on Trial*, by law professor Phillip E. Johnson (University of California, Berkeley). Johnson describes himself as a Christian and a creationist, but not a biblical literalist. His book may be one of the most significant on the evolution debate to appear in decades.

Johnson concludes that: (1) Evolution is grounded not on scientific fact but on a philosophical belief called naturalism; (2) The belief that a large body of empirical evidence supports evolution is an illusion; (3) Evolution is itself a religion; (4) If evolution were a scientific hypothesis based upon a rigorous study of the evidence, it would have been abandoned long ago; and (5) Since atheism is a *basic supposition* in the evolutionary process, it cannot be drawn as a *conclusion* from it.

Evolutionists got the cart before the horse, Johnson states. They first accepted evolution uncritically as a fact. Then they scrambled for evidence to support it, without too much success. This analysis is certainly true of human evolution. As we will show later, the concept of human evolution was well in place before any of the human fossils had been discovered.

In discussing the lack of objectivity in the interpretation of the human fossils, Johnson refers to Roger Lewin's description of the 1984 Ancestors exhibit, mentioned in chapter 1. Lewin said that the paleoanthropologists were in awe as they held their fossil ancestors in their hands. Lewin described how moving and emotional that experience was. Johnson comments:

> Lewin is absolutely correct, and I can't think of anything more likely to detract from the objectivity of one's judgment. Descriptions of fossils from people who yearn to cradle their ancestors in their hands ought to be scrutinized as carefully as a letter of recommendation from a job applicant's mother.[13]

Johnson continues:

> The story of human descent from apes is not merely a scientific hypothesis; it is the secular equivalent of the story of Adam and Eve, and a matter of immense cultural importance. . . . The needs of the public and the [secular scientific] profession ensure that confirming evidence will be found, but only an audit performed by persons not committed in advance to the hypothesis under investigation can tell us whether the evidence has any value as confirmation.[14]

Paleoanthropology's Debt to Artists

No tool has been as successful in promoting human evolution as have been the pictures and reconstructions of our ancient ancestors. Since no one has ever seen these ancient ancestors, the abilities of the artists who constructed them have been nothing short of miraculous. It gives the term "science fiction" a whole new meaning.

David Van Reybrouck has studied the pictures and drawings of fossil humans and their reconstructions, starting with those of the original Feldhofer Neandertal. Writing in the journal *Antiquity*, he states that these pictures, drawings, and reconstructions: (1) always go beyond the archaeological data; (2) always involve the speculations and prejudices of the fossil discoverers, who advise the artists; (3) always involve interpretations that are theory laden; (4) always are nonobjective but are trusted as being accurate; and (5) are used so extensively because they sell evolution so effectively. He concludes, "A good drawing is like a Trojan horse: to be rhetorically effective, its interpretation must be hidden inside."[15]

Whether or not the field of paleoanthropology has standards within its own profession, it is obvious that when communicating its message to the general public, anything goes. Evolutionists will stop at nothing in attempting to influence the public toward human evolution and against creation. From a scientific point of view, their drawings and reconstructions are outrageous. Over the years I learned of three that stick in my mind as being outrageously outrageous. The first deals with Louis and Mary Leakey.

Louis and Mary Leakey had worked at Olduvai Gorge, Tanzania, for many years. The gorge, part of the East African Rift System, had produced many stone tools and animal fossils, but no hominids. Yet Louis felt that hominid fossils had to be there. One day in 1959, Mary went out alone because Louis was ill. At a certain spot she saw teeth sticking out of the ground. Excavation revealed a large cranium having some resemblance to the South African robust australopithecines.

The stone tools found in association with the fossil led Louis to believe that this individual was a tool maker. And to Louis, tool making meant just one thing: man! Believing that they had found the first tool maker, Louis named the fossil *Zinjanthropus*, "East Africa Man." The ridiculously large molars indicated that the individual probably had lived on nuts and berries, and so the fossil became affectionately known as "Nutcracker Man."

Some of us suspected that Louis knew all along that "Zinj" was just a variant of a robust australopithecine. But the financial support Louis desperately needed to continue his work does not come from the discovery of fossil primates. It comes from finding human ancestors. The long financial association the Leakeys had with the National Geographic Society began at this time. Telling of the discovery of "Zinj" in *National Geographic*, September 1960, Louis began his report, "The teeth were projecting from the rock face, smooth and shining, and *quite obviously human* (emphasis added)."[16]

It was not long before the Leakeys began to find the remains of another type of fossil individual. This type was a far better candidate for human ancestry. Louis began to realize that "Zinj" really was just a super-robust australopithecine, and it is now known as *Australopithecus boisei*. What Louis claimed was "obviously human" turned out to be obviously nonhuman.

What I found so outrageous in the original *National Geographic* article was an artist's painting of what "Zinj" might have looked like. It was quite a piece of work. Whereas our eyes are about midway between our chin and the top of our head, there was hardly any head at all showing above "Zinj's" eyes. He had virtually no brains. His cranial capacity (brain size) is calculated at 530 cc (cubic centimeters), compared to yours and mine, which average about 1450 cc. To refer to a primate having a brain size of 530 cc as a human capable of making stone tools is an insult to all of us. Under Leakey's guidance, however, the skill of the *National Geographic* artist was amazing. "Zinj" looked like a philosopher who could debate Immanuel Kant—and win! No evolutionist ever protested this outrageous and unscientific portrait. Such is the power of the artist in promoting evolutionary theory. But few realize that this is not science. This is Madison Avenue propaganda.

My second illustration of outrageous evolutionist artistry concerns a "parade" of fifteen figures, artists' drawings, used to demonstrate human evolution. The parade starts with erect-walking protoapes, then apes, and then goes all the way up to modern humans. This parade first appeared in the Time-Life Nature Library series book *Early Man*, by F. Clark Howell (University of California, Berkeley). The book was originally published in 1965, with revisions appearing in 1968 and 1973. The parade was on a 36-inch foldout on pages 41 to 45 in the original edition.

This parade has been one of the most successful tools ever used to promote human evolution. It constituted powerful visual "proof" for human evolution that even a small child could grasp. It was a masterpiece of Madison Avenue promotion. There were few social studies classrooms and school library bulletin boards where this parade was not prominently displayed. Because of its graphic power, it is still indelibly etched into the minds of billions of people worldwide.

It is not that more recent fossil discoveries have revealed that the parade was inaccurate. No, the truth is far worse. The parade was a fake when it was first published. That is not just my personal opinion based upon my evaluation of the fossil evidence. If one reads *Early Man* carefully, the book itself reveals that the parade is fiction. Further, the author, F. Clark Howell, and the editors of Time-

Life associated with him were fully aware that the parade was manufactured evidence for human evolution.

That visually powerful parade was so successful in advancing human evolution because it received a far greater distribution and viewing than did the *Early Man* book. Worldwide mailings for advertising purposes were made of the particular pages featuring the parade. The posting of these pages in classrooms and libraries meant that far more people saw the parade than possessed the book. Perhaps less than five percent of those who had the book actually read it, but they would have seen the parade. Thus, the visual image of the parade sold the concept of human evolution even though the book revealed that the parade was fictitious.

The entire chart was outrageous. But the most outrageous part was that the parade started with erect-walking protoapes and apes. Evolutionists knew that these protoapes and apes were not bipedal (walking on two feet). In fact, these fossil apes were dated long before evolutionists believed that bipedalism had evolved. An explanatory note in the text of the book read, "Although protoapes and apes were quadrupedal, all are shown here *standing* for purposes of comparison (emphasis added)."[17]

After all these years, I still am amazed at that statement. First, these fifteen forms were not *standing*. They were *walking* across the pages from left to right. Some of them have one foot in the air as they walk. Second, if a first-grade child were asked to compare an ape or a chimpanzee with a modern human, the most obvious comparison for the child would be that apes and chimpanzees walk on all fours (quadrupedal) and humans walk on two legs. Are evolutionists that ignorant? Of course not. This is raw propaganda—brilliant propaganda, but raw nonetheless. Yet no evolutionist protested this gross lack of scientific objectivity.

My third illustration of outrageous evolutionist artistry involves a man named John Gurche. John is a special kind of artist. He sculpts the heads of ancient hominids. He was recently involved in a twelve-year project, completed in about 1996, to create seven heads of ancient hominids for the new Hall of Human Origins at the Smithsonian's National Museum of Natural History in Washington, D.C. His heads include a *Homo erectus*, a Neandertal, an *Australopithecus afarensis*, and an *Australopithecus africanus*.

I have only seen pictures of his *Homo erectus* head. It is fashioned on a cast of a female *Homo erectus* fossil, designated KNM-ER 3733, in the Kenya National Museum, Nairobi. Gurche's *Homo erectus* head is unique. I have a number of artists' drawings of what *Homo erectus* might have looked like. All of them appear quite human. They should, because *Homo erectus* was truly human and should not be called *Homo erectus* but *Homo sapiens*. But Gurche's *Homo erectus* head appears much more apelike than any of the others. He tells us why. "It's necessary to look at a feature in both great apes and in humans if you're going to extrapolate for the hominids. You can't have it work just for humans or just for apes because ancient hominids don't fit neatly into either category."[18]

Gurche tells us that he took measurements of major muscles, minor muscles, glands, fat bodies, and cartilage of both modern humans and great apes (undoubtedly chimpanzees) and extrapolated those measurements in determining the face and head measurements for his *Homo erectus*. Yet there is no evolutionist on the face of the earth who believes that modern humans evolved from modern chimpanzees. Even if we assumed that human evolution is true, those measurements would have to come from that unknown transitional form, not modern chimpanzees. Further, where would *Homo erectus* lie on the line from that unknown transitional form to modern humans? Would *Homo erectus* be about halfway between the two? Three-quarters of the way? In other words, even if we assumed that human evolution was true, Gurche's measurements are worthless, and his *Homo erectus* head is science fiction.

We can't say, "Gurche is just an artist." He is a trained anthropologist. He travels all over the world to discuss his reproductions with the most authoritative paleoanthropologists. His work is sponsored by the most respected natural history museum in the nation—the Smithsonian, supported by the United States government and American taxpayers. The evolutionary establishment endorses his work even though it is imagination, not science. It is endorsed because millions of people will see these heads each year and will be impressed as to how much we know about our ancestors.

Hero Myths and Human Evolution Stories

Misia Landau (Boston University) published a book on a most unusual subject.[19] Landau had been studying the stories of human evolution beginning with those by Charles Darwin, Thomas Huxley, Ernst Haeckel, Sir Arthur Keith, Grafton Elliot Smith, and later stories by authorities in the mid- and late-twentieth century. She recognized remarkable similarities between those human evolution stories and the folktales and hero myths of literature. Her central theme is that "paleoanthropology is story-telling." The similarities she noted are as follows:

1. The initial condition of the hero. In most folktales and hero myths, the hero is initially leading a rather safe and untroubled life in humble conditions. In many of the stories he is a bit different from the other creatures, perhaps smaller and weaker. In the story of human evolution, the hero is a nondescript primate, helpless and defenseless, living in the trees.

2. There is a change in circumstances. In most folktales and hero myths, the hero is dislodged from his home for some reason. Perhaps he goes on a trip or starts on a new adventure. In the story of human evolution, the hero leaves the trees and begins life on the ground, perhaps because of a change in the environment, the ability to be bipedal, or the development of a larger brain.

3. The hero is tested. In most folktales and hero myths, the hero meets situations that would deter him or enemies who would impede his progress. All of these incidents are to build or test his character. In the story of human evolution, the hero is tested by predators, a harsh climate, or a closely related species. There is the idea of self-destiny, and the hero seems to "make himself" into more of a human. "Indeed, the tests are specifically designed for that purpose: to bring out the human in the hero."[20]

4. Transformation by a beneficent power or donor. In most folktales and hero myths, a donor or beneficent power appears, often in nonphysical form, and gives the hero a magical agent—a cloak, a ring, or a sword—to help the hero overcome his physical deficiency. In the story of human evolution, the hero, through natural selection, acquires increased intelligence, tools, discrimination, or, as Jared Diamond describes the movement of modern humans out of Africa, a "Magic Twist" of mutations that enabled them to acquire innovation.[21]

5. The hero is tested again. In most folktales and hero myths, the hero is again tested by various circumstances or enemies to prove his achievement. In the story of human evolution, the test involves the rigorous climate of Ice Age Europe, as well as the elimination of the Neandertals and all other humans throughout the world who were not fully evolved. Thus, our hero proves that he has achieved full human status. This is, in essence, the Out of Africa Model of human evolution, the most popular model today.

6. Having accomplished great deeds, the hero succumbs to pride and is destroyed. In many of the folktales and hero myths, there is the fatal irony of the hero destroying himself as a result of his success. Almost all of the stories of human evolution end with what has become an almost obligatory warning of what we humans could do to ourselves if we are not responsible. Richard Leakey's last book, *The Sixth Extinction*, is devoted entirely to that theme.[22]

Misia Landau's insight is that the theme of almost all folktales and hero myths, the transformation of the hero from weakling to lord, is the very same as the story of human evolution—modern man emerges as the hero against many obstacles. For many years we have been told by evolutionists and religious liberals that the Bible is full of folklore and myths. It seems that if we really want folklore and myth, the place to go is not to the Bible but to the stories of human evolution in *National Geographic, Time, Discover,* and our high school and college textbooks. In truth, human evolution is a fairy tale for grown-ups.

A widely held myth is that paleoanthropologists are able to speak with the same authority as other scientists. The reality is that the quality of their information and their interpretation of that information are open to serious challenge. As Carl Sagan said, "Not all scientific statements have equal weight."

CHAPTER 3

Dead Reckoning

HAVE YOU EVER WONDERED how many hominid fossils have been discovered to date? It's a very important question, because when we deal with historical evidence, quantity makes for quality. The total number of hominid fossil individuals is universally assumed to be quite small, since paleoanthropologists have for years complained about the lack of fossil material.

The first surprise—if not shock—comes when we learn that the actual number of fossil individuals discovered to date is very difficult to obtain. There does not seem to be a central clearinghouse, a publication, or a data bank to which the interested person can go to get that information.

One publication attempted to give that information as of 1968–1976. It is the three-volume *Catalogue of Fossil Hominids* edited by Kenneth P. Oakley, Bernard G. Campbell, and Theya I. Molleson and published by the British Museum (Natural History), now called the Natural History Museum, London. Volume 1 deals with fossil hominids from Africa (the second edition of that volume was published in 1977); volume 2 deals with Europe and the USSR and was published in 1971; volume 3 covers the Americas, Asia, and Australasia and was published in 1975. The *Catalogue* was intended to serve as a reference for information on the fossil hominids. Its scholarship and authority are beyond reproach. It answers the question of approximately how many hominid fossils had been discovered up to 1969–1976.

Unfortunately, nothing as complete as the *Catalogue* has been published since 1977. And this work is extremely difficult to find even in the largest university or city libraries. In their introduction, the editors of the *Catalogue* stated that it was their intention to keep all of the volumes updated. Although Oakley died in 1981, other persons at the museum are adequate for the task. However, the Natural History Museum, London, informed me that they were not aware of plans to update the volumes. I find this surprising. This is an area where the *Catalogue* stands unique, interest in the hominid fossils is increasing, and recent discoveries have been unparalleled.

Ian Tattersall and Jeffrey Schwartz are engaged in a project of bringing the hominid fossil discoveries up to date. The first volume, *Extinct Humans,* was published in 2000.[1] A second volume, *Extinct Non-Humans,* may not be published for some time because of the difficulty in gaining access to the newer discoveries (see chapter 1). Assuming that the second volume will be like the first one, these volumes have beautiful photography and are quite thorough in the fossils they cover, but they are not exhaustive in their coverage. Hence, one still would not get an idea as to the total number of fossil individuals discovered to date.

For some unexplained reason, the *Catalogue* does not tabulate the total number of fossil hominid individuals that it covers. (An individual may be represented by just one tooth or all the way up to a complete skeleton.) According to my personal count, the total number of hominid fossil individuals, including all categories of the genus *Homo* and the genus *Australopithecus,* listed in the *Catalogue* are as follows:

Africa—1,390 fossil individuals discovered through 1976.

Europe and the USSR (now known as the CIS)—1,516 fossil individuals discovered through 1970 (but the figure for France, one of the most prolific fossil areas, goes only through 1969).

The Americas, Asia, and Australasia—1,092 fossil individuals discovered through 1974.

A grand total of 3,998 (approximately 4,000) fossil hominid individuals have been discovered as of 1969–1976. This is a surprisingly large amount of evidence dealing with human origins.

Several comments about this figure are in order. Because of the many different ways the hominid fossils have been catalogued, it is probable that in some cases different numbers have been assigned to different fossil bones of the same individual. In seeking to arrive at a "body count" of the hominid fossils, there is no way to avoid this inflation of the numbers. However, several factors would compensate for this inflation:

(1) Editors of the *Catalogue* recognized that there were gaps in their coverage of the fossil material. They appealed to their readers to help them track down unreported fossil discoveries. However, some legitimate fossils that would strengthen the creationist position were omitted.

(2) In my tabulations, I always used the minimum figures in cases where exact fossil populations were unknown. For instance, for Peking Man, the Lower Cave material indicates a minimum of forty individuals and a maximum of forty-five individuals. At Krapina, Yugoslavia, the fossil material indicates a minimum of fourteen individuals and a maximum of twenty-eight individuals. I have always used the minimum figures, although the probability is that in many cases a larger number of individuals were actually present.

(3) In cases where miscellaneous human fragments are mentioned, I have always counted these as just one individual unless it is obvious from the remains that more than one individual is involved. For instance, three femora (thigh bones) would indicate at least two individuals and possibly three.

(4) In evaluating fossil remains, when there is no duplication of bones it is usually assumed that just one individual is involved. I have tallied it as such. Although this is a logical inference, it cannot be proven. In some cases, more individuals could be involved.

I believe that my conservative approach in tallying the fossils listed in the *Catalogue* compensates for the possible inflation that I spoke of, and that the figure of approximately four thousand fossil individuals is a realistic appraisal of the total number discovered worldwide as of 1969–76. This is an immense amount of material.

By 1976, over two hundred individuals were classified as Neandertals, and over one hundred were classified as *Homo erectus*. This means that in over three hundred of these fossil individuals, enough material had been recovered to be diagnostic of these categories.

Much of this fossil material is only thousands of years old. As we go further back in time, the amount of fossil material drops off significantly. Evolutionists say that because evolutionary change occurs so slowly, the older fossils are more significant, and these are the ones that are in short supply. Yet it is the more recent fossils that effectively falsify the concept of human evolution, specifically a recent *Homo erectus* (the Cossack skull), our alleged evolutionary ancestor, who may have been alive and well just a few hundred years ago.

In light of the richness of the hominid fossil record, it is difficult to understand why the public has been exposed to statements from authorities about the small number of hominid fossils that have been discovered. About forty original fossils were brought to the American Museum exhibit in 1984. Ian Tattersall, one of the organizers of the exhibit and a curator of the museum, is quoted as having said, "When this exhibit is assembled, we'll have more than half of the most complete specimens in the human fossil record under this roof."[2] Because Tattersall is a responsible scientist, I suspect that he was misquoted. As it stands, the statement is so patently wrong as to be absurd. Even if he had said "more than half of the most important fossils," the statement would still be false. A large number of very important fossils were not brought to the exhibit.

Well after the *Catalogue* was published documenting approximately four thousand individual fossil hominids, we would read statements like the following from leading authorities. Boyce Rensberger was a senior editor of a journal published by the American Association for the Advancement of Science, one of the most respected scientific organizations in our nation. Directly under the title of his article, "Bones of Our Ancestors," is the caption, "In all the world there are only a few dozen. But these rare fossils attest to the evolutionary odyssey that created the human species."[3]

Richard Leakey excused the many mistakes that had been made in interpreting the human fossil record by saying, "I think this was inevitable by virtue of the fact we had so little material."[4] We read in *NewScientist*, "The entire hominid collection known today would barely cover a billiard table,"[5] and in *Science*, "The primary scientific evidence is a pitifully small array of bones from which to construct man's evolutionary history. One anthropologist has compared the task to that of reconstructing the plot of *War and Peace* with 13 randomly selected pages."[6]

In light of statements such as these, it is not surprising that the perception is far different from the reality. The public is unaware of the rich harvest of hominid fossils we now possess. Although some of the myths I discuss in this book are not the fault of the evolutionists, this one clearly is. The evolutionists have gone to their public wailing wall and lamented the tragic lack of human fossils.

More recently, evolutionists have been lamenting the lack of earlier fossils, especially the australopithecines, as they attempt to determine the number of australopithecine species and which line is best suited as the human ancestral line. This question is still not resolved and may never be. The fact that we did not evolve from the australopithecines is irrelevant. Evolutionists will always propose their favored scenario.

Since all paleoanthropologists worth their salt know about the *Catalogue* published by the British Museum as well as the many fossils discovered since its publication, the question arises, Are they lying? No. While they are certainly not telling the truth, in their own minds they are not lying. They are speaking a different language, and the public has not learned to translate it. Their professional colleagues understand the language. They have made the translation. To comprehend the field of paleoanthropology and its literature, you, too, must learn to translate. When workers in this field speak of the scarcity of the human fossils, they are actually saying, "Although there is an abundance of hominid fossils, the bulk of them are either too modern to help me or they do not fit well into the evolutionary scheme. Since we all know that humans evolved, what is so perplexing is the difficulty we are having in finding the fossils that would clearly demonstrate that fact."

A very common myth today is that not many hominid fossils have been discovered. The reality is that by 1969–1976, approximately four thousand hominid fossil individuals had already been unearthed. The period since that time has seen the most intensive and successful search for hominid fossils in the history of paleoanthropology. No one knows exactly how many have been found to date. However, by my own reckoning, a conservative but very rough estimate, *the total number of hominid fossil individuals discovered to date is between seven and eight thousand.*

CHAPTER 4

Monkey Business in the Family Tree

THE MOST PROFOUND question a human being could ask is, "What is truth?" Pontius Pilate asked that question two thousand years ago. However, two problems are associated with the search for truth. It is not enough just to ask the question. Pilate did that. We will never know if he asked the question in sincerity or in cynicism. But it doesn't matter, because Pilate had not resolved the deeper problem. The problem is: "How will I recognize truth if and when I see it?" Since Pilate had no idea what he was looking for, he did not recognize truth when the one who is the Truth stood before him. To know what truth is gives us direction in our search. Otherwise, our search will be aimless and fruitless.

The God revealed in the Bible claims not only to be the Truth but to be able to reveal truth to those who seek him. If a person forsakes the biblical God, his options in searching for truth are quite limited. "Science" is one of those options claiming to be truth—not in the ultimate sense, but science claims to be able to lead us down a path closer to truth. One reason science is said to lead us toward truth is that it is self-correcting.

The late Carl Sagan was a man of deep faith in the ability of science to lead us into truth. Speaking at an AAAS convention in San Francisco, he said:

> Science is also self-correcting. . . . The history of science is full of cases where previously accepted theories and hypotheses have been overthrown to be replaced by new ideas which more adequately explain the data. . . . This self-questioning and error-correcting aspect of science is its most striking property, and sets it off from other areas of human endeavor such as . . . theology.[1]

Anthropologist Vincent Sarich (University of California, Berkeley) states that he is a humanist. He also has deep faith in the ability of science to lead us into truth. He repeated a concept that he has stated many times.

> We have a faith game and a science game. Really, these are both faith games. Both are human constructs which are made to make sense out of a very complicated world. Science is a very peculiar kind of new faith, which assumes that faith itself can be self-correcting. Other faiths do not have any kind of self-correcting mechanism built into them. They are dogmatic. People doing science can be dogmatic, but they forget what science really is. It is a continuously self-testing and self-correcting faith. The faith comes in the idea that it is a way in which we can generate increasing understanding of the world around us.[2]

There is no question that this self-correcting aspect of science is hailed by those who would make science a superior worldview over biblical faith, with its belief in the inerrancy of the Word of God and its "Thus saith the Lord!" Christ claims that he is "the way and the truth and the life" (John 14:6). The Bible claims to be the Word of God in an ultimate sense. These claims are offensive to some. They either are not interested in ultimate truth, do not like the biblical teaching of ultimate truth, or do not believe that it is possible for humans to know ultimate truth.

We could ask several questions: (1) How does one know that there is no ultimate truth? (2) How does one determine that Christ's words are not ultimate truth? (3) What criteria does one use to evaluate Christ's words? The Christian has such a criterion. He believes that Jesus Christ by his resurrection from the dead validated all his claims. Anyone who has power over death commands ultimate power and traffics in ultimate truth.

It is not necessary to make science and biblical faith an either-or situation, but Sagan, Sarich, and others have chosen to do so. There is something romantic about the thought of a scientist searching for truth. However, like the steamship company slogan that says "Getting there is half the fun," many scientists say that they are not interested in finding truth. They enjoy the search. They don't particularly care if they arrive or not. It is not unusual to hear scientists say that the Bible would destroy science, because there's no point in doing science if you already have the truth—probably one of the most absurd statements ever uttered as an excuse for rejecting God's Word.

A basic question needs to be asked, but I have never heard anyone ask it. To justify science as a superior worldview, Sagan cites situations in the history of science where the self-correcting mechanism has worked. However, the question is not whether this self-correcting mechanism has worked once, twice, a hundred times, or a thousand times. The basic question is, How efficient is science as a self-correcting mechanism? or What is the batting average of science in this area? or Out of the total number of mistakes made in science, how many have been corrected?

When we put the question this way, it is obvious that there is no way of knowing the total number of mistakes made in the history of science. Nor do we know how many uncorrected errors exist in science today. If we knew about them, they would be corrected. Hence, it is impossible to know how efficient this self-correcting element in science is. But if there is no way to determine its effectiveness, then we can never know if trusting science to lead us into truth is a very wise worldview or a very foolish one. We all agree that, according to its methods, science could be somewhat self-correcting. But we are not living in a perfect world.

Behind the self-correcting aspect of science is the idea that when a scientist feels he has discovered something unique or innovative, he must publish both his results and his methodology. His work is not only subject to a process known as peer review, but it is eventually exposed to the entire scientific community for evaluation. This sounds like a very healthy and purifying process. Ideally, it is. However, in actual practice it breaks down.

Scientists simply do not have the time nor the money to check up on the research of other scientists. Scientists in the academic community are busy with their teaching assignments, their graduate student supervision, and their own research programs. They are driven by the publish-or-perish attitude prevalent today. It simply does not benefit them in any way—no fame or fortune—to confirm or falsify the work of someone else. Scientists in industry have a bottom-line mentality. They must be productive in the areas in which their company specializes. They have no time to check out the work of other scientists just for the fun of checking them out or to prove that science really is self-correcting. There are exceptions, but in practice this is normally the case.

There is a touch of irony in the fact that science works in the very opposite way that people think it works. The self-correcting aspect of science implies a self-policing action by scientists, a checking up on one another. In actuality, scientists have demonstrated an incredible faith and trust in the work of their fellow scientists. They tend to accept that work at face value without much investigation at all. A number of recent scandals have developed in which fraudulent medical research was incorporated into medical practice and the fraud was discovered far down the line. The exposure of error usually occurs only when the effects are very obvious. If the error is not obvious, it can be perpetuated almost indefinitely. One of the most amazing illustrations of faith is the trust that scientists put in their fellow scientists who work outside their own fields of expertise.

The question still is, Is science really self-correcting? One of the areas of science with which I am familiar is paleoanthropology. I can testify that in this area the track record for self-correcting by scientists is very poor. Many illustrations could be used, but I will confine myself to two of the most notorious: the faulty reconstruction of the Neandertal skeleton from La Chapelle-aux-Saints and the Piltdown Man fraud.

The Faulty Neandertal Reconstruction

In 1908, a nearly complete skeleton of a Neandertal-type individual was found buried ritualistically in the floor of a small cave near the village of La Chapelle-aux-Saints, France.[3] The site is in the famous Dordogne River Valley, carved into a limestone plateau in southwestern France, about 120 miles east of Bordeaux. The area is known for the rock shelters and caves the groundwater has carved into the limestone. These caves and rock shelters have yielded other Neandertal fossils—Le Moustier and La Ferrassie—as well as the Cro-Magnon fossils from Les Eyzies.

Marcellin Boule, the famous paleontologist at the National Museum of Natural History in Paris, was called upon to reconstruct the man from La Chapelle-aux-Saints. This was the most complete Neandertal skeleton found in western Europe up to that time. Once it was reconstructed, the world could see what a Neandertaler looked like. Not for a moment did Boule believe that the Neandertalers, with their low, wide skulls and their sloping foreheads, deserved a place in the direct ancestry of humans—certainly not in the direct ancestry of the French. He felt that the Neandertalers were a withered side branch of the family tree or a backward evolutionary group that became extinct, leaving no issue. Piltdown Man was one of Boule's proofs that modern humans with their high-domed skulls dated much further back than the Neandertalers. In Boule's mind, the tall, high-domed Cro-Magnon people, the ones responsible for the beautiful cave paintings in France and Spain, were the true ancestors of modern humans.

Boule, who made the first detailed description of the bones of the Neandertalers, emphasized what he felt were simian (apish) features in the

La Chapelle-aux-Saints skeleton and played down the human features—based on his preconceived ideas of evolution. Although there was evidence that the vertebrae were severely deformed because of arthritis and rickets, Boule ignored the pathological evidence. He claimed that the spine lacked the curves that enable modern humans to walk erect. He placed the head on the neck in an unbalanced position, thrust far forward, so that the individual probably would have sprained his neck had he looked at the sky.

Boule also decided that this individual could not extend his legs fully but walked with a bent-knee gait. He made the foot only slightly arched, resting on its outer edge, with toes pointing in. Hence the man would have walked like an ape, pigeon-toed. Boule formed a wide separation between the big toe and the other toes, making the big toe similar to an opposable thumb—such as monkeys and apes have. Under these conditions, if Neandertal Man walked at all, he would have looked like a shuffling hunchback. His center of gravity was located so far forward of his center of support that he probably would have fallen flat on his face.

Using casts of the inside of the La Chapelle-aux-Saints skull, Boule felt that the brain of Neandertal, although larger than the average brain size of modern man (1620 cc vs. 1450 cc), resembled the brain of the great apes in organization. Boule concluded that Neandertal was closer to apes than humans in brainpower, had only a trace of a psychic nature, and had only the most rudimentary language ability—possibly not much more than a series of grunts.

The stone tools found in the same cave confirmed to Boule this primitive condition. Boule wrote:

> The uniformity, simplicity and rudeness of his stone implements, and the probable absence of all traces of any pre-occupation of an aesthetic or of a moral kind, are quite in agreement with the brutish appearance of this energetic and clumsy body, of the heavy-jawed skull, which itself still declares the predominance of functions of a purely vegetative or bestial kind over the functions of the mind.[4]

Boule's report on the Neandertalers was published between 1911 and 1913 in a series of articles in three volumes of the *Annals de Paleontologie.* This twisted view of the Neandertalers dominated the world for forty-four years until William L. Straus (Johns Hopkins Medical College) and A. J. E. Cave (St. Bartholomew's Hospital Medical College, London) published their paper in 1957 on the reexamination of the Neandertals. Attending an anatomy conference in Paris in 1955, Straus and Cave decided to take a look at the La Chapelle-aux-Saints skeleton. They immediately recognized that there were some very serious problems with the reconstruction. Their study revealed that the Neandertals, when healthy, stood erect and walked normally as modern humans do. Exit *Homo neanderthalensis.* Enter *Homo sapiens neanderthalensis.* (As we will see later, however, the Neandertalers are again in danger of loosing their full human status.)

Thanks to Boule, the Neandertalers have had a bad reputation from the start, one they did not deserve. It is now admitted that their differences from modern humans are rather superficial. Their low, wide cranium and heavy browridges caused people to think of them as "savage," even though there is nothing in the anatomy of a person to indicate his morality, behavior, or degree of culture. Since the average cranial capacity of the Neandertalers was thought to be almost 200 cc higher than the average for modern humans, that should have helped their image. But thanks to Boule's prejudice, it did not.

We wish we could say that Boule was rather unsophisticated and that he made sincere mistakes. That would be the kinder explanation. However, he was world renowned for his abilities. It is difficult to escape the conclusion that his errors were deliberate. Yet an evolutionist could argue that: (1) The entire evolutionist establishment should not be held responsible for the misguided mistakes of one man, and (2) Since the errors were discovered and corrected, does that not prove that science really is self-correcting?

The answer is not quite that simple. During the time that the mistakes went undetected, the "savage-caveman" idea was being used worldwide as strong evidence for human evolution. The word *Neandertal* is still virtually synonymous with *brute.* Until recently, it would have been easy to find a children's book in almost any schoolroom where a picture of Neandertal was displayed as one of the major evidences for human evolution. We have the right to ask, Did no evolutionist look at this reconstruction for forty-four years? Whether the answer is yes or no, the implications for evolutionist scholarship are not complimentary.

If evolutionists did look at the reconstruction and did not notice the obvious errors for forty-four years, that is hardly a mark of sound scholarship. If they did not look at it for forty-four years while using it as evidence for evolution, that is hardly an illustration of the self-correcting nature of science. It is not just a matter of mistakes by one man, Boule. It is a case of the entire evolutionist community allowing those mistakes to persist for forty-four years and failing to correct them. When it takes scientists forty-four years to correct very obvious mistakes, it is hardly fair to call that a successful case of self-correction.

However, the situation does not end there. Not only did it take forty-four years for the original mistakes regarding Neandertal to be corrected, it took the Field Museum of Natural History in Chicago, one of the great natural history museums of the world, another twenty years to correct their own Neandertal display. How long it took other museums I do not know.

During that period, I visited the Field Museum perhaps fifteen times. It was not until the mid-1970s that the Field Museum removed their old display of the apish Neandertals and replaced them with the tall, erect Neandertals that are there today. What did they do with the old display? Did they throw it on the trash heap where it belonged? No. They moved the old display to the second floor and placed it right next to the huge Apatosaurus dinosaur skeleton where more people than ever—especially children—would see it. They labeled it "An alternate view of Neandertal." It was not an alternate view. It was a *wrong* view. So much for the self-correcting mechanism in science as far as Neandertal is concerned.

The Piltdown Man Hoax

The second illustration of paleoanthropology notoriously failing to be self-correcting is the famous Piltdown Man hoax. This is probably the best-known "whodunit" in all of science, and the mystery of "whodunit" is still unsolved.

Piltdown, England, is the least likely place in the universe where one would expect something momentous to happen. Forty miles south of downtown London and about twenty miles west of the site where the Battle of Hastings took place in 1066, Piltdown is a tiny hamlet in east Sussex in a once-forested area known as the Weald. Here, at a small gravel pit used for the repair of roads, Charles Dawson "discovered" Piltdown Man.

Dawson was a solicitor—a lawyer who represents the Crown in matters of the public interest. Quite knowledgeable of geology, he had made some significant fossil discoveries, was an honorary collector for the British Museum, and had been invited to be a member of the prestigious Geological Society. For all of this, he was still regarded as an amateur by professional geologists.

In 1908, a workman at the Piltdown gravel pit handed Dawson a portion of a human skull (parietal) that the workman said looked to him like a coconut. In late 1911, Dawson found several more pieces of the skull in the gravel refuse heaps beside the pit.

In early 1912, Dawson took his finds to Arthur Smith Woodward at the British Museum. His finds, Dawson believed, would rival Heidelberg Man (Mauer Mandible), which had been discovered in Germany in 1907. Smith Woodward was interested. A number of human fossils had been discovered on the Continent, and the British were embarrassed by their lack of respectable fossil ancestors. The time was ripe for Piltdown.

Studying at nearby Hastings was the thirty-one-year-old French Roman Catholic paleontologist-priest, Pierre Teilhard de Chardin, destined to become famous for his relationship with both Piltdown Man and Peking Man. Later he would become known as the formulator of a pantheistic evolutionary theology that was first condemned by the Roman Catholic Church but is now embraced by many Roman Catholics. Teilhard's "theology" has come into its own under the banner of the New Age Movement.

In early June of 1912, Dawson, Woodward, and Teilhard began a series of digs at the Piltdown pit. More pieces of the skull were found, and Teilhard discovered a fragment of an elephant molar. Then the lower jaw of Piltdown Man was discovered, together with more fossils of mammals. In December, Dawson and Woodward officially announced to the Geological Society of London the discovery of the earliest Englishman, *Eoanthropus dawsoni*, affectionately known as Piltdown Man.

The skull of Piltdown Man was quite large and quite "modern" in shape (morphology). However, the lower jaw was very "primitive" or apelike. Some doubted the association of the two fossils. Paleoanthropologists said that what was needed to resolve the strange association was the canine tooth. The following August the canine tooth was "discovered" at the feet of Teilhard as he sat on a gravel refuse pile beside the pit.

The Piltdown pit also produced fossil bones of elephant, mastodon, rhinoceros, hippopotamus, beaver, and deer. Primitive tools and crudely flaked flint stones (eoliths) were found. The last fossil ever to come out of the pit was a portion of a fossil elephant femur that had been worked to look for all the world like a bat used in the game of cricket. Some suspect this was the hoaxer's attempt to reveal that the whole thing was a prank. But the English bent on discovering their "roots" missed the clue, and they considered the bone some sort of unknown primitive tool.

Piltdown did have its skeptics. So the hoaxer supplied the one thing that would convert most of the skeptics: a second discovery at a different location that tended to confirm the original discovery. Enter Piltdown II in early 1915 in another gravel pit a few miles from Piltdown I. No one knows for sure the location of that pit where Dawson discovered Piltdown II. Dawson died in 1916, the year before the discovery was revealed by Woodward. Amazingly, at Piltdown II the exact pieces came to light that were needed to identify it with Piltdown I and thus confirm Piltdown I.

Evolutionists now like to boast that not everyone accepted Piltdown. Technically, they are correct. There were a few, such as Weidenreich and Hrdlicka, who did not accept Piltdown. But the vast majority of paleoanthropologists worldwide did accept Piltdown as legitimate, especially after the confirming discoveries at Piltdown II.

Evolutionists complain that creationists make too much of Piltdown. Frankly, it is not necessary for creationists to do so. It is not the frauds that expose the weaknesses of human evolution. It is the legitimate fossils that clearly falsify human evolution, as the later chapters of this book demonstrate. Furthermore, one does not have to be an evolutionist to recognize that Piltdown really was a dirty trick. Fraud is found in every area of human activity, including religion. We do not hold the entire evolutionist community responsible for the Piltdown fraud, which was committed by only one or a very few persons.

What we do hold the evolutionist community responsible for is its continual claim that science is self-correcting when in fact it is not. Piltdown demonstrates that it is not. The Piltdown fossils were discovered between 1908 and 1915. It was not until 1953, thirty-eight to forty-five years later, that Kenneth Oakley, Joseph Weiner, and Wilfred Le Gros Clark discovered that Piltdown Man was a fraud. The British Museum then issued a statement to that effect. That is hardly a case of efficient self-correcting.

There is the possibility that the skull itself was legitimately found in the pit. Radiocarbon dating determined it to be from 520 to 720 years old. Piltdown Common had been used as a mass grave during the great plague of AD 1348–49. The skull bones were quite thick, a characteristic of more ancient fossils, and the skull had been treated with potassium bichromate by Dawson to harden and preserve it. This was a common practice for the treatment of fossils at that time.

The other bones and stone tools had undoubtedly been planted in the pit and had been treated to match the dark brown color of the skull. The lower jaw was that of a juvenile female orangutan. The place where the jaw would articulate with the skull had been broken off to hide the fact that it did not fit the skull. The teeth of the mandible were filed down to match the teeth of the upper jaw, and the canine tooth had been filed down to make it look heavily worn.

It was only in 1982 that both the mandible and the canine tooth were determined conclusively, by collagen reactions, to be those of an orangutan.[5] Whoever put the canine tooth in the pit likely knew that the mandible was also that of an orangutan. Orangutans are found today only in Borneo and Sumatra.

Some of the mammalian bones probably came from other areas of England, but the mastodon molar is thought to have come from Tunisia and the hippopotamus molar from the island of Malta. Some of the flints in the pit may also have come from Tunisia. It is obvious that whoever perpetrated the hoax either had traveled extensively or had access to some exotic fossil and archaeological collections.

Some researchers feel that the hoax was not sophisticated, and that Dawson was the culprit. Others feel that it was a very professional hoax calling for someone having far more expertise than Dawson. Still others say that the fraud was brilliantly conceived but poorly executed. Sophisticated or not, it cannot be denied that the hoax worked very well.

Many feel that Dawson was in some way involved. The British Museum has documented other "discoveries" by Dawson as being fakes. Because of the feeling that Dawson lacked the expertise to pull it off by himself, some investigators think in terms of Dawson and a "Significant Other"—either Dawson and "Other" in collaboration, or Dawson used by "Other" as a vehicle for "Other's" deception.

Nonetheless, there are those who remove Dawson from the list of suspects altogether. Ronald Miller in his book *The Piltdown Men* said that Sir Grafton Elliot Smith of the British Museum was the hoaxer.[6] Charles Blinderman in *The Piltdown Inquest* fingers Lewis Abbott, a jeweler and amateur geologist.[7] Stephen Jay Gould (Harvard University) believes (as I do) that the Roman Catholic priest Pierre Teilhard de Chardin was definitely involved.[8] The most surprising suspect, suggested by John Winslow, is Sir Arthur Conan Doyle, creator of Sherlock Holmes.[9] Doyle, who had been trained as a medical doctor, lived just a few miles from the Piltdown pit.

The amazing thing about the Piltdown hoax is that at least twelve different people have been accused of perpetrating the fraud. All of these suspects had the expertise to do it, all had access to the materials involved, all had opportunity to do it, and all had motives for doing it. It is frustrating to know that at this late date we shall never know with certainty who committed what has been called the most successful scientific hoax of all time.

However, it is not necessary to know who perpetrated Piltdown to know that if science were really self-correcting, Piltdown should have been uncovered long before it was. Like Boule's reconstruction of the Neandertal skeleton from La Chapelle-aux-Saints, there were elements about it that were quite obvious. The file marks on the orangutan teeth of the lower jaw were clearly visible. The molars were misaligned and filed at two different angles. The canine tooth had been filed down so far that the pulp cavity had been exposed and then plugged. If science were really self-correcting, the Piltdown fraud should have been discovered soon after it was committed, rather than thirty-eight to forty-five years later.

Why was the Piltdown hoax so successful? A big-brained ancestor was what evolutionists were expecting to find. Sir Grafton Elliott Smith had predicted that a fossil very similar to Piltdown would be found. This is why he is one of the suspects. If the australopithecines had not come into favor as the preferred evolutionary ancestors of humans, and Piltdown had not become an embarrassment because it no longer fit the scenario, the fraud might still be undiscovered, and Piltdown might still be considered a legitimate fossil.

There is a touch of irony in the Piltdown story. When the Piltdown fossils were brought to the British Museum, plaster casts were made of them for display to the public. However, museums seldom indicate that the fossils on display are replicas, and most people thought that they were seeing the real thing. Yet while people thought that they were seeing actual fossils of their evolutionary ancestors, they were looking at fakes. But more than just looking at fakes, they were looking at *fakes of fakes.*

The literature produced on Piltdown was enormous. It is said that more than five hundred doctoral dissertations were written on Piltdown. The man most deceived was Sir Arthur Keith, one of the greatest anatomists of the twentieth century. Keith wrote more on Piltdown than anyone else. His famous work, *The Antiquity of Man*, centered on Piltdown. Keith put his faith in Piltdown. He was eighty-six years old when Oakley and Weiner called on him at his home to break the news that the fossil he had trusted in for forty years was a fraud.

Keith was a rationalist and a pronounced opponent of the Christian faith. Yet in his *Autobiography* he tells of attending evangelistic meetings in Edinburgh and Aberdeen, seeing students make a public profession of faith in Jesus Christ, and often feeling "on the verge of conversion."[10] He rejected the gospel because he felt that the Genesis account of creation was just a myth and that the Bible was merely a human book. It causes profound sadness to know that this great man rejected Jesus Christ, whose resurrection validated everything he said and did, only to put his faith in what proved to be a phony fossil.[11]

The widespread myth is that science is self-correcting, and because of this it is a superior worldview. In reality, science is not adequately self-correcting and for very practical reasons cannot be self-correcting in any meaningful way.

CHAPTER 5

Looks Aren't Everything

WE HAVE ALL HEARD of that famous fossil "Lucy." This three-foot-tall australopithecine individual was found by Don Johanson in Ethiopia in 1974. Forty percent of her skeleton was recovered. Since she was believed to be more than three million years old, her completeness was most unusual. At three million years, a paleoanthropologist expects only a few bits and pieces.

Why was she called Lucy? Although she was considered to be a female because of her diminutive size, the main reason was that at the time she was found, the loudspeaker at the base camp was blasting out the Beatles' song "Lucy in the Sky with Diamonds."

From their evaluation of Lucy and fossils like her, Don Johanson and Tim White decided in 1979 that Lucy was our oldest-known direct ancestor. You and I are said to have evolved, over a period of three million years, from something that was very unlike us in size, appearance, and mental ability.

While there is currently much controversy regarding our family tree, many evolutionists still agree with Johanson and White that you and I, modern humans, are genetically related to Lucy. We have Lucy's genes in us. About five million mutational events have occurred in our genes in that three-million-year period to account for the differences, but Lucy's modified genes they are.

In a public forum, anthropologist Vincent Sarich was asked why Lucy isn't here today. He replied, "Why isn't Lucy here? That's simple—because we are. She evolved into us. That's not any problem at all."[1]

Evolution always deals with populations: the collective gene pool. So when we ask why Lucy isn't here today, we are asking why there are no small, erect, chimplike animals living today that are like Lucy. And the evolutionist's answer is, "Lucy isn't here because we are."

Since contemporary evolutionists are committed to the idea that a population of chimplike animals known collectively as Lucy (*Australopithecus afarensis*) evolved into us, there cannot be any Lucys around. Whether they hold to the slow and gradual view of evolution (phyletic gradualism) or to the newer model of long periods of stability and short bursts of rapid evolutionary change (punctuated equilibria), the result is the same. Why isn't Lucy here? Because we are. Why isn't *Homo erectus* here today? The answer is the same: Because we are. From Lucy, human evolution involved that long trail of three million years and five million mutational events, but here we are.

W. W. Howells (Harvard University) explains how the process takes place in each of the two evolutionary models:

> In "phyletic gradualism," change is viewed as gradual and general over the species. . . . In "punctuated equilibria," the apparent discontinuity, seen so often in a paleontological succession, is not simply the artifact of a gap in the record but is real. The process of change is not species-wide but results from allopatric speciation [speciation in some other place]. A subspecies, ideally a peripheral isolate of the old species, becomes the new form in some significant respect and replaces populations of the old by migration. Thus the main body of the species does not undergo the gradual change to a new species.[2]

In the gradual model, the entire Lucy population would change into some form of *Homo habilis*, and that population would change gradually into *Homo erectus*. The *Homo erectus* population would change gradually into early *Homo sapiens* (or into Neandertals), and they would change eventually into us. (This scenario is a bit more clouded today than it was a few years ago.) Should there be some isolated groups along the way who did not inherit those superior mutated genes, eventually they would die out, because they would be less fit to survive in a very competitive environment.

In the newer punctuated-equilibria model, a small portion of the Lucy population, probably on the edge of the species range, obtains some favorable mutations. This small "advanced" Lucy population, to use Howells's words, "replaces populations of the old [species] by migration." This replacement of the older population is accomplished by their death, since they are the less fit and the advanced Lucy population represents the more fit. This process takes place again through some form of *Homo habilis, Homo erectus,* early *Homo sapiens,* and on up to modern humans. A very famous statement by geneticist Richard Goldschmidt about a reptile laying an egg and a bird hatching out of it has given a false impression of punctuated equilibria. Punctuated equilibria is concerned basically with evolution on the species level, not with the higher categories.

These two models have their differences. However, in both cases the time element—about three million years—is the same. The total number of mutational events needed to bring about these changes—approximately five million—is the same. And in each case those groups who did not take part in that evolutionary process must be eliminated through death. In the evolutionary process, the less fit die as the more fit survive. The more fit survive because they are better able to compete for the limited food supply, or they reproduce in greater numbers, or both. In other words, in both models, for species A to evolve into species B, species A must precede species B in time. Furthermore, after species A has evolved into species B, any species A remnants must soon die. It is thus basic to evolution that if species B evolved from species A, that species A and species B cannot coexist for an extended length of time.

The "survival of the fittest" has a flip side. It is the death of the less fit. For evolution to proceed, it is as essential that the less fit die as it is that the more fit survive. If the unfit survived indefinitely, they would continue to "infect" the fit with their less fit genes. The result is that the more fit genes would be diluted and compromised by the less fit genes, and evolution could not take place.

The concept of evolution demands death. Death is thus as *natural* to evolution as it is *foreign* to biblical creation. The Bible teaches that death is a "foreigner," a condition superimposed upon humans and nature after creation. Death is an enemy, Christ has conquered it, and he will eventually destroy it. These respective attitudes toward death reveal how many light-years separate the concept of evolution from biblical creationism.

It is possible to determine whether the concept of human evolution is a scientific theory or a philosophy. If it is a scientific theory, it must be capable of being falsified. Since human evolution is alleged to be a historic process, the evidence for it or the falsification of it must come from the fossil record. For instance, if *Homo erectus* people persisted long after they should have died out or changed into *Homo sapiens,* the concept of human evolution would be falsified. If one could show that fossils indistinguishable from modern humans existed long before they were supposed to exist (according to the process of evolution) this also would falsify the concept.

It is the burden of this book that both of these falsifications can be demonstrated. If human evolution is truly a scientific theory, the fossil record shows that it has been falsified. The fact that the evidence is ignored or disguised indicates that the concept of human evolution is a philosophy that is perpetuated in spite of and independent of the facts of the human fossil record.

If we humans evolved from a small chimplike animal like Lucy, it is obvious that we had to pass through a number of stages on this long evolutionary journey. We are classified as *Homo sapiens*. Not only are we said to have come from some form that was not our species, we are said to have come from some form that was not even our genus. This form, Lucy, is called *Australopithecus afarensis* (southern ape from Afar, Ethiopia). Because of very recent fossil discoveries, however, some evolutionists are going back to the older view that our immediate nonhuman ancestor was *Australopithecus africanus*. Evolutionists then propose a sequence going to some form of *Homo habilis* (handy man), then to some form of *Homo erectus* (erect man), then to early *Homo sapiens* (early wise man), and then to *Homo sapiens* (wise man).

In theory, the progression appears rather tidy. In actuality, it is very untidy. First of all, it may surprise the reader to learn that there is no clear-cut, universally accepted scientific definition for any of these categories, including *Homo sapiens*. While there is some consensus on these categories, there is enough uncertainty to make for some healthy confusion. Anyone who works in this area must be prepared for it.

There is also the question of how long it takes (if one assumes that evolution is true) for a new species to evolve. George Gaylord Simpson estimated that it took a quarter of a million years for a species to evolve within a genus. However, the present sequencing of the hominid fossils would imply that it could take longer. Lucy is dated at about 3 million years. *Homo habilis* is dated from 2 to 1.5 million years. *Homo erectus* is dated roughly at 1.6 to 0.4 million years, with *Homo sapiens* in the present. That sequencing implies that it could take up to 1 million years for a new species to fully develop in human evolution (assuming that there were no species intermediate between those creatures).

Another way evolutionists have attempted to answer the question of how long it takes for a new species to evolve is to estimate the rate of gene flow in a population. Recent estimates of the time one advantageous gene would take to disperse throughout hominid populations in the Pleistocene Epoch are 20,000 generations or 400,000 years.[3] Since the evolution of one species from another would require many favorable genetic mutations (the existence of a "favorable" mutation has yet to be conclusively demonstrated), it is obvious that evolution requires vast periods of time even on the species level and even if several "advantageous" genes were being dispersed throughout the population at the same time. Time thus becomes the key ingredient in the evolutionary process.

We have the right to expect, if evolution was true, that the hominid fossil record would faithfully follow the time and morphology sequence set forth by evolutionists. Since humans are supposed to have evolved from something very different from what they are today, we have a right to expect that very modern-looking fossils would not show up in Lucy times and that primitive or archaic fossils would not embarrass the evolutionist by showing up in modern times. We also have the right to expect that if a significant number of fossils are so rude as to show up at the wrong time, the evolutionist would be honest enough to admit that his theory has been falsified. In actuality, many fossils have been that rude. And evolutionists have been less than intellectually honest.

Evolutionists work their own special magic on nonconformist fossils. With the waving of a magic wand, *Homo erectus* fossils can become *sapiens* or Neandertals and *Homo sapiens* fossils can become australopithecines. To us, this is a serious matter of intellectual integrity. The evolutionist does not see it that way. To him, evolution is true. Hence, fossils must be interpreted accordingly. I will give a few examples of the type of magic-wand waving to be found in the scientific literature.

The Dating of the Taung Fossil

In 1924, Professor Raymond Dart, anatomist at the University of the Witwatersrand in Johannesburg, South Africa, acquired a skull that had come from a lime works at Taung. Dart immediately recognized that the skull was something unique. After cleaning and studying it, he announced to the world that he had discovered our evolutionary ancestor. It was the skull and endocranial cast of an extinct primate child that Dart named *Australopithecus africanus.* The skull was so well-known in the 1920s that it became the object of a joke. If a man had a blind date, he was asked, "Is she from Taung?"

Although Dart's assessment was initially met with hostility and rejection, the eventual exposure of Piltdown Man as a hoax and the discovery of adult australopithecine fossils at other locations in South Africa turned that rejection into almost universal acceptance. By 1960, it would have been difficult to find any public-school book touching on human origins that did not have in it a picture of the Taung skull. That popularity has remained. The fossil received much publicity in 1984, the sixtieth anniversary of its discovery. Pictures of Taung are still found in most books dealing with human origins.

Until Lucy was discovered in late 1974, Taung, the type specimen of *Australopithecus africanus,* was considered our oldest direct evolutionary ancestor. Although dating the South African fossils has always been a nasty problem, Taung was generally considered to be between two and three million years old. That age seemed appropriate for it as our evolutionary ancestor.

In 1973, South African geologist T. C. Partridge dropped a bomb. His investigations revealed that the cave from which the Taung skull had come could not have formed prior to 0.87 Mya[4] ("Mya" means "million years ago"; "ya" means "years ago"). That meant that the Taung skull could be at most only three-quarters of a million years old.

Since it could take up to a million years for the hominids to evolve from one species to another, to go all the way from australopithecines to modern humans in only three-quarters of a million years was out of the question. Further, true humans were already on the scene in Africa at 0.75 Mya. Karl W. Butzer (University of Chicago) clearly saw the problem when he wrote:

> If the Taung specimen is indeed no older than the youngest robust australopithecines of the Transvaal, then such a late, local survival of the gracile [a term used to describe the *A. africanus* fossils] lineage would seem to pose new evolutionary . . . problems.[5]

Anatomist Phillip V. Tobias (University of the Witwatersrand, Johannesburg) pointed to the real problem: "The fact remains that less than one million years is a discrepant age for a supposed gracile australopithecine in the gradually emerging picture of African hominid evolution."[6]

Here was a problem that, if allowed to persist, could jeopardize the concept of human evolution. It was time for the evolutionist to wave his magic wand. A. J. B. Humphreys wrote:

> One point that needs investigation at the outset, however, is the question of the identification of the Taung skull as *Australopithecus africanus*. The possibility that the Taung skull might represent *Homo habilis* or a more advanced creature than *A. africanus* . . . certainly deserves some consideration in view of this younger date.[7]

Phillip Tobias suggested another way the wand might be waved: "Because its brain and its dental characters would exclude it from *H. erectus*, it must seriously be considered whether the Taung child is not a late surviving member of *A. robustus* or *A. cf. robustus*."[8] Tobias then made this amazing confession: "Although nearly 50 yr have elapsed since its discovery, it is true to say that the Taung skull has never yet been fully analyzed and described."[9]

A basic obligation upon the one who discovers a new fossil is to intensively study it and publish descriptions of the study for the benefit of the scientific community. To fail to do this is akin to dereliction of duty. Dart did not do it. Tobias, Dart's successor at Witwatersrand and his successor as custodian of the Taung skull, did not do it. Yet for fifty years the public was told that Taung was our evolutionary ancestor. Considering that it has been with us since 1924, we can question if any fossil has been pictured more in the popular press than Taung. Now we were told that the appropriate analysis and description of the fossil had never been done. In spite of that fact, without further study (and only to relieve the embarrassment of the revised date) the suggestion was made that perhaps Taung was closer to humans (*H. habilis*) or not in the human lineage at all (*P. robustus*). Looks, obviously, aren't everything.

In 1974, Johanson discovered Lucy. Other material followed. Paleoanthropologists began to focus on the new material from Ethiopia, and the problem of the awkward dating of the Taung skull faded into the background. When Johanson and Tim White revised the hominid family tree in 1979, *A. afarensis* (including Lucy) replaced *A. africanus* (including Taung) as our direct nonhuman evolutionary ancestor. The dating of Taung thus became a nonissue. The *A. africanus* forms were moved to the australopithecine branch of the family tree, becoming the link between Lucy and the robust australopithecines.

This comfortable arrangement was severely jolted by the discovery in 1985 of the famous "black skull," KNM-WT 17000.[10] Australopithecine phylogeny is now in disarray. Dated at 2.5 Mya, the "black skull" seems to have more in common with Lucy and the robust australopithecines. *A. africanus* became the odd man out. Many evolutionists are now moving *A. africanus* (including Taung) back into the human line, between Lucy and *Homo habilis*.

What happened to Partridge's work showing that Taung could be only about 0.75 million years old and hence could not possibly be our evolutionary ancestor? His work has not even been addressed, let alone answered. Instead, it is being disparaged. A recent work discussing Taung states, "An ill-founded attempt at geomorphological dating in the early 1970s suggested an age of less than 870 k.y. [870,000 years] for the hominid."[11]

That "ill-founded" date was supported by a date of about 1 Mya by thermoluminescence analysis of calcite and uranium-series dates of 942,000 ya and 764,000 ya on limestone.[12,13] But when the evolutionist waves his magic wand, both date and taxon become plastic so that evolution might be served. Richard G. Klein (Stanford University) doesn't really face the issue, but he is at least honest when he writes, "A date for Taung of 2 Mya [million years ago] or more may seem most reasonable, but the argument is obviously circular and the true age remains uncertain."[14]

Because Taung is again a candidate for human ancestry, Tobias has withdrawn his suggestion that Taung might be a robust australopithecine. Most workers now consider it to be *Australopithecus africanus*. However, the full analysis and description of the Taung skull still has not been published, and the dating problem raised by Partridge continues to be ignored.

THE KANAPOI ELBOW FOSSIL

One of the most flagrant cases of wand waving involves a fossil found at Kanapoi, southwest of Lake Rudolf (Turkana) in northern Kenya. This fossil, known as KNM-KP 271, is the lower end of a left upper arm bone (distal end of the humerus). It was found in 1965 by Bryan Patterson (Harvard University) and is in an excellent state of preservation. The most recent dating of the fossil gives it an age of 4.5 Mya.[15] At the time it was virtually the oldest hominid fossil ever found—older than Lucy and all of the australopithecines. The question is, What is it?

To answer the question of the identity of KNM-KP 271, Patterson and W. W. Howells used the method of computer discriminate analysis. They compared KNM-KP 271 with the distal ends of the humeri of a modern human, a chimpanzee, and the only other similar fossil they had at the time: *Australopithecus (Paranthropus) robustus*, from Kromdraai, South Africa. Seven different measurements from each of the four samples were fed into the computer. Patterson and Howells published the results of their study in *Science*, 7 April 1967. "In these diagnostic measurements, Kanapoi Hominoid 1 [the original name given to the fossil] is strikingly close to the means of the human sample."[16]

After stating that their computer analysis revealed the Kanapoi humerus to be *strikingly* close to modern humans (they were not comparing it to ancient humans but to modern humans) they made the rather shocking conclusion that KNM-KP 271 "may prove to be *Australopithecus*."[17] (They meant the gracile *A. africanus* form.) Their reasoning was that the upper end of the humerus of *A. africanus* is quite similar, based on visual assessment, to that of modern humans. Hence, they assumed that the lower end was similar also, even though they did not have the lower humerus portion of *A. africanus* for comparison with KNM-KP 271. The real reason for this strange departure from their data comes out later.

Further computer analysis of many more measurements revealed even more dramatically the similarity of KNM-KP 271 to modern humans. Henry M. McHenry (University of California, Davis) wrote, "The results show that the Kanapoi specimen, which is 4 to 4.5 million years old, is indistinguishable from modern *Homo sapiens*."[18]

Regarding KNM-KP 271, David Pilbeam (Harvard University) states:

> Multivariate statistical analysis of the humeral fragment aligns it unequivocally with man rather than with the chimpanzee, the hominoid most similar to man in this anatomical region. Professors Bryan Patterson and F. Clark Howell [sic], the describers of this fragment, believe that it represents *A. africanus* rather than *A. robustus*.[19]

McHenry pointed out that multivariate analysis is an excellent diagnostic tool for this portion of primate anatomy. "The hominoid distal humerus is ideal for multivariate analysis because there are such subtle shape differences between species, particularly between *Homo* and *Pan*, which are difficult to distinguish in a trait by trait (univariate) analysis."[20]

In her studies on the australopithecines and the living primates, Brigette Senut (Museum of Man, Paris) found reason to challenge some of the earlier diagnostic work on the humeri in primates:

> In the field of comparative anatomy of primate postcranial material, it is striking that the forelimb (especially the humerus) has been much less studied than the hindlimb, and only by relatively few authors. Thus, it appears that these hominids, usually considered as similar to modern man in humeral morphology (Broom et al., 1950) exhibited quite unique features.[21]

Let me review what we have said thus far. We are dealing with the oldest respectable hominid fossil ever found up to that time.[22] The fossil represents a part of the anatomy where it is relatively easy to discriminate between humans and the other primates, both living and fossil. The appropriate diagnostic tools have been used to evaluate the fossil. The results show unequivocally (to use Pilbeam's term) that the fossil is indistinguishable from modern humans, not just fossil humans. Yet in their original report Patterson and Howells go against their own empirical evidence and suggest that the fossil represents *Australopithecus africanus*. Why?

Further study strengthened the fact that KNM-KP 271 should be ascribed to modern humans. Yet every textbook or journal article from 1967 until about 1994 that mentions the Kanapoi fossil calls it *Australopithecus africanus*. Why? We might assume that because so much fossil material has been discovered in East Africa since 1965, the appropriate *A. africanus* fossil has been discovered that confirmed the original evaluation of Patterson and Howells. As far as I have been able to determine, that is not the case.

Why, then, the universal insistence until recently that this fossil is *A. africanus* and not *Homo sapiens*? Howells, writing in 1981, fourteen years after the fossil was first ascribed to *A. africanus*, gives us the reason:

> The humeral fragment from Kanapoi, with a date of about 4.4 million, could not be distinguished from *Homo sapiens* morphologically or by multivariate analysis by Patterson and myself in 1967 (or by much more searching analysis by others since then). We suggested that it might represent *Australopithecus* because at that time allocation to *Homo* seemed preposterous, although it would be the correct one without the time element.[23]

It is obvious that looks aren't everything. Even though KNM-KP 271 is shaped exactly like *Homo sapiens*, the time element is wrong. What determines that? The concept of human evolution. The concept of human evolution decrees that it is impossible for true humans to have lived before the australopithecines—even though the fossil evidence would suggest otherwise—because humans are supposed to have evolved from the australopithecines.

The absurdity continues. In 1994, Dr. Meave Leakey and her associates announced the discovery of *Australopithecus anamensis*, which they claim is our oldest direct ancestor, older than Lucy. Among the fossils in the *Australopithecus anamensis* assemblage they have included KNM-KP 271, redated at more than 3.5 Mya. Admitting that this fossil is *Homo*-like, they have linked it with another *Homo*-like fossil, a tibia, as evidence for the first bipedal hominid. (See the chapter "The Pretend Humans.")

According to the basic principles of the philosophy of science, a theory must be falsifiable if it is a legitimate scientific theory. How could the theory of evolution be falsified? Supposedly it would be falsified if fossils are found that are woefully out of order from what evolution would predict. Many such fossils have been found. KNM-KP 271 is just one of them. However, evolutionists ignore the morphology of fossils that do not fall into the proper evolutionary time period. They wave their magic wand to change the taxon of these fossils. Thus it is impossible to falsify the concept of human evolution. It is like trying to nail jelly to the wall. That evolutionists resort to this manipulation of the evidence is a "confession" on their part that the fossil evidence does not conform to evolutionary theory. It also reveals that the concept of human evolution is a philosophy, not a science.

To the evolutionist there is but one primary fact in the universe: evolution. Everything else is just data. The value of this data does not depend upon its intrinsic quality but upon whether or not it supports evolution and its time scale. Good data is that which supports evolution. Bad data is that which does not fit evolution, and it is to be discarded. It is time to ask the paleontological community, "At what point does philosophical bias in the interpretation of the human fossil material become intellectual dishonesty?" The interpretation of KNM-KP 271 from Kanapoi justifies that question.

Was the original owner of that Kanapoi elbow bone a true *Homo sapiens*? I do not know. I was not there. Neither was Bryan Patterson or William Howells. There is no way at this point that anyone can prove it. That is beyond the scope of science. What we can say with confidence is that all of the scientific evidence indicates it was a modern human and there is no scientific evidence to the contrary. Science can go only that far. What's wrong with telling a student that? The problem is that there are metaphysical (read *theological*) implications to all of this. Evolution implies a naturalistic, mechanistic origin of things.

That one of the oldest human fossils ever found—skimpy as it is—reveals that man was virtually the same 3.5 million years ago (on the evolutionist time scale) as he is today suggests that humans appeared on the scene suddenly and without evolutionary ancestors.

We are continually told how bad it is to mix religious implications with science. But we don't mix them. Those implications are already there. Evolutionists act as if creationists have been trafficking in something very dark and sinister. How many young minds have we led astray by telling them that there is a creator?

Just for the fun of it, let's grant the evolutionist his point. Let's assume that there is something very bad about telling a student that one of the oldest human fossils ever discovered supports special creation. As bad as that might be, I can think of something even worse. That would be to tell a student that one of the oldest hominid fossils ever discovered, KNM-KP 271, supports evolution—because that would be a lie.

SECTION II

HUMAN EVOLUTION
A House of Cards

Don't trust *anyone*—not even your best friends.

—MARY LEAKEY TO NEWLY ARRIVED PALEOANTHROPOLOGIST JON KALB

It's always a bad combination when you get hominid fever on top of testosterone poisoning.

—FEMALE PALEONTOLOGIST WORKING IN KENYA

INTRODUCTION TO SECTION II

Revisionist History

WE HAVE ALL HEARD about various attempts to rewrite history—to make fact fiction and fiction fact. Communists have fine-tuned it into an art form. Others have also done it. There are some Germans (and others) who claim that the Holocaust never happened. Japanese children, I was told while lecturing in Japan some years ago, are taught that World War II didn't start because Japan attacked Pearl Harbor. Instead, the United States attacked Japan. And we all know that American history has been revised to remove all vestiges of our biblical and Christian heritage.

However, the greatest and most far-reaching attempt at revisionist history has never been called by that term. Nonetheless, it is revisionist history just the same. And it has been quite successful. This is the attempt by evolutionists to rewrite human history.

Having no human fossils to support his theory, Darwin used other details of nature to weave a philosophical scenario of our origins. It was revisionist history intended to replace Genesis 1 and 2. An unknown person (or persons) used a human skull and the lower jaw of an orangutan to fabricate a fake human ancestor, Piltdown Man. It was a brilliant case of revisionist history that was almost universally accepted for forty years. A prejudiced Marcellin Boule advanced the cause of Piltdown Man as our ancestor while articulating the first skeleton of a Neandertal to make him look as apelike as possible. He was wrong on both counts. But it took about forty years for us to realize that he also had indulged in revisionist history. Eugene Dubois hid Wadjak Man while promoting Java Man as our evolutionary ancestor. He, too, was a successful revisionist historian. And so it goes on. The human fossils are still being forced into a philosophical framework in which they do not fit. In this area, revisionist history never ceases.

Join me in a flight of fantasy. Think about a world in which evolution has never been injected into the human thought-stream. What would it be like if humans were responsible to the God who created them? Would it be different from our present world of violence and moral bankruptcy in which each person, thinking he has an animal origin, is a law unto himself? Politicians seeking to unseat an incumbent like to ask, "Are we better off today than we were a few years ago?" Certainly we are better off technology-wise than we were before Darwin. But technology has nothing to do with evolution. The proper question to ask is, "Are we better off as a society morally and spiritually than we were without Darwin?" You answer the question.

The concept of human evolution involves every moral issue. Anyone who thinks that "evolution is just a scientific theory" is living in a make-believe world or on another planet. Every sin condemned by the Bible is condoned by evolutionists as being an appropriate evolutionary adaptive behavior for some group. A recent article in a science journal states, "Nearly every type of same-sex activity found among humans has its counterpart in the animal kingdom. Homosexual behaviour is just as 'natural' as heterosexual behaviour" (*NewScientist*, 7 August 1999, p. 33).

In the Bible, those who gave a false spiritual or moral message contrary to what God has clearly said were called false prophets. The Old Testament penalty for being a false prophet was to be stoned to death. Being stoned to death is a rough way to go. If that penalty seems too severe or harsh, let me give an illustration. Suppose a medical doctor gave you a diagnosis that he knew was false according to the best medical procedures. The result was that you failed to get the proper medical treatment and you died. Would we not all agree that that would be a crime of the highest order?

However, those who knowingly give a false moral and spiritual diagnosis, contrary to the Bible, are dealing not just with the body but with the soul—a soul that could be separated from God forever. By his resurrection from the dead, Jesus Christ proved that he has power, information, and knowledge far beyond anything you and I have. He said, "Do not be afraid of those who kill the body but cannot kill the soul. Rather, be afraid of the One who can destroy both soul and body in hell" (Matt. 10:28). God's attitude toward false prophets is a bit different from our own.

The concept of human evolution is not just revisionist history. It is far more serious than that.

This section tells how this revisionist history, this house of cards, came to be a part of the human thought-stream.

CHAPTER 6

With a Name like Neandertal, He's Got to Be Good

Praise to the Lord, the Almighty,
The King of Creation!
O my soul, praise Him,
For He is thy health and salvation!
All ye who hear,
Now to His temple draw near;
Join me in glad adoration!

IF YOU ATTEND CHURCH regularly, the probability is high that you have sung this beautiful hymn within the past year. It is found in most hymnals. The fact that there is a relationship between this hymn and the first Neandertal fossil proves that truth is indeed stranger than fiction.

In the late 1600s, an evangelical (Lutheran) theologian and school rector, gifted in poetry and hymn writing, took long walks in the country near Hochdal, Germany. As he strolled, he composed hymns and sang them in praise to God. One of his favorite spots was a beautiful gorge through which the Dussel River flowed, about ten miles east of Dusseldorf. He strolled in this one valley so often that it became identified with him and was eventually named after him. His name was Joachem Neander, and the valley became known as the Neanderthal—the Neander Valley (*tal*, or *thal* in Old German, means "valley," with the *h* being silent).

Almost two hundred years later, this valley was owned by Herr von Beckersdorf. As the owner quarried limestone in the valley for the manufacture of cement, his workmen came across some caves in the side wall of the gorge. One cave, known as the Feldhofer Grotto, had human bones in the soil of its floor. Because the prime interest of the workmen was to quarry the limestone, what probably had been a complete skeleton was largely destroyed. Only the skullcap, some ribs, part of the pelvis, and some limb bones were saved. The year was 1856. The first Neandertal had been discovered.[1]

Although the bones were obviously human, they looked different. Beckersdorf took the bones to a science teacher, J. K. von Fuhlrott, who was also president of the local natural history society. Recognizing the antiquity and the extreme ruggedness of the bones, Fuhlrott felt that they were probably the remains of some poor soul who had been a victim of Noah's flood.

Von Fuhlrott, in turn, invited Hermann Schaafhausen, professor of anatomy at the University of Bonn, to study the bones. He agreed that the bones represented an ancient race of humans, perhaps barbarians who lived in Europe before the arrival of the Celtic and Germanic tribes.

Eventually the bones came to the attention of Rudolf Virchow, a professor at the University of Berlin. A brilliant man and a true scientist, Virchow is recognized as the father of pathology. Virchow questioned the antiquity of the bones. He felt that they belonged to a modern *Homo sapiens* who had suffered from rickets in childhood and arthritis in old age and who had received several severe blows to the head. As we shall see later, Virchow's diagnosis may be as valid today as when he first made it.

However, William King, professor of anatomy at Queen's College, Galway, Ireland, read an evolutionary history into the bones. It was he who eventually gave them their first scientific name, *Homo neanderthalensis*. The name is significant. King believed that the bones represented a person so primitive that he did not belong to the same species as modern humans. Hence, Neandertal was placed in the genus *Homo* but in a separate species.

Controversy surrounded the fossils. In 1886, two skeletons were found in a cave near Spy (pronounced "shpee"), Belgium. The Spy fossils made it obvious that the original Neandertal was not a freak but did represent some form of ancient human. Found with the human skeletons at Spy were the remains of cave bear, mammoth, and woolly rhinoceros.

In 1908, the first reasonably complete skeleton of a Neandertal type was found. The very faulty reconstruction of it by Marcellin Boule has been described

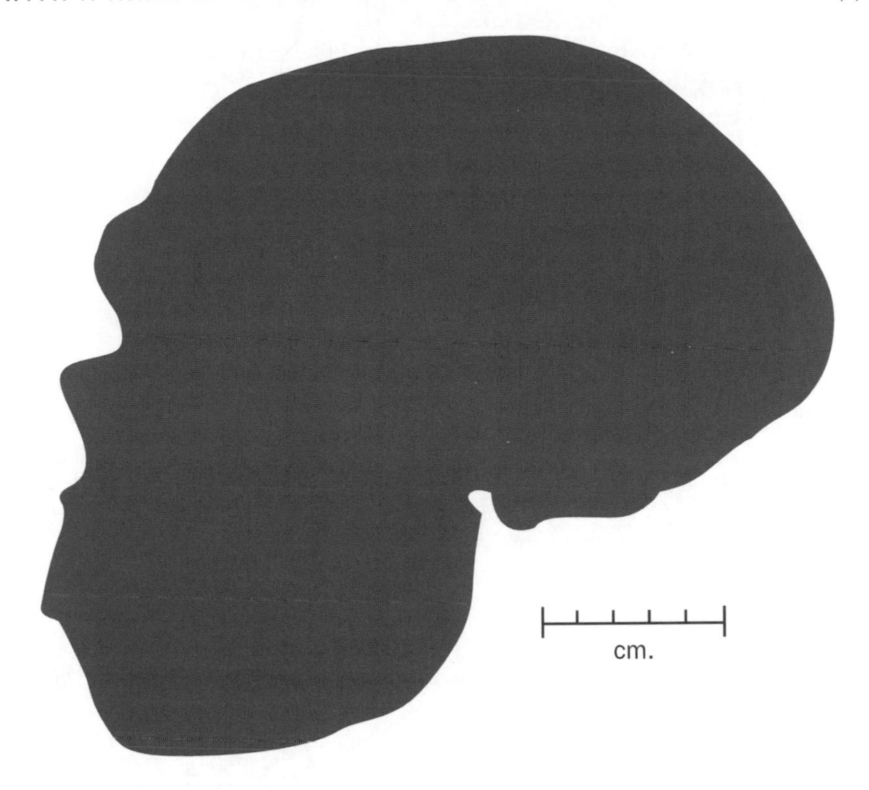

Neandertal Skull
from La Chapelle-aux-Saints, France

in chapter 4. This faulty reconstruction is responsible for some false views of Neandertal that exist in the popular mind to this day. However, a proper view of Neandertal began to emerge in the 1950s. Although the Neandertals are among the most recent of fossil humans, and although remains of more than 475 Neandertal-type individuals have been recovered, they are still the most enigmatic humans from the past—especially to the evolutionist.

Because of the richness of the Neandertal fossil record, we do have a general idea of what they looked like. There is a distinct Neandertal morphology: (1) large cranial capacity, the average being as large as the average for modern humans; (2) skull shape low, broad, and elongated; (3) rear of the skull rather pointed, with a bun; (4) large, heavy browridges; (5) low forehead; (6) large, long faces with the center of the face jutting forward; (7) weak, rounded chin; and (8) postcranial skeleton rugged with bones very thick.

The typical Neandertal does differ somewhat from the typical modern human. But the two also overlap. In fact, there should never have been a question about Neandertal's taxonomic status. When the first Neandertal was discovered, even "Darwin's bulldog," Thomas Huxley, recognized that Neandertal was fully human and not an evolutionary ancestor. Donald Johanson, in his book *Lucy's Child*, writes:

> From a collection of modern human skulls Huxley was able to select a series with features leading "by insensible gradations" from an average modern specimen to the Neandertal skull. In other words, it wasn't qualitatively different from present-day *Homo sapiens*.[2]

Although Neandertals are often presented as being inferior to modern humans, Neandertal authority Erik Trinkaus (Washington University, St. Louis) writes:

> Detailed comparisons of Neanderthal skeletal remains with those of modern humans have shown that there is nothing in Neanderthal anatomy that conclusively indicates locomotor, manipulative, intellectual, or linguistic abilities inferior to those of modern humans.[3]

The evidence indicates that the Neandertals were people of incredible power and strength—far superior to all but the most avid bodybuilders of today. Trinkaus continues:

> One of the most characteristic features of the Neanderthals is the exaggerated massiveness of their trunk and limb bones. All of the preserved bones suggest a strength seldom attained by modern humans. Furthermore, not only is this robustness present among the adult males, as one might expect, but it is also evident in the adult females, adolescents, and even children.[4]

Valerius Geist (University of Calgary) says:

> Neanderthal was far more powerful than modern humans. Whereas archeologists can experimentally duplicate the wear pattern on tools such as were used by people from the Upper Paleolithic (the people that followed Neanderthal . . .), the wear patterns on Neanderthal's tools cannot be duplicated. We do not have the strength to do it. Neanderthal's skeleton reflects a supremely powerful musculature.[5]

Conflicting Views of Neandertal

Beyond their incredible physical strength, almost everything else about the Neandertals is the subject of intense debate. It is hard to imagine a greater contrast than the two pictures of the Neandertals presented by Geist and by

Jared Diamond (University of California, Los Angeles). Geist pictures the Neandertals as the mightiest of hunters, deliberately provoking and attacking the largest game animals: "Neanderthal's kill patterns, slanted heavily to large-bodied grazers and carnivores and almost devoid of small game, are beyond comparison with any modern hunting culture."[6]

Although stone-tipped spears have been discovered at Kebara and Qafzeh Caves, Israel,[7] Geist suggests that the Neandertals may also have had to engage in close-quarter hunting. While one hunter would divert the animal's attention, one or more other hunters would rush in and attack the sides of the animal, where vital internal organs were more accessible and could be damaged more easily. If the animal had long hair, such as mammoths, woolly rhinos, steppe wisents, and horses, one hunter might actually attach himself to the body of the prey, holding onto the long hair with his powerful grip. With the animal thus distracted, another hunter, using handheld tools, would rush in and crush the lumbar vertebrae, slash open the chest cavity, or cut the tendon of the hind leg.

Geist explains how this hunting activity is related to the unique shape of the Neandertal skull.

> If great strength, agility, and precision and speed of bodily movements were required for such a hunting technique, those parts of the brain controlling motor functions in the hunter had to be greatly developed. Neanderthal possessed a massive cerebellum and motor cortex compared to modern humans. This pulled the brain case rearward, creating an occiput that reached farther rearward than in modern humans, explaining, in part, the large, long, low brain case and bun-shaped occiput of the Neanderthals.[8]

It is obvious that this kind of precision would be impossible without a quick and fluent language capability. While the Neandertals may not have been as culturally sophisticated as the people who followed, Geist concludes that the Neandertal people were not primitive but were the most highly specialized of all the humans of the past.

However, Jared Diamond presents a contrasting picture of the Neandertals.[9] He claims that the Neandertals were rather ineffective hunters, their tools and weapons were nothing to write home about, and although their average brain size was about equal to ours, they obviously were not as smart as we are. Their brains were not "wired" as well as ours. The Neandertals, Diamond claims, lacked art, needles for sewing, boats, long-distance overland trade, and, most of all, the precious human quality of innovation. He refers to them as "humans, and yet not really human."[10]

The magnitude of the confusion about the Neandertals can be seen from the fact that while Geist presents them as the mightiest of hunters and Diamond presents them as ineffective hunters, John Shea (Harvard University) presents the

Neandertals as basically vegetarians.[11] Considering the number of large animal fossils found in association with the Neandertals, Shea's view is absurd.

Diamond suggests that in various ways the Cro-Magnon people caused the extinction of the Neandertals, because, in the long run, brains always win over brawn. Geist, however, sees it the other way around. "It was probably only after the Neanderthals' extinction that modern people could colonize the land they once roamed over, for they must have been fighters of stunning abilities, for whom Upper Paleolithic people were no match."[12]

The Neandertal Problem

Ever since Darwin, evolutionists have sought to discover the path by which humans arose from their alleged primate ancestors. This, of course, remains the crucial issue in human evolution. In recent years, a second matter has also attracted major attention: the path by which our own species, *Homo sapiens*, arose. Squarely in the middle of this second issue sits "The Neandertal Problem."

The Neandertal problem is primarily the evolutionists' problem. Simply put, evolutionists don't know where the Neandertals came from or where they went. Just a few years ago, their beginnings were said to extend back only as far as 200,000 years.[13] Now, thanks to a striking discovery in a cave in Spain (see chapter 20), the Neandertals may go back 800,000 years on the evolutionist time scale. Equally mysterious is their alleged rapid disappearance at about 34,000 ya.

The Neandertal problem, however, is only secondarily the question of Neandertal's pedigree out of an assumed *Homo erectus* stock and his alleged disappearance at the beginning of the Upper Paleolithic Age. The major problem is Neandertal's relationship to two contemporary populations, one seemingly more modern in morphology and the other more archaic, all three living at the same time. The charts at the end of the book list these fossils on the evolution time scale. (The reader should bear in mind that the author does not subscribe to the evolution time scale.) The morphological spectrum represented in the charts is the type of genetic variation in the human family that the creation model would predict. Hence, the creation model explains this data better than does the evolution model.

The older view regarding the Neandertals is called the Neandertal phase of human evolution. It saw the Neandertals in the mainstream of the evolutionary process, moving from *Homo erectus* to an early phase of *Homo sapiens* to Neandertal and on to modern humans. This was the view of one of the early leaders of American anthropologists, Ales Hrdlicka, and was later held by Franz Weidenreich. A modern version of this view—that the Neandertals evolved into modern humans—is the Multiregional Continuity Model, held by Milford Wolpoff and Alan Thorne (Australian National University). A variation of this view is that the Neandertals were absorbed by more modern *Homo* populations through gene migration and hybridization. Either of these views means that modern humans have Neandertal genes in their makeup.

The most popular view today considers the Neandertals to be an isolated side branch on the family tree. It is the African Eve or the Out of Africa Model of human evolution. It holds that the Neandertals were exterminated in one way or another by more modern humans who invaded Europe and the Near East from Africa. Thus, the main branch of human evolution passed by the Neandertals. In this view, modern humans would not be the genetic descendants of the Neandertals. The "African Eve" view is a manufactured and politically correct position designed to escape the racism inherent in human evolution. It will be considered in detail later in the book.

Since anatomically modern humans existed in Europe, Africa, and the Far East at the same time as the Neandertals, the Neandertal problem is still very much an unresolved problem in contemporary paleoanthropology. Like the Cheshire cat that disappeared while its grin remained, Neandertal has disappeared, but his grin remains to taunt evolutionists.

It is commonly believed, especially in the African Eve Model, that the Neandertals became extinct—for whatever reason—between 30,000 and 35,000 ya. The Neandertal remains from Saint-Césaire, France, dated at 36,300 ya, are considered to be the most recent Neandertals.[14] This limit (*terminus ad quem*) is very rigidly maintained even though most of the Neandertal remains are poorly dated. The reasons for distancing modern humans from the Neandertals are philosophical. Since the Neandertal problem is still unsolved, the evolutionist must keep his options open. If he eventually decides that the best solution is to derive modern humans from a Neandertal stock, he must allow enough time for that to happen. Even 30,000 years or so is not enough time for the two evolutionary mechanisms—mutation and natural selection—to work their transforming magic.

However, there is evidence that the Neandertals persisted long after their alleged demise. The Neandertal skull known as Amud I from Upper Galilee, Israel, was found as a burial just below the top of layer BI. If Amud I was buried into layer BI, it follows that he cannot be older than layer BI but could be younger. *The radiocarbon date for Upper BI is just 5,710 ya.* Michael Day (Natural History Museum, London) states, "These dates are believed to be too 'young' as the result of contamination by younger carbon."[15] While it is certainly true that younger carbon compromises a radiocarbon date, this is also the standard excuse given whenever a radiocarbon date is too young to fit the system. Day gives no evidence that young carbon was present. It is understood by evolutionists that if a radiocarbon date is too young to fit the evolutionary scenario, that is proof enough that the sample was contaminated, since a "good" date would unquestionably fit the scheme. In dating Amud I, it is bad enough that uranium/ionium growth gives a date of only 27,000 ya and uranium fission-track gives a date of only 28,000 ya, with a margin of error of almost 10,000 years. Anything is better than a date of 5,710 ya for a Neandertaler.

Some of the Shanidar, Iraq, Neandertal material from layers C and D gives radiocarbon dates as low as 26,500 ya,[16] and the Neandertal Banyolas mandible, found near Girona, Spain, gave a radiocarbon date of 17,600 ya. After recording this date, obtained by the UCLA radiocarbon laboratory, *Radiocarbon Journal* made the following statement:

> Comment: for a Neanderthal, present date is too recent. The possibility of more modern travertine contaminating older travertine to yield a more recent composite date, or the relocation of an ancient mandible into travertine is open.[17]

Possibilities are given for the too recent date, but no physical evidence is cited to indicate that these possibilities are valid. The arbitrary assertion that the date is too recent for a Neandertal apparently settles the matter.

If there is any legitimacy to these recent dates for Neandertal, it could mean that Neandertal, like his smaller edition known as *Homo erectus*, persisted until quite recently. That would be additional evidence that the differences between Neandertal and anatomically modern humans had nothing to do with the evolutionary process. For evolutionists, the Neandertal problem remains unsolved.

A Nonevolutionary Explanation for Neandertal Morphology

Whether the Neandertals were in the main line of human descent, or whether they were a side branch that led to extinction, the evolutionist believes that the somewhat different Neandertal morphology was the result of the evolutionary process. The two evolutionary mechanisms are mutation and natural selection; mutations supply the raw material (new information) upon which natural selection can work. Special creation and evolution are thus mutually exclusive. If God by special creation supplied the genetic information that accounts for the existence of humans, then evolution is not necessary. If random mutations are able to supply new information upon which natural selection works to produce humans out of a nonhuman stock, then the concept of special creation is not necessary. Mutations, which are mistakes, cannot create new genetic information. Just as only an intelligent human can create new information in a computer program, so only God has created information in our genetic program.

The evolutionist improperly introduces other mechanisms into the alleged evolutionary process, such as the founder principle, geographic isolation, and genetic recombination. While these are legitimate processes, they are not evolutionary processes. They do not create unique new genetic information. Nor do these processes discriminate between special creation and evolution. They would apply in either case. The evolutionist smuggles these nonevolutionary mechanisms into the evolutionary process even though they have nothing to do with evolution. These processes do account for variation, but they cannot produce evolutionary changes that result in increased complexity; that would demand the creation of entirely new genetic information.

It is impossible for the evolutionist to demonstrate that the Neandertal morphology was the result of mutation and natural selection. That is only a dogmatic assertion that is part of his belief system. If evidence could be supplied to show that the Neandertal morphology could be achieved by means other than mutation and natural selection, the concept of human evolution would be called into question, and the concept of special creation would be strengthened. This is what we propose to do.

Over the years, the scientific literature has suggested a number of conditions—geographical, environmental, pathological, cultural, and dietary—that could produce a Neandertal-like morphology. Richard Klein writes:

> The forward placement of Neanderthal jaws and the large size of the incisors probably reflect habitual use of the anterior dentition as a tool, perhaps mostly as a clamp or vise. Such para- or nonmasticatory use for gripping is implied by the high frequency of enamel chipping and microfractures on Neanderthal incisors, by nondietary microscopic striations on incisor crowns, and by the peculiar, rounded wear seen on the incisors of elderly individuals. Similar, though less extensive damage occurs on the teeth of Eskimos, who also tend to use their anterior jaws extensively as clamps.
>
> Biomechanically, the forces exerted by persistent, habitual, nonmasticatory use of the front teeth *could account in whole or in part* for such well-known Neanderthal features as the long face, the well-developed supraorbital torus, and even the long, low shape of the cranium. Massive anterior dental loading could further explain the unique Neanderthal occipitodmastoid region which perhaps provided the insertions for muscles that stabilized the mandible and head during dental clamping (emphasis added).[18]

In two paragraphs, Klein has given a plausible nonevolutionary explanation for most of the unique features of Neandertal morphology. Just as the hands of a blacksmith develop calluses as a result of the unique wear and stress they are subjected to, so the facial and skull morphology of the Neandertals could be the result of the unique stresses their jaws and teeth were subjected to when used as tools. Klein also states, "The long, low shape of the Neanderthal cranium with its typically large occipital bun probably reflects relatively slow postnatal brain growth relative to cranial vault growth."[19]

In a statement cited earlier in this chapter, Geist also gave a plausible nonevolutionary explanation, based on Neandertal's prowess as a hunter, for his unique skull morphology:

> If great strength, agility, and precision and speed of bodily movements were required for such a hunting technique, those parts of the brain controlling motor functions in the hunter had to be greatly developed. Neanderthal possessed a massive cerebellum and motor cortex compared to modern humans. This pulled the brain case rearward, creating an occiput that reached farther rearward than in modern humans, explaining, in part, the large, long, low brain case and bun-shaped occiput of the Neanderthals.[20]

Klein also recognized the effect geographic isolation could have had on the development of the Neandertals when he wrote that "some of the European mid-Quaternary fossils clearly anticipate the Neanderthals, while like-aged African and Asian ones do not. Clearly, the implication is that the Neanderthals were an indigenous European development."[21]

Health factors can be reflected in the skeleton, including a vitamin D deficiency resulting in rickets. J. Lawrence Angel (Smithsonian Institution) writes, "Pelvis and skull base tend to flatten if protein or Vitamin D in diet is inadequate."[22] This was the diagnosis of Rudolf Virchow, "the father of pathology," when he examined the first Neandertal discovery. He was overruled by those who favored an evolutionary interpretation. In 1970, Francis Ivanhoe published in *Nature* an article entitled, "Was Virchow Right about Neandertal?"[23] He presented a strong case, based on diagnostic evidence, that the Neandertals were really modern humans who suffered from rickets.

Another possible explanation of the Neandertal morphology is disease, especially syphilis. D. J. M. Wright (Guy's Hospital Medical School, London) observed, "In societies with poor nutrition, rickets and congenital syphilis frequently occur together. The distinction between the two is extremely difficult without modern biochemical, seriological, and radiographic aids."[24]

Based on his examination of the Neandertal collection at the British Museum, Wright found a number of features in the Neandertal's morphology compatible with congenital syphilis. These conditions are seen in both adult and child skulls. Wright specifically mentioned the original Neandertal skullcap as well as the Gibraltar II, Starosel'e, and Pech de l'Aze Neandertal remains.

To suggest that all of the above factors contributed to the Neandertal morphology would be overkill. It is well within reason, however, to suggest that one of them or several in combination are responsible for the group of individuals known collectively as Neandertals. Obviously, to perform controlled experiments on humans today to determine which factors would produce the skeletal features of the Neandertals would be both immoral and impossible. But in contrast to the lack of rigorous scientific evidence that mutation and natural selection could produce these effects, there is a sizable body of scientific data that suggests one or more of the above-mentioned factors would constitute a reasonable and nonevolutionary explanation for the Neandertal morphology.

A remarkable situation has recently transpired. For 150 years, it has been assumed that the site of the original Neandertal discovery, the Feldhofer Grotto, was destroyed in the mining of limestone. However, a team from Germany's Office for the Preservation of Archaeological Monuments, using 150-year-old maps, rediscovered the original cave in 1997. In it they found 36 human remains, some fitting the original Neandertal fossils. A date of 40,000 ya was obtained on one of the fossils by radiocarbon. Fossils of modern humans were found there also, dated at 44,000 ya.

Their report continues, "In addition to the fossils, thousands of stone tools and flakes from tool production have been found, representative of both Mousterian and Aurignacian lithic technologies. Burnt and cut bones of non-cave-dwelling animals at the site seem to show evidence of food preparation."[25] A *Nature* article adds that "this is evidence of food preparation and cooking, indicating that the Neanderthals belonged to a settlement."[26]

Thus the original Neandertal fossils testify that the Neandertals were contemporaries with modern humans and were fully modern culturally as well. The 4,000-year difference in dating is of no consequence. A recent Institute for Creation Research breakthrough furnishes scientific proof that the earth is young and that the radiometric dating methods are invalid. (See the section on "The Creationist Dating Revolution.")

When Joachem Neander walked in his beautiful valley so many years ago, he could not know that hundreds of years later his name would become world famous, not for his hymns celebrating creation but for a concept that he would have totally rejected: human evolution.

CHAPTER 7

JAVA MAN

The Rest of the Story

THE STORY OF THE group of fully human fossils called *Homo erectus* by evolutionists begins with Java Man. Java Man! Breathes there a human with soul so dead who has not heard of Java Man? Java Man is like an old friend. We learned about him in grade school. They called him the ape-man and told us that he was our evolutionary ancestor. The drawings of that beetle-browed, jaw-jutting fellow were quite convincing. In fact, the vast majority of people who believe in human evolution were probably first sold on it by this convincing salesman. Not only is he the best-known human fossil, he is one of the only human fossils most people know.

The story of Java Man has been told many times. Before the turn of the century, a Dutch anatomist, Eugene Dubois, went to the Dutch East Indies (now Indonesia) in search of the "missing link" between apes and humans. In 1891, along the bank of the Solo River in Java, he found a skullcap that seemed to him to have a combination of human and ape features. A year later, about fifty feet away, he found a thigh bone (femur), very human in appearance, that he assumed belonged with the skullcap. Dubois named his "transitional form," which is now dated by evolutionists at about half a million years, *Pithecanthropus erectus* (Erect Ape-Man). But to the general public he will always be known affectionately as Java Man.

Because Java Man was found so long ago, because another like him (Peking Man) was not found until the 1920s, and because he was made to fit the evolutionary picture so beautifully, Java Man is virtually synonymous in the popular mind with human evolution.

In the past, it has been a frustrating task to research Java Man. Basic information is in very short supply. Historians of science have paid little attention to Dubois or his methodology. I had long felt that an authoritative biography of Dubois would help answer some of the basic historical questions, but no suitable work was available in either Dutch or English.[1] However, the situation has been somewhat corrected. The work by Bert Theunissen, *Eugene Dubois and the Ape-Man from Java*, published in the Netherlands, brings to light information that has hitherto been unavailable to most researchers.[2]

My conclusions on Dubois and Java Man are as follows: (1) Java Man is not our evolutionary ancestor but is a true member of the human family, a postflood descendant of Adam, and a smaller version of Neandertal; (2) Dubois seriously misinterpreted the Java Man fossils, and there was abundant evidence available to him at that time that he had misinterpreted them; (3) The evolutionists' dating of Java Man at half a million years is highly suspect; (4) More modern-looking humans—possibly including Wadjak Man—were living as contemporaries of Java Man; and (5) Java Man was eventually accepted as our evolutionary ancestor in spite of the evidence because he could be interpreted to promote evolution. The historical and scientific questions regarding Java Man are as legitimate today as they were when the fossils were first discovered. This chapter will cover the problems concerning the geology and dating of Java Man.

Accurate dating is essential to the proper interpretation of a fossil. Since Dubois claimed that Java Man was *the* missing link between apes and humans, he had to show that it dated at the appropriate time when a certain ape stock was allegedly evolving into humans. If Java Man was rather recent in date, as may well be the case, it could not serve as an evolutionary transitional form because modern humans were already on the scene.

Dubois claimed that the skullcap and the femur came from a rock stratum known as the Trinil layer, named after a nearby village in central Java. He believed that these rocks were below what is known as the Pleistocene-Pliocene (Tertiary) boundary. Dubois was convinced that "real" humans evolved later in the Middle Pleistocene. Hence his dating of Java Man was quite appropriate for a missing link. His interpretation was not exactly straightforward, however, as the man who later found other Java men, G. H. R. von Koenigswald, tells us:

> When Dubois issued his first description of the fossil Javanese fauna [animals] he designated it Pleistocene. But no sooner had he discovered his *Pithecanthropus* than the fauna had suddenly to become Tertiary [older than the Pleistocene]. He did everything in his power to diminish the Pleistocene character of the fauna. . . .
>
> The criterion was no longer to be the fauna as a whole, but only his *Pithecanthropus.* Such a primitive form belonged to the Tertiary! Dubois' view . . . did not go uncontested. But there was no getting at him until he had described his whole collection and laid all his cards on the table. That was why we all had to wait for a study of his finds, and to wait in vain.[3]

The first problem with the geology and dating of Java Man is that Dubois was not qualified to make those determinations. In 1877, Dubois entered the University of Amsterdam as a medical student. He soon lost interest in becoming a practicing physician and majored in anatomy instead. After graduation he taught anatomy and worked in an anatomical laboratory. In 1884, he also qualified as a medical doctor. Years later he was quoted as saying, "I am always described as a doctor. But I am an anatomist."[4] Actually, he was qualified to be both, but the point is that he was not a geologist. He had an interest in geology and paleontology as a child and may have studied them a bit in high

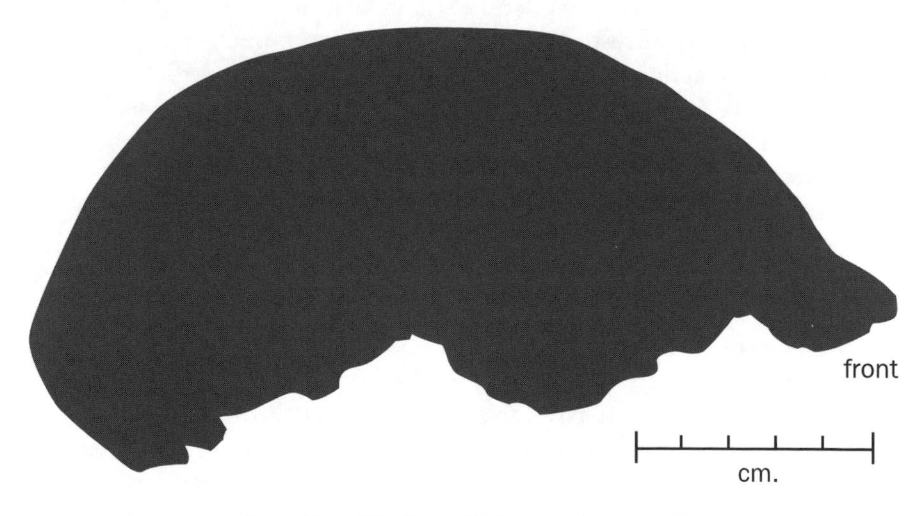

Java Man Skullcap (Pithecanthropus I)
Homo erectus

school. But there is no record of Dubois having any formal training in geology or paleontology before he went to Java.

It is true that in 1899 the University of Amsterdam appointed him Professor Extraordinary of Geology, that in 1907 it made him a full professor, and that he taught in that area until 1921. As far as can be determined, however, all of his formal training in geology was done after he returned from Java, not before. While in Java, Dubois was an amateur geologist at best.

Not only was Dubois not qualified to determine the date and location of the Pliocene and Pleistocene deposits in Java, but the geology and paleontology of Java were virtually unknown. Dubois himself said that the Dutch East Indies were practically virgin territory in this regard.[5] The entire science of geology was still in its infancy, and most of the geological work that had been done was done in Europe. Even there, the date of the Pleistocene/Pliocene boundary had been estimated to be all the way from 100,000 ya by Hein to 800,000 ya by Lyell. If geology was hardly a precise science in Europe, it certainly was not precise in Java. Fifty-seven years after Dubois made his discoveries, the famous paleoanthropologist Alan Houghton Brodrick, writing in 1948, said that the stratigraphy of the Trinil beds still was not clear.[6]

If Dubois's lack of expertise in geology, together with a general lack of geological information about Java, isn't enough of a problem, another factor complicates the picture: the way Dubois collected his fossils. Today it is understood that when an important fossil is discovered, the utmost care is taken to document the precise rock layer in which the fossil is located. This is absolutely imperative for both relative and absolute dating of the fossil. The inability to locate a fossil in its geological context is enough to disqualify it from serious consideration.

Dubois had gone to the Dutch East Indies as an army doctor and was then given permission to search for fossils. To help him in his work, Dubois was assigned two corporals from the engineering corps, who acted as supervisors, and fifty forced laborers. Dubois was usually at his headquarters, but he made periodic visits on horseback to the digging sites. He also maintained written contact with the engineers about the progress of the work. When fossils were found, they were sent to Dubois for preparation and provisional identification. Dubois himself did not uncover any of the important fossils ascribed to him, and he never saw any of them in situ (except Wadjak II). He was entirely dependent on his two engineers to determine the position of the fossils in the deposits—engineers who knew even less about geology than he did.[7] (Brodrick questions that Java Man came from the Trinil beds, suggesting that the fossils came from higher up in the rock layers, which would make them more recent in age.[8]) By today's standards, these fossils would have been disqualified. Yet they became for many years the primary evidence for human evolution.

Perhaps it was Dubois's lack of expertise in geology, together with the fact that he was not on the scene when the fossils were excavated, that is responsible for a peculiar characteristic of his reports: a lack of geological information. His reports always centered on the anatomical aspects of his fossils, which is not surprising, since he was an anatomist. When the skullcap and the thigh bone were discovered, Dubois spread the word officially and widely. He gave a rather detailed preliminary description of the fossils themselves, but he gave only the briefest account of the locality and the geological circumstances surrounding the discovery. His dating of *Pithecanthropus* was based on the mammalian fossils found with *Pithecanthropus*, but his proof consisted of only sketchy tidbits of information.

It wasn't until late 1895, after he had returned from Java, that he displayed the first profile drawing and maps of the excavation site. The scientific community was continually frustrated because of his lack of geological information. Dubois had promised that he would publish a comprehensive work on *Pithecanthropus erectus* and the Javanese fossil fauna. It was many years before the report on *Pithecanthropus* was published, but he never did produce the promised report on the Javanese mammalian fauna.

Up until 1900, Dubois had been very active in promoting Java Man as the missing link and had allowed full access to the fossils. After 1900, he withdrew completely from the public debate for twenty years, published very little about the fossils, and refused to allow anyone to see them. The reason usually given for this behavior is that Dubois wanted Java Man to be accepted as *the* missing link. Because of the initial controversy over his interpretation, he retaliated by refusing access to the fossils. From what we know of Dubois's personality, this explanation is possible.

Theunissen suggests another explanation.[9] Dubois was very proud, deeply suspicious, and fiercely independent. In 1899, Strasbourg anatomist Gustav Schwalbe produced a significant publication on the skullcap that far overshadowed anything Dubois had done. Dubois must have sensed that if he allowed continued access to the fossils, others would do the work and get the recognition that rightfully should be his. Hence he began the embargo on allowing others to see the fossils. However, Dubois continued to drag his feet. In 1907 and 1908, he published some rather superficial articles on some of the Javanese fossil fauna. His main reason for publishing seems to be that another expedition was working at Trinil at that time, the Selenka-Trinil expedition, and Dubois did not want them to steal his thunder. It was not until 1924, thirty-three years after the skullcap had been discovered, that Dubois published a definitive paper on it. Two years later he published a major paper on the thigh bone.

Why Dubois procrastinated on matters so vital to the scientific community cannot be answered with finality. Part of the answer might be found in a comment by one of his students: "Dubois had the habit of just lifting a corner of the veil of a scientific conception, but he was loath to settle down and work it out thoroughly."[10]

PLEISTOCENE EPOCH

The Pleistocene Epoch is the geological term for the period of the Ice Age or Ages. It is allegedly the time when the bulk of human evolution took place. The Pleistocene is followed by the Holocene Epoch and is preceded by the Pliocene Epoch. Pleistocene dates have been the subject of much debate, especially on the lower end. The date of the Pleistocene/Pliocene boundary was agreed upon by the geological community in about 1980. The generally accepted dates are as follows:

Holocene Epoch	Present to 10,000 ya
Upper Pleistocene	10,000 to 100,000 ya
Middle Pleistocene	100,000 to 700,000 ya
Lower Pleistocene	700,000 to 1.7 Mya
Pliocene Epoch	1.7 to 5.5 Mya

("ya" means "years ago"; "Mya" means "million years ago")

CHAPTER 8

Java Man

Keeping the Faith

GEOLOGICAL PROBLEMS WERE NOT the only ones swirling around the skullcap of Dubois's Java Man. There was also the problem of identifying that unusual individual. Was it a human, an ape, a man-ape, or an ape-man? No one knew. To explain the controversy at the turn of the twentieth century, I must first transport you back in time.

To step back into the nineteenth century would cause us culture shock in many ways. One of the shocks would come when we realized what was commonly believed about creation. Before evolution became popular, creation was the accepted scientific model of the universe and of humans. However, it was a type of creationism that few of us would recognize and no biblical creationist would endorse.

The concept was known as the Great Chain of Being, patterned not after Moses but after Plato.[1] According to this concept, the Almighty had created a great ladder or chain of living things, from single-celled organisms all the way up to humans, each organism being a bit more complex than the one below it. All of nature fit into this ascending organizational scale. Like the keys on a piano, each organism was discrete but a bit higher in the organization and more complex than the one below it. Just as there are no missing keys on a piano, there could be no spaces or gaps in this ladder (no extinction). This Great Chain of Being looked much like an evolutionary progression, but it was static. Each organism was created by the Almighty in its particular slot and did not evolve upward.

Since any imperfections in nature would be a reflection upon the work of the Almighty, it was further believed that nature was perfect. This idea was obviously in conflict with the biblical account of the fall of man into sin (Genesis 3) and its results on nature. Further, this theory implied that extinction was impossible because that would indicate that the Almighty did not take proper care of his creatures after he had created them. These ideas were in conflict with the Noachic flood in Genesis 6–9, where the whole purpose of the flood, as a judgment from God, was to destroy the world as it then was (2 Peter 3:6). It is obvious that in the Great Chain of Being we are dealing not with biblical concepts but with pagan Greek philosophy.

With the discovery of fossils that showed the existence of creatures in the past that were different from those we know today, some people believed that those fossils were either naturally formed objects, fakes, or "sports" (objects placed in the earth by the Almighty to fool us or to test our faith). Later, when it could no longer be denied that fossils were legitimate remains of past life, some of which was extinct, some people rejected the whole concept of creation. This misunderstanding of the biblical account of creation helped pave the way for the acceptance of evolution as the only viable alternative.

It is not unusual for people to begin with a wrong idea of what the Bible teaches, reject that view, and then reject the entire Bible because "the Bible is unscientific." This is what Darwin did. Most people today believe that Darwin disproved biblical creationism and proved evolution. The Darwinian Revolution, one of the most significant revolutions of all time, is generally thought to be the establishment of the concept of evolution on a solid, empirical base. Not so. In the words of Harvard biologist Ernst Mayr, the Darwinian Revolution was actually a philosophical revolution from a theistic worldview to a worldview in which God was not involved in any way.[2]

Darwin did not reject biblical creation; he knew nothing about it.[3] Even though he studied for the ministry at Cambridge, it is obvious from his writings that he did not have a clue as to what the Bible actually taught regarding special creation. Darwin heavily criticized special creation in *The Origin of Species*. He claimed that the imperfections of nature demonstrated that a wise and all-powerful God could not have done such a sloppy job. He seemed oblivious to the fact that those imperfections of nature are the result of the fall and that the world is not now the way God originally made it.

Darwin's abysmal ignorance of what Genesis teaches is seen in that not until he was fifty-two years old, two years after he published *Origin*, did he realize that the much-ridiculed date of 4,004 BC for creation was not a part of the text of Genesis. It was instead the work of Archbishop James Ussher, who lived from 1581 to 1656.[4] (Ussher's date for creation was flawed because he did not take into account some gaps in the Genesis chronologies.) Darwin first rejected an unbiblical, philosophical creationism and then rejected biblical creationism as well. It is a classic case of throwing out the baby with the bathwater. The philosophy of nature he then developed had as its cardinal principle that it is "unscientific" to believe in supernatural causation. Darwin's purpose was to "ungod" the universe.

The concept of the Great Chain of Being by its very nature became a "setup" for evolution. All one had to do was to change that static chain to a dynamic one, with the forms gradually evolving upward from one into another, and one had the basic evolutionary scenario. That philosophical preparation explains why evolution was accepted so rapidly after the publication of Darwin's *Origin*.

The Great Chain of Being was responsible for even more mischief. It allowed for the endorsement of slavery. When the nations of Africa and the East were opened up and world trade routes developed, western Europe learned about the many "savage" tribes that inhabited large portions of the earth. The differences in culture and language of these "savages" were proof to the chauvinistic western Europeans that these strange peoples were inferior races. The "savages" were fitted into the Great Chain of Being above the apes and below the Europeans. There was no evolutionary significance in that placement. Europeans believed that the Almighty had created the "savages" as true humans but as inferior races. Hence, since the Almighty had created them as inferior races, it was proper for the superior races of western Europe and the United States to keep them in the place that had been ordained for them by the Almighty. Some even went so far as to claim that the Almighty created these inferior beings without souls, to be used by the superior races much as they would use domestic animals.

To justify this outrageous idea, some even appealed to the Bible. They claimed that there was a curse on Ham, one of the sons of Noah, and that Ham was the father of the blacks (Gen. 9:25–27). Even a superficial reading of those verses, however, shows that God's judgment was not on Ham but on Canaan. Ham had four sons (Gen. 10:6): Cush, Mizraim, Put, and Canaan. Cush (Ethiopia) is the biblical term for the black race. Mizraim is the common Hebrew word for Egypt. Put refers to Libya. Canaan was the father of the Canaanites who settled in Palestine and in North Africa around ancient Carthage. God's judgment was upon the Canaanites (who were white) and was subsequently carried out. God has the right, as God, to judge nations as well as individuals. He has done so in the past, and he will do so in the future, but it is done for deeds, not race. There is absolutely no basis in the Bible for racism. The Bible clearly declares that all humans are made in the image of God (Gen. 1:26, 27; 9:6), and Paul declares that "From one man he made every nation of men" (Acts 17:26).

It was against this background of the Great Chain of Being and its pagan idea of inferior races that the first Neandertal discoveries and Java Man were interpreted at the turn of the twentieth century. Many, if not most, in the anthropological community saw the Java Man skullcap as truly human.[5] Those anthropologists did not know then of the tremendous genetic variability that exists in humans as we know it today. Yet they did not consider Java Man to be a transitional form. They considered him to be truly human and able to reproduce with any other member of the human family. The only qualification was that they considered the owner of that skullcap to be racially inferior, in the same way that they characterized Africans, Southeast Asians, and native Australians at that time.

Many anthropologists at the turn of the twentieth century also noted the remarkable similarity in shape between the Java Man skullcap and the Neandertal skulls, which would indicate that Java Man and the Neandertals, although differing in size, represented the same race of humans. Up until his return from Java, Dubois had ignored the Neandertal remains because he felt they were pathological. With the discovery of more Neandertal material, Dubois was forced to reconsider them, and even he admitted the close resemblance between his own Java Man and the Neandertals.

One of Dubois's earliest papers on Java Man was presented in 1895 in Dublin, shortly after his return from Java. It was entitled, "On *Pithecanthropus erectus*: a Transitional Form Between Man and the Apes." In it Dubois emphasized what was to be his theme for the rest of his life: Java Man was not a human being, nor was he an ape; he was a true intermediate or transitional form possessing features of both humans and apes.

Sir Arthur Keith, the famed Cambridge University anatomist, was asked to comment on Dubois's paper. He replied that the chief question to be settled was whether or not the skullcap was human. In answering that question, one had to determine the criterion of a human skull versus an ape skull. To his mind there were two basic differences: first, the very large cranial capacity of human skulls as compared to ape skulls, and second, the large muscular ridges and processes, connected with the chewing apparatus, which ape skulls have compared to human skulls. On both points Keith declared that the Java Man skullcap was distinctly human.[6] The cranial capacity of the anthropoid apes never exceeds 600 cc and averages about 500 cc. On the other hand, the cranial capacity of Dubois's Java Man was estimated at 1000 cc, which is well within the range of modern humans.

Sir William Turner, also responding to Dubois's paper, said that since the Java Man thigh bone had a pathological growth on it, possibly the skullcap was pathological also. He pointed out that in the Edinburgh University Museum is the cast of a microcephalic woman with a frontal flattening very much like that of Java Man.[7]

The main question about the skullcap of Java Man was, Is it human? However, there was no question about the nature of the thigh bone, found a year after the skullcap and fifty feet from it. From the time of its discovery, virtually every authority except Dubois felt that it was indistinguishable from the modern human femur. The great question on the femur was, Did it belong with the skullcap? It seemed far too modern in morphology to be associated with the rather archaic shape of the skullcap.

It was here that Dubois's weakness in geology and the shortcomings of his methodology put him on the defensive. He did none of the actual digging, nor was mapping done at the time of excavation. All quadrant maps and diagrams were made after the fact. In Dubois's first report of his find to Dutch authorities in August 1892, he stated that the femur was located ten meters from the place where the skullcap had been found a year earlier. In early September he changed this figure to twelve meters. His official report later that same month, and his official publication on *Pithecanthropus*, mentions fifteen meters. In a much later 1930 publication, he again mentions twelve meters.[8]

One of the most amazing facets of the Java Man saga is this: Throughout the twentieth century, the skullcap and the femur were presented to the public together as Java Man, our evolutionary ancestor, by evolutionists. Yet the association of the skullcap with the femur has *always* been questioned by the most respected evolutionary anatomists from the time of Java Man's discovery until today. It is just one of many illustrations of the fact that evolutionists will use whatever "proof" possible to sell evolution to the general public, regardless of its scientific authenticity.

In 1938, Franz Weidenreich described several femoral fragments of Peking Man. (Both Peking Man and Java Man are now called *Homo erectus*.) Whereas the skulls of Peking Man and Java Man were quite similar, the Peking Man femora differed from the Java Man femur in the very places where the Java Man femur was similar to modern humans. Since the association of the Peking Man skulls and femora was undisputed, Weidenreich concluded that the Java Man femur was not a true *Homo erectus* femur but was instead a modern one.[9]

A recent assessment of the Java Man femur comes to the same conclusion. Michael Day and T. I. Molleson compared the Java Man femur, the Peking femora, and the femur known as Olduvai Hominid 28 (OH-28) found by Louis Leakey in Olduvai Gorge, Tanzania, in unquestioned association with other *Homo erectus* material. They state that OH-28 and the Peking Man femora, although truly human, are much more similar to each other than either is to the Java femur. Their conclusion is that OH-28 and Peking Man represent a *Homo erectus* anatomy, whereas the Java femur is more modern.[10]

Here, then, is the problem faced by evolutionist paleoanthropology. If the Java skullcap and femur actually belong together, then it is difficult to maintain a species difference between *Homo erectus* and *Homo sapiens*. The distinction would be an artificial one, and it would compromise these fossils as evidence for human evolution. If, on the other hand, the skullcap belongs to *Homo erectus* and the femur belongs to *Homo sapiens*, it shows that these two forms likely lived together as contemporaries. It likewise removes these fossils as evidence for human evolution, because fluorine analysis indicates that the fossils are the same age.[11]

A recent suggestion is that the femur came from a much younger stratigraphic horizon than did the skullcap.[12] But this naive attempt at time separation—one hundred years after the fact—reveals the awkwardness evolutionists feel regarding these fossils. This suggestion comes not from the physical evidence at the scene but from a transparent attempt to salvage Java Man as evidence for human evolution.

The Java Man skullcap and femur are evidence that the distinction between *Homo erectus* and *Homo sapiens* is an artificial one, that these two forms are both truly human, and that they lived as contemporaries. The differences attributed to evolution are instead evidence of the wide genetic variation found in the human family.

For the rest of his life, Dubois maintained that the skullcap and the femur belonged together and that *Pithecanthropus* was unique. When fossils similar to his Java Man were found, Dubois rejected the evidence out of hand. He labored hard to find tiny areas where his fossils differed from anything else that had been found so as to defend the uniqueness of his discovery.

G. H. R. von Koenigswald wrote of Dubois, "On this point he was as unaccountable as a jealous lover. Anyone who disagreed with his interpretation of *Pithecanthropus* was his personal enemy."[13] Von Koenigswald should know. He made the unfortunate mistake of finding a number of fossils very much like Dubois's original Java Man in—of all places—Java.

Both evolutionists and creationists report that toward the end of his life Dubois renounced *Pithecanthropus* as the transitional form and declared that it was just a giant gibbon. Although this report is not true, Dubois himself is responsible for the misunderstanding. To distance *Pithecanthropus* from similar fossils that were later discovered in both Java and China, Dubois so emphasized what he believed to be gibbonlike characteristics of *Pithecanthropus* that others thought he was saying that *Pithecanthropus* was just a giant gibbon. However, to the end of his life Dubois "kept the faith," believing that his beloved *Pithecanthropus* was uniquely *the* missing link.[14]

CHAPTER 9

Wadjak Man

Not All Fossils Are Created Equal

CHANCES ARE YOU HAVE NOT heard of Wadjak Man (now spelled Wajak). Whether or not that is a significant omission in your life, you alone must decide. If you decide that it is, you can blame Eugene Dubois for your ignorance of Wadjak.

A former president was called the "Teflon President" because the adversities that would have tarnished the popularity of a lesser man did not seem to stick to him. Dubois was certainly the Teflon Paleoanthropologist. Neither his sins of commission nor his sins of omission seemed to stick to him. One of his sins of commission was his handling of Wadjak Man. It was so serious that it should have jeopardized his credibility as a scientist. Instead, his fellow scientists excused it.

When Dubois went to the Dutch East Indies, he first went to Sumatra, another large Indonesian island northwest of Java. He found animal fossils there, but no hominids. He was already planning to search in Java when he heard of a remarkable discovery that confirmed the wisdom of those plans. B. D. van Rietschoten, a Dutch mining engineer who was looking for marble, had discovered a fossil human skull near the village of Wadjak, on the south coast of east Java, in 1888. Van Rietschoten sent the skull to the curator of the Natural Science Museum in Batavia (now Jakarta), and he in turn sent it on to Dubois, who was delighted. The fossil is now known as Wadjak I. Upon his arrival in Java, Dubois found a second skull at the same site. Known as Wadjak II, it is not quite as complete as Wadjak I. Later Dubois found fragments of a skeleton there.

The same lack of precision plagues the geological details of this site as it did that of *Pithecanthropus*. Theunissen calls the site a rock shelter and later calls it a cave.[1] Von Koenigswald wrote that the fossils were found in a cave opened up in the Wadjak marble quarry.[2] But Sir Arthur Keith wrote that the fossils were found in one of the terracelike deposits on the limestone bluffs on the shoreline of an ancient freshwater lake.[3] Marble is a metamorphosed limestone, so it is not unusual for limestone and marble to be found together. But the accounts do seem to be contradictory. If they can be reconciled, the solution is not readily apparent. Carleton Coon has stated, "As the site has since been destroyed by quarrymen, the exact date may never be known."[4] And it is the date that is of vital importance.

The Wadjak skulls, found before Dubois discovered *Pithecanthropus*, are truly those of modern man, morphologically speaking. Whereas *Pithecanthropus* had a cranial capacity of 1000 cc, the Wadjak skulls have cranial capacities of 1550 cc and 1650 cc respectively. Although Dubois called them *Homo wadjakensis*, there has never been any doubt that they were modern humans and should have been classified as *Homo sapiens*.

The Wadjak skulls were found in 1888 and 1890. Dubois brought them back from Java in 1895 and kept them sequestered in his home in Haarlem, Holland. He made no public announcement about these fossils until May 1920, thirty years after they were found. The motivation then for Dubois's revelation of his Wadjak discoveries was Stuart A. Smith's publication of a monograph on Talgai Man, claiming that he had discovered the first "proto-Australian." Dubois's massive ego could not let that claim go unchallenged. He unveiled the Wadjak skulls and said that he had discovered the first "proto-Australian" years before Smith did.[5]

Why did Dubois keep his discoveries at Wadjak a secret for thirty years? Most evolutionist writers who mention Wadjak do think that Dubois's actions were a bit strange. Theunissen is the most casual about it.[6] He says that Dubois went to the Dutch East Indies to find the missing link. Since the big-brained Wadjak was obviously not the missing link, Dubois was not that interested in the fossils. He did not even remove the hard sediment that still partly covered the fossils until 1910. By that time, Dubois's view of Neandertal had changed, and Wadjak became worth studying.

Wadjak presented another "window of opportunity" for Dubois, says Theunissen. Marcellin Boule was busy moving the Neandertals out of man's direct ancestry because they were too "primitive" for their advanced date. Dubois had already claimed to have discovered the missing link. By revealing Wadjak and emphasizing their advanced features over Neandertal, Dubois could claim credit for having discovered a second direct human ancestor, although a more modern one. The time was thus ripe to tell the world about Wadjak.

Theunissen has done well in explaining why Dubois revealed Wadjak, but he has not explained why Dubois hid Wadjak in the first place. Sir Arthur Keith let the cat out of the bag. After he described Dubois's strange behavior, he wrote:

> We cannot question his honesty; the Wadjak fossil bones were discovered under the circumstances told by him. There can be no doubt that if, on his return in 1894, he had placed before the anthropologists of the time the ape-like skull from Trinil side by side with the great-brained skulls of Wadjak, both fossilised, both from the same region of Java, he would have given them a meal beyond the powers of their mental digestion. Since then our digestions have grown stronger.[7]

Keith had a way with words. He was saying in a gracious way that if Dubois had revealed the Wadjak fossils at the time he revealed *Pithecanthropus*, his beloved *Pithecanthropus* would never have been accepted as the missing link. Dubois was well aware of that fact. There is evidence that Wadjak was approximately the same age as *Pithecanthropus*, so to sell *Pithecanthropus*, Dubois had to hide Wadjak.

Wadjak was not the only occasion when Dubois manipulated fossils to protect his claim for *Pithecanthropus*. Earlier he had found a jawbone fragment at Kedoeng Broeboes, Java. Jawbone fragments are sometimes difficult to diagnose, and Dubois did have problems with it. He first ascribed it to the genus *Homo*. But when he wrote his initial description of *Pithecanthropus erectus* in 1894, he recognized that he had a problem. Theunissen describes how he handled it:

> He did make a very glancing reference to the jaw fragment from Kedoeng Broeboes which he had earlier attributed to the genus *Homo*, now suggesting that this fossil might also have belonged to Pithecanthropus. It is fairly easy to fathom

> the reason for this change of opinion. It "must have belonged" to Pithecanthropus because Dubois believed that the Kedoeng Broeboes sediment layers were as old as those at Trinil. The existence of a *Homo* fossil in layers of the same age as those in which the ape-man had been found, would have considerably weakened his argumentation that *Pithecanthropus erectus* was a transitional form between ape and Man. It was obvious, however, that his doubts about the jaw fragment had not dissipated, since he omitted any description or illustration; we can imagine that he was anxious not to present his critics with an easy target.[8]

To preserve the uniqueness of *Pithecanthropus* as the missing link, Dubois had to make sure that no fossils of more modern morphology could be assigned to the same stratigraphic level or given the same date. He did this by (1) changing the assignment of the Kedoeng Broeboes jawbone from *Homo* to *Pithecanthropus*, while at the same time withholding any description or illustration of it so that no one could challenge his assignment, and (2) hiding Wadjak for thirty years so that no one would know he had found such fossils until *Pithecanthropus* had been thoroughly established in the human thought-stream as our evolutionary ancestor.

In 1982, creationist Duane Gish debated C. Loring Brace on the campus of the University of Michigan. Gish had stated that Dubois's secrecy regarding Wadjak "can only be labeled as an act of dishonesty." Brace responded:

> Dubois did publish preliminary accounts of his Wadjak material in 1889 and 1890 before his Trinil discoveries were even made, and he recapitulated these in print in 1892 before becoming involved in what he correctly realized was the far more significant *Pithecanthropus* issue. If there is a question of honesty involved, it has nothing to do with Dubois.[9]

Yes, there is a question of honesty involved. Dubois was deceptive. Technically, Brace told the truth. But he also gave a very false impression.

It is possible to lie by telling the truth. It is done often. Suppose a man owes you one hundred dollars. Because you need the money, you call him to find out when he can pay you. His wife answers the phone and tells you that he is out. You take that to mean that he is unavailable. You don't know that he is standing just outside the front door of his house so that his wife can "honestly" say that he is "out." She justifies herself in that she technically told the truth. But she really lied, because she intended that you would think that "out" meant "unavailable." She lied by telling the truth.

Brace said that Dubois published accounts of the discovery of Wadjak in 1889, 1890, and 1892. Brace knew full well that the two thousand people attending that debate would take "publish" to mean that Dubois had made an announcement to the world, or at least to the anthropological community, about

Wadjak. When someone publishes about a fossil discovery, that is normally what is meant. However, this is not what happened. This publishing was nothing more than Dubois's quarterly and annual reports to the director of education, religion, and industry of the Dutch East Indies government, the department that had granted him permission to work in Java. These were bureaucratic reports and were not intended for the public or for the scientific community. They may not have been read even by the bureaucrats. Did Brace know that? Of course he did. Trivia buffs will recognize that the dates of Dubois's government reports are probably some of the least significant facts in the history of the universe. Anyone who knows those dates certainly knows what kind of reports they were. If they had been public reports, why would Keith and other anthropologists have made excuses for Dubois's keeping Wadjak a secret?

Dating the Wadjak Skulls

The mystery is still not solved. What was there about Wadjak that caused Dubois to keep it a secret for thirty years? Allow me to take you on a behind-the-scenes excursion. We are going to undertake a project: the dating of Wadjak Man. We are going to do it, moreover, by pretending that we are evolutionists. I have long felt that if I were going to critique evolution, I owed it to my readers, as well as to my own integrity, to learn to think like an evolutionist, to honestly try to understand where he is coming from. I have tried very hard to get inside the brain of the evolutionist and understand his system. I do not claim to have succeeded. I claim only to have honestly tried.

As "evolutionists" we recognize an awesomeness, a pagan elegance in unaided nature's capacity to produce the incredible variety of living things we see about us. There is a logic to evolution that cannot be denied. It is not difficult to imagine that the more complex forms we see about us came from simpler forms by gradual change over time. We are familiar with some of the thousands of experiments that tend to support evolution. In just looking at nature, it seems obvious that evolution is true, because things can be arranged in an evolutionary pattern. The fact that some doubt evolution is frustrating. It is hard to deal with people who deny the obvious and will not open their minds.

Thanks to Darwin, we know that it is unscientific to appeal to supernatural forces to explain the universe. Living in a scientific age, we must oppose with all our might those who would turn back the clock and corrupt science by trying to inject religion into it under the term *scientific creationism.* The very idea of scientific creationism is a contradiction in terms. If creationists only understood the scientific method, they would see the light. It is obvious that we evolutionists have done a very poor job of teaching science. We must do a better job in the future, for science itself is at stake.

Our immediate task as evolutionists is to date the Wadjak fossils. To start, we consult Michael Day. He writes of Wadjak, "In view of the degree of mineralization of the skulls and the modern fauna, the earliest possible date is probably late Pleistocene."[10]

We notice several aspects of Day's remarks. By "late Pleistocene" he probably means the Upper Pleistocene, although his term is a bit vague. When he refers to "the earliest possible date," he means that the fossil could be younger than late Pleistocene but certainly not older. He mentions two points upon which his evaluation is based: the mineralization of the skulls and the modern fauna. By modern fauna he means that most, if not all, of the fossil animals, especially mammals, found in association with the skulls are living today. If many of the fossil animals found in association with human fossils are extinct, this seems to indicate a long intervening time span by evolutionary standards. This is then referred to as older fauna.

Dubois recognized that the Wadjak skulls were highly mineralized, or fossilized,[11] which would imply that the fossils did have a bit of age, although fossilization rates are dependent upon a number of conditions. Thus a date of middle Upper Pleistocene, about 50,000 ya, would certainly be a reasonable assessment, and a few evolutionists have given Wadjak that date.

Day further states that the fossil fauna associated with the skulls "does not differ significantly from that found in modern Java."[12] This could argue for a date more recent than 50,000 ya. In the *Catalogue of Fossil Hominids*, the date is given as "late Upper Pleistocene or early Holocene."[13] Since the Holocene is the more recent epoch after the Pleistocene, that would indicate that Wadjak could be dated at about 10,000 ya. This is the generally accepted date of Wadjak. Thus Wadjak is considered to be too recent to have played any part in human evolution, and it is for this reason that Wadjak is virtually unknown. It simply is not an important fossil as far as evolutionists are concerned.

Michael Day and the editors of the *Catalogue*—Oakley, Campbell, and Molleson—are extremely competent scholars. We can safely place our faith in their date for Wadjak. The editors of the *Catalogue* list their resources for their dating of Wadjak: Dubois and von Koenigswald. We place our faith in the evaluation of Oakley, Campbell, and Molleson. They place their faith in the accuracy of Dubois and von Koenigswald.

Since there is a bit of fuzziness about the date of the Wadjak fossils, we could wish that they had been found after the advent of the radiometric dating methods. If these methods had been used on the Wadjak fossils, there would be no doubt about the date. We could trust the results. That is why they are called absolute dating methods. We "evolutionists" have heard that creationists challenge the very foundations of the radiometric dating techniques, but creationist writings are such a confused mixture of science and religion that they are not to be trusted. Besides, creationists have this "infallibility of the Bible" thing that locks them into a preconceived philosophical framework. They are not free to accept data that conflicts with the Bible. We "evolutionists" are not constrained by some rigid philosophical framework. We are free to go with the facts wherever they may lead.

Back to Wadjak. It seems that we can safely conclude that the date for the Wadjak skulls is about 10,000 ya and that our project is complete. We have consulted many other authoritative scholars, and there is almost universal agreement on that date for Wadjak.

The *Catalogue* does mention that found among the fossil fauna at Wadjak was a species of tapir, *Tapirus indicus.* A tapir is a hoofed mammal that looks like a pig but is considered to be more closely related to the rhinoceros. It is a bit curious that of all the fossil fauna found at Wadjak, this is the only one mentioned in the *Catalogue.* Yet nothing is said about its being significant. If it is not significant, why is it mentioned? If it is significant, why wouldn't they give at least a hint as to the reason?

Wadjak seems like such an open-and-shut case that it is very easy to let the tapir item pass by. We are pretending to be evolutionists, however, and evolutionists, by definition, are scholars. So, almost for the fun of it, we decide to check it out. After much research and a feeling that we are wasting our time, we come across a 1951 article by American Museum of Natural History anthropologist Dirk Albert Hooijer, in which he writes, "*Tapirus indicus,* supposedly extinct in Java since the Middle Pleistocene, proved to be represented in the Dubois collection from the Wadjak site, central Java, which is late—if not post—Pleistocene in age."[14]

Hooijer clearly states that Wadjak is late or post-Pleistocene—that is, about 10,000 ya. Yet a species of tapir that became extinct in the Middle Pleistocene was found by Dubois among the fossil fauna. Now we understand why Dubois hid the Wadjak fossils rather than reveal them and try to explain away *Tapirus indicus.* Fossil forms that are believed to have become extinct within a specific time frame can be important indicators in dating fossil assemblages.

A question comes to mind. If *Tapirus indicus* was found in the Wadjak faunal assemblage, and if this species of tapir is believed to have become extinct in the Middle Pleistocene, why couldn't Wadjak be Middle Pleistocene? In spite of Dubois's original claim that *Pithecanthropus* was of Pliocene age, it is now universally placed in the Middle Pleistocene. That would mean that the big-brained Wadjak skulls and smaller-brained *Pithecanthropus* skullcap could be approximately the same age. Perhaps we have just answered our own question. Wadjak and *Pithecanthropus,* both in the Middle Pleistocene, would not fit the evolutionary scenario well at all; in fact, it would be evidence against it. We certainly understand now why Dubois hid the Wadjak skulls for thirty years.

Since you and I still endeavor to think like evolutionists, we recognize that we have a problem. Is there a solution? We think of the vast array of evidence and experiments that support evolution, and of the fact that virtually the entire scientific community endorses it as true. It seems incredible that one tiny fact should threaten such a vast array of evidence. Certainly if there were serious challenges to the fact of evolution, these would be published in the accepted scientific literature. But we do not read of any. Of the two possibilities—that evolution is wrong or that this one fact is wrong—it seems more logical that this one little fact is wrong. Let's start there.

We can explore a number of possibilities for error in the matter of *Tapirus indicus.* (1) Is it possible that this species of tapir did not become extinct in the Middle Pleistocene but actually lived into the Upper Pleistocene or even into the Holocene before becoming extinct, and that fossils of this tapir from these time periods have not been discovered yet? The answer obviously is yes. An argument based on extinction is really an argument from silence, and that type of argument always carries with it a bit of risk. (2) Although Dubois was an anatomist, he would not have been as familiar with tropical animals, and the tapir is a tropical animal. Is it possible that the fossils in question belonged to another species of tapir, or to another type of animal altogether, and that Dubois misinterpreted them? Once again, the answer is yes. (3) Is it possible that the fossils of this particular tapir came from another locality, and that they were put in the Wadjak collection by mistake, by Dubois or someone else? Yes. There may even be other possibilities for a legitimate mistake to have been made regarding these fossils. Since there are so many possibilities for mistakes, and since evolution is so well established, it seems best to go with evolution rather than with the negative implications of this one little questionable fact.

After mentioning the discrepancy in the Wadjak fauna, Hooijer comments on the faunal dating of the fossils in Java:

> It would seem to be only logical to assume that, upon further accurate monographic studies of the various elements to the Pleistocene fauna of Java, the faunal differences between the Lower, Middle and Upper divisions will dwindle more and more, and the picture of the evolution of the fauna will become less and less cataclysmic.[15]

Later Hooijer writes, "Thus, it seems that in Java this method of Pleistocene chronology fails. The method has certainly part of the truth, but in Java it is completely overshadowed by the epeirogenic movements [earth uplifts and downwarps] of much greater magnitude."[16]

Hooijer is saying that the differences in the fossil fauna of the Upper, Middle, and Lower Pleistocene of Java are not that great. This, together with the fact that most of the fossils upon which the faunal dating in Java is based were collected by hired nationals who had absolutely no knowledge of geology, causes Hooijer to say, "The finds of Early Man in Java, therefore, cannot be exactly dated, and the only thing we can honestly say is that they are Pleistocene in age, and most probably neither lower-most nor late Pleistocene."[17]

It is shocking to learn that our problem is far greater than just the dating of Wadjak. Between 1936 and 1941, von Koenigswald found a number of fossils similar to Java Man, now all classified as *Homo erectus.* Some were found in the Trinil beds at Sangiran, some in the lower Djetis beds, and some in the higher Ngandong beds. All of these were found before the advent of radiometric dating (although some of the beds have been dated radiometrically well after the fact). All of these Javanese human fossils were dated according to the alleged evolution of the fossil fauna. Based upon this, much has been made of human evolution within the taxon *Homo erectus* in Java. It is a favorite subject for displays in museums, such as San Diego's Museum of Man. Yet all of us evolutionists know that the Javanese fossils are very poorly dated. When we are faced with such uncertainties, perhaps the best thing to do is to go with the one thing we know for sure, the fact of human evolution—and interpret the fossils accordingly.

The problem is that you and I do not have the time, the money, or the expertise to go to Java to check these things out to our own satisfaction. Even if we did go to Java, it would do no good. The Wadjak site has long since been removed for marble. We have no choice but to put our faith in the skill, accuracy, and evaluation of the experts. Since we know that evolution is true, it seems logical to take the position on the Wadjak fossils that is most consistent with it. We thus accept the date of 10,000 ya for Wadjak.

We have a passing thought. We found a little detail that seemed to militate against Wadjak being late Upper Pleistocene or Holocene. We ruled against it on philosophical grounds. Is it possible that in much of the evidence that we lean on in support of evolution, there were also anomalies that researchers missed or ignored? No, we feel sure that such discordant items would have been published. However, in our report on Wadjak, we will ignore the tapir incident. Our report is more for the public, and the public likes direct answers without a lot of uncertainties and conditions. Too much of that makes it look as if we "evolutionists" don't know what we are doing.

We return now from our imaginary experience as evolutionists. I have attempted to describe the process an evolutionist goes through in dating Wadjak. The newer radiometric dating methods do not clarify the picture. Because scientists refer to radiometric dating as absolute dating, they grant an assumption of precision and accuracy to those methods that they do not deserve. They are usually considered as independent verifications of the age of something. Since the age of *Pithecanthropus I*, the original Java Man, is usually given in the literature as about 500,000 ya, it might be of interest to know how that figure was obtained. Kenneth Oakley, referring to the category of radiometric dating known as A-3 (Absolute 3 category), wrote:

> A3 Dating: The age of a specimen in years inferred by correlation of the source bed with a deposit whose actual age has been determined. Thus the original *Pithecanthropus I* remains from river gravel at Trinil, Java, can be dated as c 500,000 years old on the basis of the K/Ar age of leucite in volcanic rock found elsewhere in Java but containing Trinil fauna (von Koenigswald, 1968).[18]

Thus, the radiometric age of Java Man is not an independent confirmation of the faunal date, but is directly related to the estimates regarding the Trinil fauna. In light of Hooijer's conclusions about the unreliability of dating by the Trinil fauna, the radiometric date for Java Man—almost universally accepted—is worthless. This heightens the possibility that Wadjak and *Pithecanthropus* could be approximately the same age.

I'm sure that an evolutionist would cry "foul" on several of my points in the dating of Wadjak. First is the point that in many situations he goes on faith rather than by facts. In spite of his protests, it is true. We personally check out and examine only a tiny portion of the things we accept as fact. Everything else we accept on faith, which really means we trust in the credibility and accuracy of someone else.

The second point is my contention that he evaluates facts according to his theory or his philosophy. He believes it is the other way around, that his facts have developed his theory. He is mistaken. But I dare not criticize him for doing it, lest I condemn everyone. The old idea that theory always comes from facts is universally believed, even though it is totally inaccurate.

As early as 1935, the Austrian philosopher of science (later at the University of London) Sir Karl Popper demonstrated in his monumental work *The Logic of Scientific Discovery* that scientists do not work according to the so-called scientific method and that they could not work that way even if they wanted to.[19] To say you can start with observations but without a theory is absurd. Scientists simply do not go around collecting observations and data indiscriminately and then trying to fit them into theories. They must start with some theory or concept. This then gives them direction in the collecting of data.

If this idea still sounds strange to you, let me assure you that it is indeed the way science works. Perhaps another illustration will help. Writing about a conversation with Albert Einstein, German physicist Werner Heisenberg, who had been emphasizing the importance of observations in the formulation of scientific theory, recalls his surprise at Einstein's response:

> "But you don't seriously believe," Einstein protested, "that none but observable magnitudes must go into a physical theory?"
>
> "Isn't that precisely what you have done with relativity?" I asked in some surprise. "After all, you did stress the fact that it is impermissible to speak of absolute time, simply because absolute time cannot be observed; that only clock readings, be it in the moving reference system or the system at rest, are relevant to the determination of time."
>
> "Possibly I did use this kind of reasoning," Einstein admitted, "but it is nonsense all the same. Perhaps I could put it more diplomatically by saying that it may be heuristically useful to keep in mind what one has actually observed. But on principle, it is quite wrong to try founding a theory on observable magnitudes alone. In reality the very opposite happens. *It is the theory which decides what we can observe*" (emphasis added).[20]

Obviously the context of that quotation was not a discussion of human fossils. It makes no difference. From physics all the way to anthropology, theory strongly influences facts. That is the way science works because that is the way people work—evolutionists, creationists, and everyone in-between. But no one ever says it. Why? Because it sounds so wrong. It sounds so wrong that we don't even like to think it. But it is not wrong. It is the way we operate. It is the only way we can operate. The only deception is self-deception when we try to tell ourselves that we don't operate that way—that we, above all others, are objective and without bias. In theory, facts determine theory; but in fact, theory determines facts. Ultimately, *everything* is philosophy—or theology.

Meanwhile, it does not seem fair that Dubois's Java Man enjoys worldwide fame while Wadjak Man lives in virtual obscurity. But it does serve to prove that not all fossils are created equal.

CHAPTER 10

The Selenka-Trinil Expedition

A Second Opinion

DUBOIS WAS CERTAIN that he had found what he set out to find: the missing link. But his fossils were the source of heated debate in the scientific community at the end of the nineteenth century.

First was the problem of whether the very archaic-looking skullcap and the very modern-looking femur belonged in the same category, let alone to the same individual. Did their being found fifty feet apart constitute "being together" as Dubois claimed? The bones could just as well have been the result of chance association in flood deposits on the bank of the Solo River.

Second, the nature of *Pithecanthropus* was also a mystery. Was he a human, an apish man, a mannish ape, or an extinct primate with no relationship to humans at all? The limited information did not allow a definitive answer. The Wadjak skulls would have helped, but no one knew about them at the time.

Third was the question of where *Pithecanthropus* fit into the alleged human evolution sequence. Dubois felt that he was in the direct line of human evolution. Some scientists considered him an evolutionary dead end with no direct bearing on human origins; most felt that he was already fully human but belonged to an inferior race.

Last was the matter of his age. The geology of Java was virtually unknown, and Dubois's information was frustratingly skimpy. Dubois assigned the fossil-bearing stratum at Trinil to a much earlier geological period than later investigations determined it to be.

A Munich zoologist, Professor Emil Selenka, saw clearly that the solution to the problem of *Pithecanthropus* was not continued discussion but further exploration. Selenka was one of the leading scientists of his day. Sir Arthur Keith said that he "did more than any man of his time to advance our knowledge of the higher primates."[1] Selenka's dream was to have an expedition go to the place in Java where Dubois had found *Pithecanthropus* and search for more fossils. He was in the process of organizing such an expedition when he died. His wife successfully completed the project, which became known as the Selenka-Trinil expedition of 1907–08.

Frau M. Lenore Selenka was a professor and academician in her own right. As she undertook the expedition, she enlisted Professor Max Blanckenhorn of Berlin as her associate. Sponsorship came from the Berlin Academy of Science and from at least one academic institution in Munich. However, the bulk of the expenses for the expedition came from Frau Selenka's private purse.

Besides Selenka and Blanckenhorn, seventeen other specialists were involved in the study of the 43 large crates of fossils that were returned to Germany. These specialists produced an excellent scientific report of 342 pages, edited by Selenka and Blanckenhorn, published in 1911, and entitled *Die Pithecanthropus-Schichen auf Java.* Few, unfortunately, have ever heard of it. The report has suffered the fate decreed for all evidence that is contrary to evolution: consignment to the lower reaches of oblivion.

British creationist A. G. Tilney, who died in 1976, first made me aware of the Selenka-Trinil expedition report. Tilney served as secretary of the Evolution Protest Movement (now the Creation Science Movement) in England and was a modern language scholar. He searched over sixty libraries on the continent of Europe before he finally located a copy of the report in the Volkschulla Library in Aachen, Germany.

The report has never been translated into English. Being a modern-language professor, however, Tilney was able to study it. Unfortunately, he did not make a full translation of it before his death. My information on the expedition report comes from Tilney's pamphlet on the subject and from a review of the report by Sir Arthur Keith, published in *Nature.*[2,3]

The field of paleoanthropology has had more than its share of scandals and sloppy scholarship. In contrast, the manner in which Frau Selenka organized and executed her expedition and published its results, Keith said, "commands our unstinted praise." The thoroughness and scientific integrity of the expedition are exceeded only by the obscurity into which it has fallen. Although the purpose of the expedition was to confirm Dubois's findings, its results actually contradicted his claims.

According to Tilney, the seventeen specialists who contributed to the expedition's report were also on the scene of the excavations in Java. If that was the case, it was truly a sizable project. Keith mentioned only that "scientific investigators were sent out." If Tilney was right, then Frau Selenka was a pioneer in the multidisciplinary approach to paleoanthropology. Not until the 1970s did F. Clark Howell institute the practice of bringing a number of specialists, such as paleobotanists, paleoecologists, paleozoologists, and geologists, as well as paleoanthropologists, to a fossil site. This practice is considered a quantum leap in the study of the past. Richard Leakey, known for his organizational ability, has popularized this multidisciplinary approach in his work in East Africa. Yet Frau Selenka may have pioneered this concept in 1907, more than sixty years earlier.

In 1907, the Selenka-Trinil expedition journeyed to the banks of the Solo River, an area known as "the hell of Java," and found the stone that Dubois had used to mark the site of his original discoveries. Seventy-five national workers were hired, and barracks were built for them. A Dutch sergeant who had worked for Dubois was also employed by Frau Selenka.

Extensive mining and digging operations were required, as the fossil-bearing layer of the Trinil formation was under 35 feet of volcanic sediment. More than 10,000 cubic meters of material were removed in the search for more remains of *Pithecanthropus*. But *Pithecanthropus* was not to be found.

What they did find was more significant than what they did not find. Three of the specialists, Dr. E. Carthaus, Frau H. Martin-Icke, and Dr. J. Schuster, concluded that Dubois had seriously overestimated the age of the stratum in which *Pithecanthropus* was found. Paleontologist Martin-Icke reported that 87 percent of the gastropods found in it were of modern forms, and botanist Schuster testified that the flora was not too dissimilar from that of today. This would indicate a rather recent age for *Pithecanthropus*, and that alone would eliminate him as the missing link.

In the very same stratum in which *Pithecanthropus* was found, Dr. Carthaus discovered splinters of bones and tusks, foundations of hearths, and pieces of wood charcoal. As a geologist, he felt that the *Pithecanthropus* stratum was rather recent and that modern humans and *Pithecanthropus* (Java Man) had lived at the same time. This was another blow to the missing link status of *Pithecanthropus*.

The most striking discovery was by Dr. Walkhoff, an anthropologist. In the dry bed of a tributary of the Solo River, about two miles from Dubois's famous discovery, he discovered the crown of a human molar. The dentine within the enamel cap had been replaced by a fossilized organic matrix. Although the tooth belonged to a relatively modern human, Walkhoff concluded that its condition indicated an even greater age than the age assigned to *Pithecanthropus*. The tooth is known as the Sondé fossil.

All of the members of the Selenka-Trinil expedition were evolutionists. The purpose of the expedition was to confirm Dubois's findings of fossil evidence for human evolution. But Frau Selenka, the leader of this exemplary expedition, concluded that modern humans and *Pithecanthropus* both had lived at the same time and that *Pithecanthropus* played no part in human evolution. This is the same conclusion that would have been reached had Dubois revealed Wadjak at the time he paraded *Pithecanthropus* before the public. As it was, the Wadjak skulls were still sequestered beneath the floorboards of Dubois's home. It would be another ten years after the release of the Selenka-Trinil expedition report before the Wadjak skulls would see the light of day.

Perhaps the most remarkable part of the report was its description of the violent volcanic eruptions from nearby Mount Lawu-Kukusan and the subsequent flooding that took place in that part of Java every thirty years or so. The geologic activity was so intense that the report concluded that the volcanic Trinil sediments which contained *Pithecanthropus* were far too young to yield any information on human origins. Native traditions tell of the Solo River actually having changed its course in the thirteenth or fourteenth centuries, which could mean that the Trinil beds were only about 500 years old, not the 500,000 years old believed today. Because volcanic matter is heavily mineralized, the report states that the degree of fossilization of *Pithecanthropus* was the result of the chemical nature of the volcanic material, not the result of vast age.

The summary chapter of the report was written by Dr. Max Blanckenhorn. In it he apologized to the reader because what they had hoped would be a corroboration of Dubois's findings seemed more like a debunking of Dubois's work. He used the German word for "fruitless" to describe their failure to substantiate Dubois's claim that the famous Java Man he had discovered was our evolutionary ancestor.

It did not occur to Blanckenhorn that their work was not fruitless but had produced very positive results. It showed that humans have wide morphological variation, something that anthropologists have only recently come to appreciate. It is possible that the variation in early humans could have exceeded what variation there is today. Although the Selenka-Trinil expedition is universally considered to have failed in its primary purpose, in true scientific terms it was a resounding success. Had the wide variation in the human family been appreciated at that time, many of the later mistakes in anthropology could have been avoided.

In light of the expedition's findings, it is interesting to see how evolutionists have handled the Selenka report. With one exception, the newer works on paleoanthropology ignore the Selenka report completely. About half of the books written between 1945 and 1975 mention the expedition or the report but do so in such a way that it would be impossible for the English-reading researcher to discover what the Selenka report actually said. It is an amazing conspiracy of silence.

The Selenka report is most often mentioned for its excellent description of the fossil fauna of the Trinil deposits (a description that Dubois was expected to give but never did). All evolutionist works that refer to the geological findings of the Selenka report misrepresent those findings, which were that the Trinil beds appeared to be recent Pleistocene deposits (this, I assume, would mean Upper Pleistocene). Some writers actually claim that the expedition supported Dubois's assessment that the Trinil beds were Pliocene. Other authors state that the expedition reduced the age of the beds to Lower or Middle Pleistocene age from Dubois's original claim of Pliocene age.

Some authors report that no *Pithecanthropus* fossils were found, but they do not mention that other human fossils and artifacts were found. Other authors state flatly that no artifacts were found at all, when in truth artifacts were found.

Alan Houghton Brodrick mentions that a *Homo sapiens* tooth was found, but since he gives no details, the reader would not know that the tooth was believed by the Selenka-Trinil expedition to be older than *Pithecanthropus.*[4] On the other hand, Carleton Coon mentions the Sondé tooth as probably coming from the Trinil beds, but he is very fuzzy on the category to which it belongs (*Homo sapiens*) and does not associate it with the Selenka-Trinil expedition.[5] Once again, the reader would miss the true impact of the discovery.

Only two books reveal the true significance of the Sondé tooth discovery. The first one is a 1924 work by George Grant MacCurdy:

> The Selenka Expedition of 1907–08 . . . secured a tooth which is said by Walkhoff to be definitely human. It is a third lower molar from a neighboring stream bed and from deposits older (Pliocene) than those in which *Pithecanthropus erectus* was found. Should this tooth prove to be human, *Pithecanthropus* could no longer be regarded as a precursor of man.[6]

The other book is the recent work by Theunissen. Notice that his account does not entirely agree with MacCurdy's.

> The only discovery of (subfossil) hominid remains was that of a fully human tooth, uncovered at Sondé, some distance from Trinil. However, the age of the

> specimen was uncertain. Dubois joined in the discussion arising from this find and—naturally—refused to accept that the tooth could possibly have come from the same layer as the *Pithecanthropus* remains. *He even suggested that there was fraud somewhere* (emphasis added).[7]

The man who could have, and should have, given us a full account of the Selenka-Trinil expedition was G. H. R. von Koenigswald. He spoke German, had access to the report, was a noted paleoanthropologist, and worked many years in Java. Yet he disappoints us. He has, however, added a vital bit of information: "The spot at which the Selenka expedition pitched its camp is still easy to find. It is strewn with innumerable broken beer bottles, testifying to the expedition's thirst."[8] This was a German expedition. Did he expect them to drink Tetley Tea?

The members of the Selenka-Trinil expedition went to Java to confirm Dubois's findings. They thought they had failed. Actually, they succeeded in revealing the true nature of the human fossil record—that the human family had wide morphological diversity—and that Java Man was not our evolutionary ancestor.

I have written at length on the background and details of Dubois and his Java Man. I have done so to show that one of the conclusions of this book—that *Homo erectus* was not our evolutionary ancestor but lived as a contemporary of more modern humans—is not original with me. The evidence has existed for over one hundred years. Dubois himself had three warnings of it: (1) The Trinil femur and skullcap evidenced wide morphological diversity in a single contemporaneous population. Dubois rejected that evidence. (2) Wadjak and *Pithecanthropus* demonstrated the possibility that humans having wide morphological diversity had lived together. Dubois retaliated by hiding Wadjak for thirty years. (3) The Selenka-Trinil expedition confirmed that humans with wide morphological diversity had lived together in recent times. Dubois responded by claiming that the Selenka findings were fraudulent.

Dubois went to the Dutch East Indies to prove human evolution by discovering fossil evidence. He then undertook a campaign to sell his fossils to the world. He succeeded. The result was that he set the study of paleoanthropology back one hundred years.

Was Dubois a bad man? Perhaps yes. Perhaps no. Theunissen puts it this way: "If Dubois deceived others, he deceived himself just as much."[9]

CHAPTER 11

HOMO ERECTUS

A Man for All Seasons

JAVA MAN (*PITHECANTHROPUS I*) was the first of at least 280 similar fossil individuals that have been discovered to date. It would be impossible to exaggerate the importance of this group of fossils, known collectively as *Homo erectus.* For the evolutionist, *Homo erectus* is the major category bridging the gap between the australopithecines (which everyone recognizes as nonhuman) and the early *Homo sapiens* and Neandertal fossils (which are truly human). Thus *Homo erectus* is indispensable to the evolutionist as *the* transitional taxon. (*Homo habilis* is a flawed taxon that will be given a respectful burial in a later chapter.)

Surprisingly, *Homo erectus* furnishes us with powerful evidence that falsifies the concept of human evolution. Three questions are crucial. First, is *Homo erectus* morphologically distinct enough to warrant its being classified as a species separate from *Homo sapiens*? The evidence clearly says no. By every legitimate standard applicable, the fossil and cultural evidence indicates that

Homo erectus is fully human and should be included in the *Homo sapiens* taxon. That is the subject of the next chapter.

Second, are the *Homo erectus* fossils found in the relevant time frame so as to serve as a legitimate transitional form? The clear answer from the fossil record is again no. The *Homo erectus* charts in this book are among the most complete listings of fossils of *Homo erectus* morphology to be found in the scientific literature. When compared with the other charts in this book, they show that *Homo erectus* individuals have lived side by side with other categories of true humans for the past two million years (according to evolutionist chronology). This fact eliminates the possibility that *Homo erectus* evolved into *Homo sapiens.* That this two-million-year contemporaneousness has been largely camouflaged is a tribute to the skill of evolutionist writers.

The third question is whether or not there are adequate nonevolutionary explanations for *Homo erectus,* early *Homo sapiens,* and Neandertal morphology. The answer is yes. There is a touch of humor in the fact that the nonevolutionary factors that adequately explain this morphology are some of the same factors that evolutionists themselves use to explain *Homo erectus*-like fossils when these fossils mischievously show up in the wrong time frame. In this chapter we will discuss the time frame of *Homo erectus.*

The fossil charts in this book are central to the falsification of the theory of human evolution. Both the dates of these fossils and their assignment to specific categories are well documented from recent and reliable evolutionist sources. Many fossils have a degree of uncertainty on their dates. Only the most notorious cases are marked "date uncertain." When a fossil has a range of dates, its placement on the chart represents an average (a fossil dated between 400,000 and 600,000 ya would be placed at 500,000 ya). In some cases there is strong evidence for going against the evolutionist date, and that evidence is documented. It is important to emphasize that the uncertainties in the dating of the individual fossils are not serious enough to affect the conclusions reached in this book, because the fossil sampling is so large and the margin of uncertainty in the dating of each of the fossils is so small relative to the time scale of human evolution.

The evolutionary category to which a fossil belongs can also be a problem. For example, since the category *Homo erectus* includes a suite of morphological characteristics, it is obvious that there must be enough of the fossil individual recovered to make a proper diagnosis. Usually, the less material recovered, the more precarious the diagnosis. It is also generally true that the older a fossil is, the less of its material is preserved. While there are striking exceptions, this means that the older fossils are often the more difficult to diagnose. In the fossil charts, the assignment usually represents the most recent evolutionist scholarship. However, because evolutionists have done much fudging when a fossil is not found in its "proper" time slot, I have not hesitated to go against the grain when the evidence justifies it. In such cases, I have given ample documentation.

Homo erectus is fast becoming the human evolutionist's worst nightmare. Actually, the problem is not so much the category itself but the dates for the fossils in that category. Further, it isn't so much the problem of getting dates for them—although there is some of that—as it is accepting the dates that the results give.

There has always been evidence that *Homo erectus* was both very old and very young—spanning the entire time range of humans. About twenty years ago, however, evolutionists were unified as well as in complete denial. Milford Wolpoff, the leading advocate of the Multiregional Model of human evolution, dated *Homo erectus* from about 400,000 ya to 1.4 Mya.[1] William W. Howells, one of the founders of the African Eve Model (originally called "Noah's Ark" Model—they seem to like Bible names), dated *Homo erectus* from about 400,000 ya to 1.5 Mya.[2] They, together with almost all others, tended to ignore both the older and younger *Homo erectus* fossils. But now there are at least 140 *Homo erectus* fossil individuals dated younger than 400,000 ya and 32 dated older than 1.5 Mya.

It is ironic that it is the island of Java—the site of the discovery by Dubois of the first *Homo erectus*, the original Java Man—that is causing so much trouble. Because there have always been questions about the dating of the Java fossils, veteran dating specialist Garniss H. Curtis (University of California, Berkeley) set out to redate them using the latest methods. He dated the oldest Java beds containing the infant skull (calvaria) of the Mojokerto infant at 1.81 Mya.[3] That age was bad enough.

The real shock came when Curtis dated the Java Solo *Homo erectus* fossil skulls from the Ngandong beds. The dates ranged from 27,000 ya to 53,000 ya, almost 400,000 years younger than some previous estimates.[4] The authors of the article in *Science* then make this absurd statement: "The new ages raise the possibility that *H. erectus* overlapped in time with anatomically modern humans (*H. sapiens*) in Southeast Asia." Possibility? There is absolutely no question that *H. erectus* overlapped in time with anatomically modern humans (*H. sapiens*) in Southeast Asia—and in other parts of the world as well.

Many paleoanthropologists are having trouble accepting the new dates. Tattersall seems to accept the new Asian dates whereas Klein does not.[5,6] The problems are: (1) For *Homo erectus* to coexist with modern humans is not the way evolution is supposed to work; and (2) For *Homo erectus*, or any species, to exist for almost 2 million years without any significant evolutionary change is also not the way evolution is supposed to work.

Regarding the problem of coexistence, even a span of one million years of coexistence makes evolutionists very nervous, as seen in the following quotation by Susman, Stern, and Rose regarding two *Homo erectus* innominate (hip) bones from East Africa.

> The morphological similarity of O. H. 28 and KNM-ER 3228 supports their assignment to the same taxon *but for the large discrepancy in their ages.* The age of KNM-ER 3228 is roughly 1.5 m.y. (or greater) while O. H. 28 is dated at around 0.5 m.y. The possibility thus exists that 3228 represents the taxon *H. habilis* while O. H. 28 represents *H. erectus* and that locomotor anatomy grades subtly from one taxon to the other. KNM-ER 3228, a male, suggests considerable sexual dimorphism within *H. habilis* and that dimorphism was subsequently reduced in *H. erectus* (emphasis added).[7]

The fossil hip bones, KNM-ER 3228 and O. H. 28, are very similar and are normally assigned to *Homo erectus.* The problem is that they are dated one million or more years apart, and they show no evolutionary change over a million-year span. The solution is to wave the magic wand and turn KNM-ER 3228 into *Homo habilis* without a shred of supporting physical evidence. However, *Homo habilis* is now known to be considerably smaller than *Homo erectus* (about half the size). The solution is to call 3228 a male (without evidence), and to ascribe the large size difference in *Homo habilis* to sexual dimorphism. The implication then is that all of the other *Homo habilis* fossils (with the exception of 1470, 1481, and 1590) are female because of their small size. That assumption is also without physical evidence. A further undocumented assumption is that in evolving from *Homo habilis* to *Homo erectus,* sexual dimorphism changed a great deal, while the locomotor anatomy of the two taxa changed hardly at all. I will allow my readers to pass judgment on the validity of those assumptions. The most recent dating suggests an even larger time separation of O. H. 28 and KNM-ER 3228 hip bones—1.2 million years.

A different type of problem is the date obtained by Curtis of 1.8 Mya for the Mojokerto infant *Homo erectus.* This date causes *Homo erectus* to trespass dangerously into the private time domain of *Homo habilis,* which is dated from about 1.5 Mya to about 2.0 Mya. Nor are evolutionists comfortable in having *Homo erectus* come all the way up to 27,000 ya. Most evolutionists believe that we have *Homo erectus*-like genes in us, whether we evolved from the entire *erectus* population (neo-Darwinism) or from a small segment of that population (punctuated equilibria). With the new view that modern humans originated in Africa at about 200,000 ya, to have *Homo erectus* dated from 27,000 to 53,000 ya is unacceptable. Some have tried to solve the problem by splitting the *Homo erectus* taxon, with *Homo erectus* being strictly an Asian phenomenon and the African *Homo erectus* fossils being considered a separate and older species, *Homo ergaster*. *Homo ergaster*, not *Homo erectus*, would then be our ancestor in the African Eve Model.

The problem is not just the possible 27,000-year-old date for the Java Solo fossils. At the Upper Pleistocene site at Coobool Crossing, New South Wales, Australia, a number of fossils are said to exhibit the robust (*erectus*-like) morphology similar to other *Homo erectus* discoveries in Australia.[8] The literature also indicates that about 130 fossil individuals are found at that site and that the fossil population there is a mixture of robust and gracile individuals.[9] Since no published material indicates how many individuals exhibit this robust morphology, I have entered it on the chart as (2+) and counted it as 2, indicating that more than one *erectus*-like individual is at that site. It is likely that many more than two *erectus*-like individuals were at that location, but we may never know. Because of demands by the Australian aboriginal community, those remains were reburied in 1985.

There is a sense in which, for the evolutionist, the *Homo erectus* situation in Asia is much like the Neandertal situation in Europe; all the more so because *Homo erectus* is just a slightly smaller version of Neandertal. The youngest Neandertal fossils, other than those challenged by evolutionists, are 28,000 to 30,000 years old. Since biblically Neandertals and modern humans would be interfertile, that would be more than enough time to go from a Neandertal morphology to a modern human morphology by genetic recombination. However, 30,000 years is not enough time for the Neandertals to *evolve* into modern humans—for reasons set forth elsewhere. Therefore, the evolutionist must insist that the Neandertals became extinct without issue to protect the African Eve Model and to avoid any racist implications.

There are *at least* 78 *Homo erectus* fossil individuals dated more recently than 30,000 years ago, the youngest one being 6,000 years old. Since biblically *Homo erectus* and modern humans would be interfertile, that also would be more than enough time to go from a *Homo erectus* morphology to a modern human morphology by genetic recombination. However, 6,000 years is not enough time for *Homo erectus* to *evolve* into modern humans. Therefore, evolutionists must ignore these fossils, challenge the date, challenge the morphology and claim that they are *Homo sapiens*, or claim that they were in an evolutionary backwater and became extinct. But those 78 *Homo erectus* fossil individuals simply will not go away.

On the far end of the *Homo erectus* time continuum, *Homo erectus* is contemporary with *Homo habilis* for 500,000 years. In fact, *Homo erectus* overlaps the entire *Homo habilis* population, as the *Homo habilis* chart in this book shows. The oldest *Homo erectus* (or *Homo ergaster*) fossil is dated at 1.95 Mya. The oldest *Homo habilis* fossil is just a little over 2.0 Mya.[10] Thus the almost universally accepted view that some form of *Homo habilis* evolved into *Homo erectus* (or *Homo ergaster*) becomes impossible. In a later chapter, we will demonstrate that

Homo habilis is a flawed taxon because it is a mixture of some fossils that might be human (KNM-ER 1470, 1481, and 1590) and other fossils that are definitely not human. But even if *Homo habilis* were a legitimate taxon and 1470, 1481, and 1590 were proper members of that taxon, *Homo habilis* could not be the evolutionary ancestor of *Homo erectus* (or *Homo ergaster*) because the two groups lived at the same time as contemporaries.

In chapter 5 we discussed the process by which one species allegedly evolves into another species (regardless of whether the evolutionist believes in phyletic gradualism or punctuated equilibria). For some form of *Homo habilis* to evolve into *Homo erectus* (or *Homo ergaster*), *Homo habilis* must precede *Homo erectus* in time. Furthermore, after *Homo habilis* has evolved into *Homo erectus*, *Homo habilis* must be eliminated by death, because *Homo erectus* is supposedly the better fit of the two in the intense competition for limited resources. Yet the fossil record shows that (according to evolutionist dating) *Homo habilis* and *Homo erectus* existed side by side as contemporaries for half a million years. The fossil record also shows that *Homo erectus* lived alongside the early *Homo sapiens* and the Neandertals for the entire 700,000 years of early *Homo sapiens* history and the 800,000 years of Neandertal history, and that *Homo erectus* lived alongside modern *Homo sapiens* for two million years (according to evolutionist chronology). This does not constitute an evolutionary sequence.

When a creationist emphasizes that, according to evolution, descendants can't be living as contemporaries with their ancestors, the evolutionist declares in a rather surprised tone, "Why, that's like saying that a parent has to die just because a child is born!" Many times I have seen audiences apparently satisfied with that analogy. But it is a very false one. In evolution, one species (or a portion of it) allegedly turns into a second, better-adapted species through mutation and natural selection. In the context of human reproduction, I do not turn into my children; I continue on as a totally independent entity.

Furthermore, in evolution, a certain portion of a species turns into a more advanced species because that portion of the species allegedly possesses certain favorable mutations that the rest of the species does not possess. Thus the newer, more advanced group comes into direct competition with the older, unchanged group and eventually eliminates it through death. The older group is not able to compete successfully for the limited resources available. This competition and eventual death of the less fit is indispensable to the evolutionary process. However, in the human reproductive process, I do not compete with my child. I devote all of my resources to the survival of my child—not to his death. The analogy used by evolutionists is without logic, and the problem of contemporaneousness remains.

Terms like *Homo erectus* and *Homo habilis* are convenient terms to use in reference to groups of fossil material. But it is obvious that when evolutionists give dates for *Homo erectus* that do not fit the fossil material, or when they say that *Homo habilis* evolved into *Homo erectus*, contrary to what the fossil material shows, they are using those terms in a manipulative manner without regard for the fossil material in those categories. It is not unusual in evolutionary charts to show *Homo habilis* somewhat below *Homo erectus*, implying that *Homo habilis* is earlier in time. Klein does this several times. The reasons this manipulation is not valid are: (1) It is just an assumption that *Homo erectus* must have had evolutionary ancestors, and (2) Almost all of the fossils dated earlier than 2.0 Mya are too fragmentary for a legitimate diagnosis. Anyone who is concerned about truth in packaging ought to be concerned over the way the relationship of *Homo habilis* to *Homo erectus* is presented in the evolutionist literature. Evolutionists demonstrate a marvelous faculty for snatching fantasy from the jaws of truth.

In 1991, Bed I at Olduvai Gorge was redated using the 40Ar-39Ar laser-fusion technique. At least nine *Homo habilis* individuals have come from this bed. The reasons given for redating Bed I are that "precise age estimates have been elusive" and "its detailed chronology is largely unknown."[11] Since Bed I had already been redated several times, and the Leakeys and others have worked at Olduvai Gorge for over thirty years, the need to redate Bed I does not inspire confidence in the dating methods. Although the reasons given for redating are certainly valid ones, I suspect that there was also the hope that redating could put a bit more age on *Homo habilis* to improve its credibility as a transitional taxon. Unfortunately, there was no evidence for an older date for *Homo habilis*, and the need was expressed for even further dating work on Lower Bed I.

If the fossil evidence at the far end of the *Homo erectus* spectrum shows no evidence of its having evolved from *Homo habilis*, the evidence at the near end of that spectrum shows no evidence of its having evolved into *Homo sapiens*. The most dramatic illustration of this condition of the fossil record is the Kow Swamp, Australia, fossil material initially discovered in 1967 and published in 1972 in *Nature*.[12] (The Cohuna cranium, having the same morphology, was discovered in Kow Swamp in 1925.) Others that are included in the group known as robust, *erectus*-like Australian fossils are the Talgai cranium discovered in 1886, the Mossgiel cranium discovered in 1960, and the Cossack skull discovered in 1972. Newer Australian sites yielding robust fossils include Willandra Lakes and Coobool Crossing. The Java Solo fossils from the Ngandong Beds exhibit this same robust, *erectus*-like morphology.

These discoveries, however, are just the tip of the iceberg. At least four other locations near Kow Swamp are said to contain material of similar robust (*Homo erectus*) morphology but have not yet been explored in detail. Two of these sites are at Gunbower and Bourkes Bridge.[13] A third site is near the Murray River, and a fourth is at Lake Boga.[14]

The evolutionists' response has been interesting and predictable. Although these robust Australian fossils (Kow Swamp and Cohuna,[15] Mossgiel,[16] Talgai,[17] and WHL-50 from Willandra Lakes[18]) are said in the literature to have *Homo erectus* features, the evolutionist has waved his magic wand and called them *Homo sapiens* because of the very late date. Since *Homo erectus* long ago was supposed to have evolved into *Homo sapiens*, it is simply unthinkable that any *Homo erectus* fossils could still be around so recently. Thus, any thinking person would know that these fossils are *Homo sapiens*, no matter what they look like. The most common explanation now given by evolutionists for this *erectus*-like morphology is cranial deformation (discussed in a later chapter). We are asked to believe that evolution produced the *Homo erectus* morphology in the Lower and Middle Pleistocene, but that cranial deformation was responsible for this very same *Homo erectus* morphology after these individuals had allegedly evolved into *Homo sapiens* in the Upper Pleistocene.

While evolutionists have not yet developed a formal definition for *Homo erectus*, a suite of characteristics is generally accepted:

1. Skull low, broad, and elongated
2. Cranial capacity 750–1250 cc
3. Median sagittal ridge
4. Supraorbital ridges
5. Postorbital constriction
6. Receding frontal contour
7. Occipital bun or torus
8. Nuchal area extended for muscle attachment
9. Cranial wall unusually thick overall
10. Brain case narrower than the zygomatic arch
11. Heavy facial architecture
12. Alveolar (maxilla) prognathism
13. Large jaw, wide ramus
14. No chin (mentum)
15. Teeth generally large
16. Postcranial bones heavy and thick

Where there is material for comparison, the Kow Swamp fossils, as well as the other robust Australian fossils, fit the above description well—allowing for reasonable genetic variation. They qualify as *Homo erectus*, as the evolutionist uses the term.

The evolutionists' attempt to explain the Kow Swamp fossils (and others) by calling them the result of an isolated population that was removed from the evolutionary mainstream also fails. While most of these robust fossils were found in southeast Australia, the Cossack skull was found on the west coast of Australia, two thousand miles away, and the Java Solo people were found three thousand miles away. We are dealing with a continent-wide phenomenon. Furthermore, the Cossack skull has a maximum age of 6500 ya but a minimum age of just a few hundred years.[19] Thus, it is possible that *Homo erectus*, whom the evolutionist claims is our evolutionary ancestor, walked the earth just a few hundred years ago. He is truly a man for all seasons.

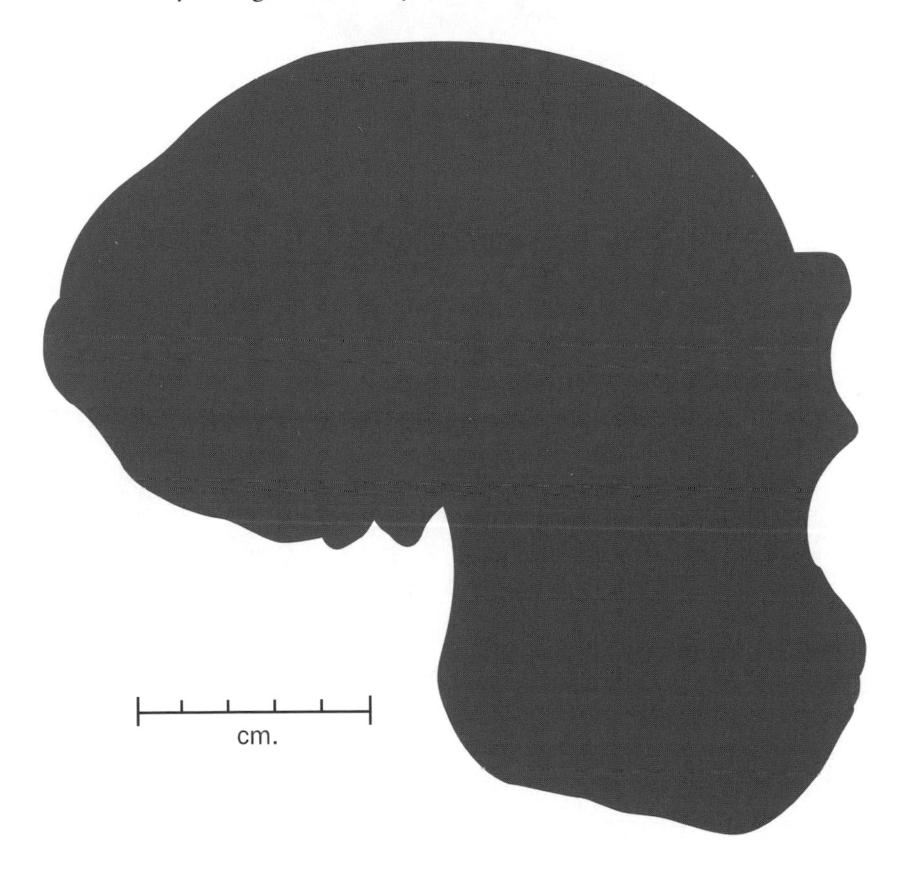

Peking Man Skull, Zhoukoudian, China
(Weidenreich reconstruction)
Homo erectus

CHAPTER 12

Homo Erectus

All in the Family

Homo sapiens to *Homo erectus*: "I'm OK. You're just so-so."

—Evolutionist version of "I'm OK; you're OK."

IS *HOMO ERECTUS* MORPHOLOGICALLY distinct enough to warrant its being classified as a separate species? There have always been evolutionists who have asked that question. Now their tribe is increasing. Michael Day, reviewing G. Philip Rightmire's 1990 book, *The Evolution of Homo erectus*, writes:

> Of the three stages we know of the evolution of man (the australopithecine ape-men, *Homo erectus* the first true men, and early *Homo sapiens* our own species) *Homo erectus* of the Middle Pleistocene would have seemed the most clearly understood and the most taxonomically stable of them all a relatively few years ago—not any more. Important new finds as well as new ways of thinking about hominid taxonomy have thrown this "species" into the same turmoil as all of the others.[1]

Day then enumerates the many questions that now embroil *Homo erectus.* One is, "Does *Homo erectus* exist as a true species or should it be sunk into *Homo sapiens*?"[2]

Today, that question has politically correct overtones. Those favoring the Out of Africa or African Eve Model, to camouflage the latent racism of human evolution in the present, tend to shift that racism to our past history. Hence, they suggest a proliferation of species in the past, including significant distinctions between *Homo erectus* and *Homo sapiens.* However, all modern humans evolved from that African Eve who was already a *Homo sapiens,* thus minimizing all racial distinctions and avoiding racism.

Adherents of the Multiregional Continuity Model believe that *Homo erectus* evolved independently into *Homo sapiens* in Asia, Africa, and Europe. Sufficient gene flow between the three areas prevented the evolution of three different species of modern *Homo.* They claim that it is impossible to discern a specific point at which *Homo erectus* became *Homo sapiens.* Hence, the only significant distinction is one of *time—Homo erectus* being older than *Homo sapiens.* Since time alone is not a sufficient basis for establishing a new species, they favor combining the two species.

Because of the political nature of today's debate, I have included older quotations to demonstrate that this is more than just a political issue. It involves a very false distinction between two legitimate portions of the human family.

Milford Wolpoff has been one of the most vocal evolutionists calling for the "sinking" of the taxon *Homo erectus* into *Homo sapiens.* He writes in conjunction with Wu Xin Zhi (Institute of Paleoanthropology, Beijing) and Alan G. Thorne (Australian National University), "In our view, there are two alternatives. We should either admit that the *Homo erectus/Homo sapiens* boundary is arbitrary and use nonmorphological (i.e., temporal) criteria for determining it, or *Homo erectus* should be sunk [into *Homo sapiens*]."[3]

They then quote Franz Weidenreich. Weidenreich wrote the original descriptions of *Sinanthropus pekinensis* (Peking Man, now *Homo erectus*) and made the very fine plaster casts of those fossils before the originals were lost. Writing in 1943, more than sixty years ago, Weidenreich recognized that this classic *Homo erectus* material was not that different from *Homo sapiens*: "It would not be correct to call our fossil '*Homo pekinensis*' or '*Homo erectus pekinensis*'; it would be best to call it '*Homo sapiens erectus pekinensis.*' Otherwise it would appear as a proper 'species' different from '*Homo sapiens*' which remains doubtful, to say the least."[4]

William S. Laughlin (University of Connecticut), in studying the Eskimos and the Aleuts, noted many similarities between these peoples and the Asian *Homo erectus* people, specifically Peking Man. He concludes his study, in 1963, with a very logical statement:

When we find that significant differences have developed, over a short time span, between closely related and contiguous peoples, as in Alaska and Greenland, and when we consider the vast differences that exist between remote groups such as Eskimos and Bushmen, who are known to belong within the single species of *Homo sapiens*, it seems justifiable to conclude that *Sinanthropus* [Peking Man] belongs within this same diverse species.[5]

In the 1970s, a number of evolutionists expressed the fact that *Homo erectus*, while a bit different, is not so different from modern humans as to warrant a separate species designation. Gabriel Ward Lasker (Wayne State University) has written:

Homo erectus is distinct from modern man (*Homo sapiens*), but there is a tendency to exaggerate the differences. Even if one ignores transitional or otherwise hard to classify specimens and limits consideration to the Java and Peking populations, the range of variation of many features of *Homo erectus* falls within that of modern man.[6]

The authors of the Time-Life book on *Homo erectus*, entitled *The First Men*, had as their consultants two eminent authorities, F. Clark Howell (University of California, Berkeley) and Bernard Campbell (University of California, Los Angeles). They comment on the postcranial anatomy of *Homo erectus*:

His bones were heavier and thicker than a modern man's, and bigger bones required thicker muscles to move them. These skeletal differences, however, were not particularly noticeable. "Below the neck," one expert has noted, "the differences between Homo erectus and today's man could only be detected by an experienced anatomist."[7]

In an article in *Geotimes*, October 1992, regarding Kenneth A. R. Kennedy (Cornell University), the author states: "Kennedy would like to bury the taxon *Homo erectus* altogether."[8]

A remarkable discovery of an almost complete *Homo erectus* skeleton dated at about 1.6 Mya occurred in Kenya in 1984. It was the first time that a *Homo erectus* postcranial skeleton was uncontested as to its association with a *Homo erectus* cranium. This skeleton brought some surprises. Designated KNM-WT 15000, it is thought to be that of a twelve- to thirteen-year-old boy. Yet he was 5'4" to 5'6" tall. Had he grown to maturity, it is estimated he could have been six feet tall. This was a new insight into the stature of *Homo erectus*, who was always assumed to have been smaller than modern humans.

Susman, Stern, and Rose give their assessment of *Homo erectus* anatomy: "Changes in locomotor anatomy from *H. erectus* to modern man were relatively minor and by earliest *H. erectus* times body size was essentially modern."[9]

The only true test of whether or not *Homo erectus* and *Homo sapiens* are members of the same species is the test of interfertility, which for obvious reasons is impossible to apply. Most Bible scholars would agree that if Adam and Eve were somehow restored to life, we would be able to produce offspring with them. The intervening years would not have affected the interfertility of the human race from its beginning. The biblical word *kind* is based upon this interfertility. We thus speak of mankind and humankind. Unfortunately, fossils do not reproduce. However, Donald Johanson expresses this opinion:

> It would be interesting to know if a modern man and a million-year-old *Homo erectus* woman could together produce a fertile child. The strong hunch is that they could; such evolution as has taken place is probably not of the kind that would prevent a successful mating. But that does not flaw the validity of the species definition given above, because the two cannot mate. They are reproductively isolated by time.[10]

Note that the major obstacle Johanson sees to a *sapiens-erectus* mating is the time element. We have already stressed that a species distinction based solely on the time element is an evolutionary concept. It is valid only if evolution is valid. The fact that Johanson believes that *erectus* and *sapiens* could mate if they were living at the same time is actually a confession, although he would deny it, that the two belong in the same species. Furthermore, the fossil record shows that *Homo sapiens* and *Homo erectus were living at the same time.*

Johanson's statement reveals the semantic confusion rampant today. If one million years would not produce enough significant genetic change to inhibit conception, then the differences between *Homo erectus* and *Homo sapiens* are not the result of evolution but instead represent genetic variation within one species. Furthermore, although I am genetically isolated from my great-grandmother because of time, this does not mean that she and I are in different species. A species distinction based primarily on time is an evolutionary necessity but absurd nonetheless. Evolutionary time is imaginary time.

When we compare the crania of *Homo erectus* with those of early *Homo sapiens* and Neandertal, the similarities are striking. My own conclusion is that *Homo erectus* and Neandertal are actually the same: *Homo erectus* is on the lower end, with regard to size, of a continuum that includes *Homo erectus,* early *Homo sapiens,* and Neandertal. The range of cranial capacities for fossil humans is then in line with the range of cranial capacities for modern humans. Modern humans have a cranial capacity range from about 700 cc all the way up to about 2200 cc.[11] This range—a factor of three—is an amazing spread and is most unusual in the biological world. It is recognized that this spread has virtually nothing to do with intelligence, because human intelligence is more dependent on how the brain is organized than on brain size alone.

The cranial capacity of *Homo erectus* goes from about 780 cc (Gongwangling, China; Lantian 2, China; and Dmanisi D22880, Georgia, CIS) to about 1225 cc (the largest Peking Man skull, Zhoukoudian X). The "classic" Neandertal cranial capacity begins at about 1200 cc (Saccopastore I, Italy) and goes to about 1650 cc (La Ferrassie, France), with a few Neandertals possibly going a bit higher. If the African early *Homo sapiens* fossils are factored in, they would fit in at the transition, with cranial capacities of about 880 cc (Salé, Morocco) to about 1480 cc (Jebel Irhoud I, Morocco). Even when these three categories are considered as a single unit, they still do not reflect the cranial capacity range of modern humans.

Homo erectus and the Neandertals are very similar in cranial morphology. In the question period following a lecture at Colorado State University by Neandertal authority Erik Trinkaus, I asked him, "Other than brain size, what are the differences in cranial morphology between *Homo erectus* and Neandertal?" His reply was, "Virtually none."[12]

In an article on *Homo erectus*, C. Loring Brace wrote, "Some scholars even treat Neandertal as late *erectus*."[13] And Harry L. Shapiro (American Museum of Natural History) earlier observed, "When one examines a classic Neanderthal skull, of which there are now a large number, one cannot escape the conviction that its fundamental anatomical formation is an enlarged and developed version of the *Homo erectus* skull."[14]

It is because of the similarities of the various categories of fossil humans that evolutionists have had great difficulty in assigning certain fossils to a specific category. The African early *Homo sapiens* fossils have been referred to in the past as "African Neandertals." Many of the Asian *Homo erectus* fossils have been termed "Asian Neandertals." A running battle is currently going on over the precise assignment of many fossils. The problem of deciding which fossils properly belong in the *Homo erectus* taxon is so vexing that Jerome Cybulski (National Museum of Man, Ottawa) laments:

> Indeed, one may well wonder whether agreement will ever be reached as to which fossils do belong to or represent the taxon, and on what morphological-cum-phylogenetic grounds fossil hominids are or are not to be regarded as *Homo erectus*.[15]

An evolutionist would claim that in an evolutionary continuum we would expect to have a number of fossils "on the line," since the transition points of the various species are arbitrary because of the nature of the evolutionary process. We agree. If evolution were true, that would be the case. However, there is another, more satisfying way in which to interpret the data: a morphological continuum that includes just one species of humans, called *Homo sapiens.* In a court of law, a case must be proven beyond all reasonable doubt. The fact that the data can be explained in at least two ways shows that the evolutionist is far from proving his case. That he *never* considers the other arrangement shows that he is far from impartial.

There is a way by which we can discriminate between the two possible explanations of the data and thereby determine which is the more likely to be correct. That is to place *all* of the relevant fossil material on a time chart according to the evolutionist dates for each of the fossil individuals and to evaluate the results as to whether the evidence favors an evolutionary or a morphological continuum. When this is done, as it is done in this book, the evidence is strongly in favor of a morphological continuum, both horizontally across species and vertically over time. The horizontal continuum shows that anatomically modern *Homo sapiens*, Neandertal, early *Homo sapiens*, and *Homo erectus* all lived as contemporaries over extended periods of time. The vertical continuum shows that as far back as the human fossil record goes, humans have not evolved from something else.

This condition is what the creation model would predict. It is what we would expect if creation were true. The evidence, in fact, is so strong for the creation model of human origins that it is extremely unlikely that any future fossil discoveries could weaken it. This is because no future fossil discoveries in the 1–4.5 Mya time period could cancel out the solid body of factual evidence that has already been accumulated. Up to now, new fossil discoveries have only strengthened the creationist position. It is understandable why evolutionist books seldom carry the type of human fossil charts found in this book. Charts of bits and pieces of the human fossil record abound in evolutionist books, but one will look in vain in an evolutionist work for a time chart that arranges all of the relevant human fossil material according to the *morphological* description of the individual fossils. In studying the latest works of Richard Klein, Ian Tattersall, and others, it is obvious that many of their dates are based upon their belief in the Out of Africa Model of human evolution, rather than on the morphology of the fossils.

The Archaeological Evidence

Archaeological evidence also demonstrates that the distinction between *Homo erectus* and *Homo sapiens* is an artificial one. Although there are limitations in the archaeological record, all of the evidences that one could reasonably expect to be discovered that would demonstrate the full humanity of *Homo erectus* have already been found in association with him.

Of the 83 localities where *Homo erectus* fossils have been found, a minimum of 40 sites have also yielded stone tools. Eight of the *Homo erectus* sites contain evidence of the controlled use of fire. Most significant is that at the oldest sites, both in East Africa and at Swartkrans, South Africa, thought to date between 1.5 and 2 Mya, *Homo erectus* fossils have been found in association with both tools and fire.[16] While it is technically impossible to prove that the stone tools and the fire at a given site were made by *Homo erectus* individuals, the sheer number of associations makes it unreasonable to believe otherwise.

Three Upper Pleistocene *Homo erectus* sites show evidence of burial, one of a cremation, one of the use of red ochre, and another of the use of bone-chopping tools. Perhaps most amazing is the discovery of a two-inch-long piece of quartzite rock that appears to be carved into a human figurine. It is 400,000 years old and was found near Tan-Tan, Morocco. Four *Homo erectus* fossil individuals, dated a bit earlier, come from Morocco. When examined under a microscope, many of the grains of the rock in the object were fractured, a condition only duplicated by using a stone hammer and flake. There were also microscopic remnants of iron and manganese oxide (chemicals used in early red pigments) on it, indicating that the object had been painted. It could be the oldest art object ever found.[17]

The Chinese *Homo erectus* were thought to be less evolved because of a lack of stone tools. Acheulean-like tools have now been found in China, dated at 800,000 ya.[18] Evidence of shelters, allegedly made by *Homo erectus,* have been found in Japan and dated at 500,000 ya.[19] We hardly dare ask the archaeological record for more evidence of the true and full humanity of the individuals having a *Homo erectus* morphology.

There is also evidence of the possible use by *Homo erectus* of seaworthy craft. An amazing discovery in 1994 on the island of Flores, Indonesia, has been reported by Michael Morwood (University of New England, NSW, Australia) and associates. At two sites on Flores, animal fossils and 14 stone tools—flakes and choppers for working wood or animal carcasses—have been found. According to zircon fission-track dating, they are 900,000 years old. The tools show signs of having been used and retouched. The animal remains found in direct association with the stone tools are large elephant, pigmy elephant, giant tortoise, crocodile, Komodo dragon, and giant rat. Although there were no human fossils, it is assumed that the tools were made by *Homo erectus.* We know that *Homo erectus* (Java Man) reached Java, which was, during the Ice Age, connected to the Asian mainland.

The remarkable thing about this discovery is that Flores is several hundred miles and three deepwater straits east of Java. The widest of these is 12 miles, and none were dry land during the Ice Age. Morwood and his associates write, "We conclude that *Homo erectus* in this region was capable of repeated water crossings using watercraft." They add that this amazing discovery, and others, "suggests that the cognitive capabilities of this species [*Homo erectus*] may be due for reappraisal."[20] Regarding the nature of the watercraft, Morwood reasons, "I think you need directed watercraft. You'd have to have some means of steering, and some means of propulsion. If you try to put a few logs together and jump on it, you're probably going to die."[21] Before this discovery, the earliest use of watercraft was thought to have been by modern humans in the colonization of Australia about 40,000 to 60,000 years ago.

One of the most popular myths of human evolution is that stone tools testify to the increasing mental and conceptual abilities of humans as they evolved. The most primitive stone tools, Oldowan, once were identified by evolutionists with *Homo habilis*, the most "primitive" of humans. Acheulean tools (named for the French site at which they were discovered) were associated with the more evolved *Homo erectus*. Neandertal was said to be associated with the even more advanced Mousterian tool kits. The most sophisticated and artistic stone tools were identified with Cro-Magnon and other relatively modern peoples. Thus, stone tools were once considered an almost independent confirmation of the evolutionary development of the human mind.

Things are different now. Almost every basic style of tool has been found with almost every category of human fossil material. Stringer and Grün write, "The simplistic equation of hominids and technologies in Europe has thus been abandoned."[22] The fallacy of the evolutionary archaeologist was to equate simple with primitive. Louis and Mary Leakey were among the first to identify the primitive Oldowan tools with the primitive *Homo habilis*, both thought to be about two million years old. However, Mary Leakey tells of discovering Oldowan-type tools in Kanapoi Valley, northern Kenya, associated with potsherds and hut circles that gave evidence of being rather recent. She wrote:

> The occurrence of an industry restricted to heavy duty tools of Lower Palaeolithic facies associated with pottery and hut circles, is an anomaly hard to explain. It may be noted, however, that a crude form of stone chopper is used in the present time by the more remote Turkana tribesmen in order to break open the nuts of the doum palm.[23]

On these "primitive" Oldowan tools, Lawrence Robbins (Michigan State University) commented, "It is interesting that these oldest of technological items were among the most successful inventions for they continued to be manufactured throughout the entire Stone Age."[24]

If these Oldowan tools were so successful and efficient that they were used throughout the entire Stone Age (Paleolithic, Mesolithic, and Neolithic), if they were the best tools for certain jobs, and if they are still the best tools for certain jobs in some parts of the world today, is it intellectually honest for evolutionists to refer to them as primitive and use them as evidence for the evolution of the human brain?

The World's First Swiss Army Knife

One of the most common Acheulean tools is the hand ax, which in the past has been almost exclusively identified with *Homo erectus*. So complete was this identification that if hand axes were found at a habitation site, it was called a *Homo erectus* site even though no *Homo erectus* fossils were found there to so identify it. Now it is known that Acheulean tools, including hand axes, have also been found with Neandertal and other *Homo sapiens* fossils.[25] Furthermore, many Asian *Homo erectus* fossils are found with tools considered to be more primitive than those of the Acheulean culture. (In Asia, it is now believed, bamboo tools were used more extensively than stone tools, which could account for stone tools being less frequent and more "primitive" there.)

The Acheulean hand ax, however, was truly used worldwide. It is found from northern Europe to southern Africa, and from the Mediterranean to India and Indonesia. It is also mystifying. Although it is called a hand ax, no one knows for sure what its use was. In shape it resembles a giant almond, pointed at one end and round on the other. The pointed end is thinner, the rounded end thicker, but overall it is rather flat like the almond. Because the rounded end is thicker, it has an eccentric center of gravity; in lengthwise cross section, it looks like a very tall and skinny teardrop. The length ranges from a few inches to well over a foot. I have seen some that were rather crudely made and others that were works of art. It has a cutting edge all around its perimeter, and as far as we know, it was never hafted (used with a handle). The assumption is that it was some type of chopper; hence its name. The problem is that since it is sharp all around, it could do as much chopping on the hand using it as it did on the object being chopped.

Eileen M. O'Brien (University of Georgia) has a better idea.[26] Her experiments led her to conclude that the hand ax was actually a flying projectile weapon, thrown discus style and used in the hunting of large game. To test this idea, she had a fiberglass replica made of one of the largest hand axes in the collection at the National Museums of Kenya, Nairobi. It was about a foot long and weighed a bit over four pounds. She then had several discus throwers practice with it. When thrown, the hand ax spun horizontally as it rose, like a discus. But when it reached its maximum altitude, it flipped onto its edge and descended that way. It landed on its knife-sharp edge 93 percent of the time, and 70 percent of the time it landed point first. The average throw was over one hundred feet, and it was usually accurate to within two yards right or left of the line of trajectory. The Olympic record for the discus, which weighs about the same as O'Brien's hand ax replica, is well over two hundred feet. O'Brien believes that ancient humans could have attacked large animals over two hundred feet away with great accuracy.

Acheulean Hand Axes

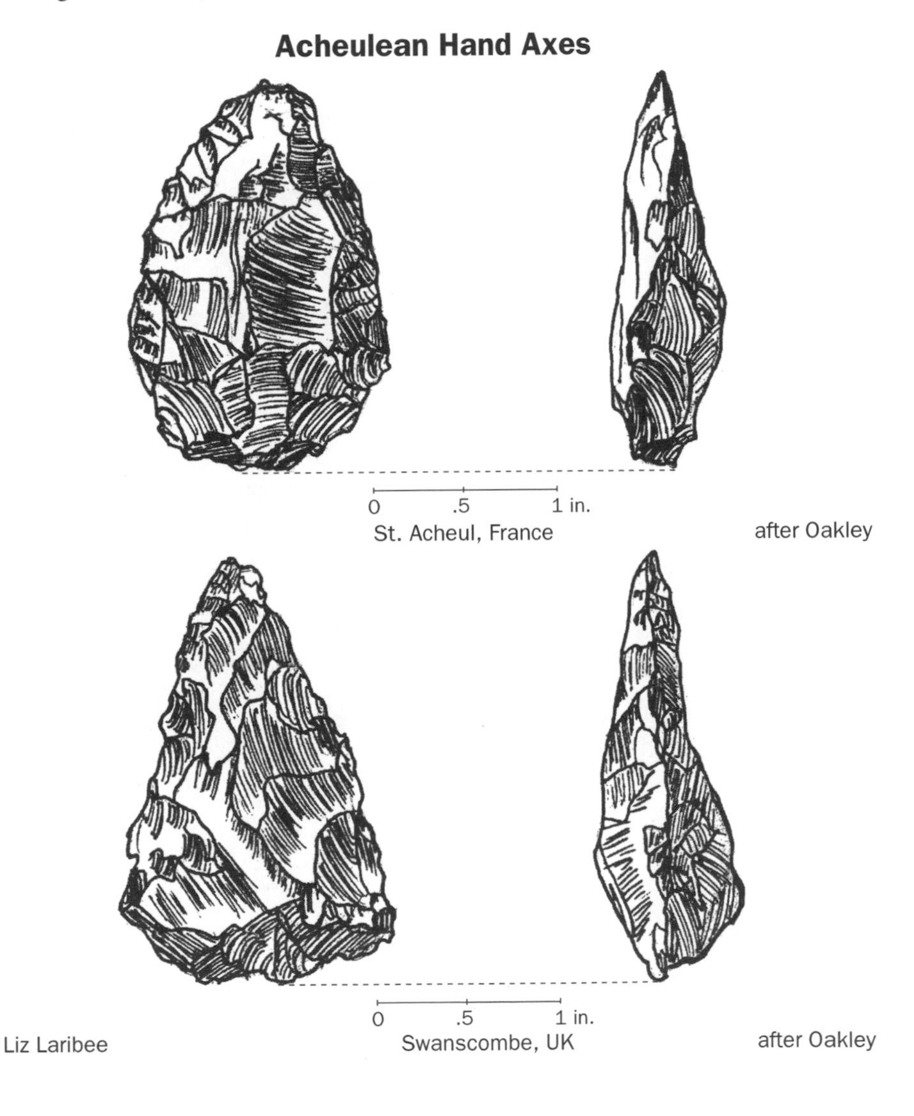

One of the puzzling aspects of hand axes is that they are found in great numbers in places that used to be streams, rivers, or lakes. It would be logical for bands of ancient humans to attack animals when they came to water. Hand axes that landed in the dirt could be retrieved. Hand axes that were overthrown and landed in the water usually could not be recovered, which might explain why we find them in those places today.

Also significant is that the Acheulean hand ax first appears in the archaeological record at about the same time as evidences of large animal kills—hippopotamus, elephant, and *Dinotherium* (an extinct elephant-like animal with large tusks in the lower jaw). Bands of ancient humans, throwing four-pound hand axes two hundred feet, could inflict heavy damage on even the largest and toughest game.

Although it was round, the discus of the ancient Greeks was unhafted, edged all around, and made of stone. To some people, the discus throw seems a strange and unlikely sport. O'Brien suggests the possibility that the Olympic discus throw is a carryover from the hand ax hunting technique of ancient humans. She asks, "Is it possible that the ancient Greeks preserved as a sport a tradition handed down from that distant yesterday?"[27]

If that is the case, that "distant yesterday" may not have been so long ago.

SECTION III

Evolution and Racism

A Fatal Attraction

The luckiest thing that ever happened to me was that my father didn't believe in God, and so he had no hang-ups about souls. I see ourselves as products of evolution, which itself is a great mystery. . . .

See, the reality is that we are genetically very unequal now.

—James Watson, codiscoverer of the double helix structure of DNA (*Discover*, July 2003, pp. 20–21)

From one man he [God] made every nation of men, that they should inhabit the whole earth.

—The historian Luke (Acts 17:26)

INTRODUCTION TO SECTION III

"So Just What Is Racism?"

CHARLES DARWIN WAS A RACIST. In all fairness to the man, we must state that about 98 percent of his fellow Englishmen were also racists—and about that same percentage of the French and the Germans. In fact, Europeans have a grand and "noble" tradition of chauvinism toward all non-Europeans. That tradition came to full flower during the period of discovery and colonization—the very time when Darwin published his famous 1859 work. Most people are unfamiliar with the full title of that work. It is: *The Origin of Species by Means of Natural Selection or The Preservation of Favored Races in the Struggle for Life.*

Two things are obvious in the full title of Darwin's book. First, since Darwin did not publish on human evolution until 1871, it is clear that he used the terms "species" and "race" interchangeably for all forms of life. Today, it is the custom to use the term "race" only for humans. Second, Darwin felt that the process of evolution was able to preserve those "favored races." In fact, it was evolution that produced those favored races in the first place.

Racism is defined as "the inherent superiority of certain races and stirs up prejudice and hatred for races said to be inferior" (Charles Winick, *Dictionary of Anthropology*, p. 449). Racism, as a concept, seems to be relatively new in our society—at least our awareness of what it entails. The *Oxford Universal Dictionary*, 1944 edition, does not even have an entry for "racism."

Racism centers around three elements. First, racism always involves differences in population *groups*. Often the differences involving racism are ethnic, tribal, cultural, or even religious. Racism is not about the differences that are found among individuals. The popular word for those differences is the term *diversity*.

Second, the crucial factor in racism is "*inherent* superiority." Throughout most of history, this "inherent" superiority was based upon some vague belief that one's own group was for some reason superior to others. Since the 1800s and the rise of evolution with its "scientific racism," the emphasis has been on *genetic* superiority. Evolution deals with mutational changes in the genes, which are the very stuff of life. Hence it is obvious that evolution is not only the cause of that alleged "inherent superiority," but, according to Darwin, evolution also preserves that "inherent superiority." When applied to humans, this "inherent superiority" of some race or group over others is properly called "racism."

Third, racism always involves prejudice and rejection—active or passive, latent or expressed. Whereas the qualities of love, acceptance, and respect always unite, racism, with its prejudice, hatred, and rejection, always divides. That is why racism is evil. A "loving racist" is a contradiction in terms. And because evolution is racist, evolution is an evil philosophy. Almost everyone focuses on the alleged "scientific" aspects of evolution. Few ever consider the moral implications and ethical consequences of evolution.

The concept that some entities are inherently superior or "more fit" is basic to evolution. Evolutionists believe that two similar entities existing in the same environment cannot coexist indefinitely. Over time, one of them will acquire some slight mutational advantage, usually in feeding, defense, or reproductive mechanisms, so that it will simply out-compete the other. It will survive, being more favored or "more fit." The other entity, being less favored or less fit, will eventually die out. In other words, evolutionists claim that nature works by what Darwin called "natural selection" to cut out the weak and thus allow the strong to proliferate.

When the concept of evolution was applied to nature generally, nothing could have seemed more benign, innocent, or obvious. It was utterly racist, but so was the audience to which it was addressed. It wasn't that no one noticed the racism. No one cared. And when, with Darwin's 1871 publication, the concept of evolution was extended to include humans, that racism seemed normal. No European, hearing reports of the various savage races throughout the world, doubted for a moment that he or she, as a European, was among those favored races of mankind. No one could deny that those "savages" were less favored. As proof, when Europeans came into direct competition with those "savages," the "savages" always lost. This is what evolution is all about: "the survival of the fittest."

Racism has been practiced throughout all of human history. However, only recently has our society become sensitized to its wrongs. It was the sheer horror of the Nazi death camps and the systematic attempt of the Nazis to exterminate the Jewish people that served to shock the world regarding racism. I do not wish to imply that evolution is the *cause* of racism. Its cause is the sin nature that all humans possess (Genesis 3). No, evolution is not the *cause* of racism. The *sin* of evolution, and I use that term literally, is that evolution gives humans an allegedly scientific *justification* for racism. Further, because evolution is devoid of any moral core, it opens the door to a host of other evils, including genocide, murder, infanticide, abortion, and euthanasia.

The entrance of sin into the human family has distorted the view we humans have both of ourselves and of others. Throughout human history, various segments of humanity have been denied full human status. These include women, slaves, various races, the mentally and physically disadvantaged, the aged, the poor, and children. Today, thanks in part to the concept of evolution, two groups that are being denied full human status are our descendants (the unborn) and our ancestors (the Neandertals and other fossil humans).

The concept of evolution is behind the racism of abortion. The philosopher and theologian Norman Geisler (Southern Theological Seminary) has pointed out that from conception on, absolutely nothing is added to the fetus except food, oxygen, and water. The entire DNA blueprint for growth toward an adult human is in place at conception. There is no other logical point at which human life could begin. It is the false concept of the "Biogenetic Law," the entirely disproved evolutionary idea that the fetus goes through all of the stages of evolution before it becomes a human, that is often used to justify taking an innocent life. Many have written, and written well, on the wrongs of racism expressed toward our descendants (abortion). In this book, we will focus on the nature of racism and how it is expressed toward our worthy ancestors, the Neandertals and other fossil humans.

This section will survey the fatal attraction of evolution and racism, first in the nineteenth and on into the twenty-first century. When, because of the Nazis, the true nature of racism was revealed, evolutionists began their attempts to camouflage the racism in evolution. The most sophisticated camouflage is found in the Out of Africa Model. Finally, with smoke and mirrors, evolutionists are even beginning to claim that evolution is responsible for our equality. The age of miracles is not over.

Just because evolution is racist does not, in itself, prove that evolution is untrue. Unfortunately for the scientific community, rigorous proof for evolution, in the normally accepted sense of the term "proof," is an impossibility. However, let us pretend that evolution could somehow be proven beyond all reasonable doubt. We would then be obliged to accept it—racism and all. We would have no option but to accept it: sadly for those whom the scientific community could show to be genetically inferior; gladly for those whom the scientific community could show to be genetically superior.

Thankfully, our Lord gave us a valuable guide for discernment and evaluation. He said, "Every good tree bears good fruit, but a bad tree bears bad fruit. A good tree cannot bear bad fruit, and a bad tree cannot bear good fruit. . . . Thus, by their fruit you will recognize them" (Matt. 7:17–18, 20). Jesus was talking primarily about people, but this applies to philosophies also.

The purpose of this section is simply to demonstrate that evolution is racist. The purpose of this entire book is to demonstrate that evolution is both racist and untrue.

We open with the famous voyage that began to give racism scientific respectability.

CHAPTER 13

THE VOYAGE

TRUTH IS INDEED STRANGER than fiction. No one would have dreamed that when Charles Darwin sailed from England aboard HMS *Beagle* on a scientific expedition around the world, it would result in an intellectual and spiritual revolution that would shake the foundations of modern society. It was December 27, 1831. Darwin was only twenty-two. Yet Darwin's voyage was far more important philosophically than was the voyage of Christopher Columbus geographically. It was on this voyage that Darwin's ideas on evolution began to crystallize and much of his "evidence" for evolution was collected.

When the *Beagle* left Plymouth, it had on board some very strange entries on its passenger manifest—the names of three former "savages" from Tierra del Fuego, that bleak land at the extreme tip of South America. These three had been taken hostage from their homeland by Robert FitzRoy, the captain of the *Beagle*, on its first trip a few years earlier. The three Fuegians on the return trip were a young woman named Fuegia Basket and two young men, Jemmy Button and York Minster. Unfortunately, another young man, Boat Memory, named after a stolen whaleboat, had died in England of smallpox.

On that earlier expedition of the *Beagle*, the ship's five-oared whaleboat had been stolen one night by Fuegians. FitzRoy responded by capturing four Fuegian hostages with the intent of exchanging them for the whaleboat. To his utter amazement, the Fuegians showed no interest in exchanging the whaleboat for their four tribesmen. The hostages also seemed to be quite happy to remain on the *Beagle*. Apparently, there were practical reasons why FitzRoy could not return the hostages to their own people without the return of the whaleboat. So he decided on an alternate plan—to take them to England.

Although misguided, FitzRoy had the best of intentions. His plan was to educate the Fuegians in England and return them to their own people as missionaries. Darwin writes that FitzRoy took them to England:

> Determining to educate them and instruct them in religion at his own expense. To settle these natives in their own country, was one chief inducement to Captain Fitz Roy to undertake our present voyage; and before the Admiralty had resolved to send out this expedition, Captain Fitz Roy had generously chartered a vessel, and would himself have taken them back.[1]

The land to which these Fuegians were being returned was, for Darwin, the strangest place he had ever seen. The Indians of Tierra del Fuego were hunter-gatherers and were thought to number about ten thousand at that time. They were considered to be among the most primitive people on earth. Ashley Montagu (Princeton University) writes that these Indians:

> Live in perhaps the worst climate in the world, a climate of bitter cold, snow, and sleet, and heavy rains a great deal of the time, yet they usually remain entirely naked. During extremely cold weather they may wear a loose cape of fur and rub their bodies with grease.[2]

The Fuegians wove baskets in the simplest manner; their huts were often nothing more than windbreaks made of branches and brush or animal skins; they made fire using flint or pyrites for small hearths; they hunted guanaco (a small deer-sized camel) with dogs; they had primitive canoes and used harpoons with stone points to spear fish, seals, and otters; they used rocks as hammers or pounders and fashioned thick end-scrapers and knives. Apart from body decorations, the Fuegians had virtually no interest in art. They had no adhesives or glue for hafting tools, no domesticated plants, no lamps, no metallurgy, no musical instruments, no needles or awls for sewing, no nets for fishing, no pottery, no rope, no long-distance trade, and no writing. They did not have the wheel, and they had no observable worship or religious practices.

Darwin's account of his four-year expedition, entitled *Voyage of the Beagle*, although virtually unknown, is a first-rate adventure classic. I quote him extensively because it is important in demonstrating his racism that we see firsthand his reaction to the Fuegians. In what could be called an overview of his impressions, he writes, "I believe, in this extreme part of South America, man exists in a lower state of improvement than in any other part of the world."[3]

Because of the degree of culture acquired in England by the three Fuegians on the *Beagle*, Darwin was not prepared for the shock of seeing the Fuegians in their native habitat. He writes:

> It was without exception the most curious and interesting spectacle I ever beheld. I could not have believed how wide was the difference between savage and civilized man: it is greater than between a wild and domesticated animal, inasmuch as in man there is a greater power of improvement. . . . Their skin is of a dirty coppery red colour. . . . Two men were ornamented by streaks of black powder, made of charcoal. The party altogether closely resembled the devils which come on the stage in plays like Der Freischutz [an opera by Weber].
>
> Their very attitudes were abject, and the expression of their countenances distrustful, surprised, and startled. . . . The language of these people, according to our notions, scarcely deserves to be called articulate.[4]

Later, Darwin describes another Fuegian tribe:

> These poor wretches were stunted in their growth, their hideous faces bedaubed with white paint, their skins filthy and greasy, their hair entangled, their voices discordant, and their gestures violent. Viewing such men, one can hardly make oneself believe that they are fellow-creatures, and inhabitants of the same world. It is a common subject of conjecture what pleasure in life some of the lower animals can enjoy: how much more reasonably the same question may be asked with respect to these barbarians! At night, five or six human beings, naked and scarcely protected from the wind and rain of this tempestuous climate, sleep on the wet ground coiled up like animals.[5]

Darwin refers to the Fuegians a number of times as "savages." He also uses terms such as their being "miserable, degraded savages," their being in "a savage state," their "savage land," and their "wild cry." He states that they are "like wild beasts" and calls them "savages of the lowest grade." Darwin continues:

> The different tribes when at war are cannibals. From the concurrent, but quite independent evidence of the boy taken by Mr. Low, and of Jemmy Button, it is certainly true, that when pressed in winter by hunger, they kill and devour their old women before they kill their dogs: the boy, being asked by Mr. Low why they did this, answered "Doggies catch otters, old women no." This boy described the manner in which they are killed by being held over smoke and thus choked; he imitated their screams as a joke, and described the parts of their bodies which are considered best to eat.[6]

Darwin sums up his feelings regarding them:

> They cannot know the feeling of having a home, and still less that of domestic affection; for the husband is to the wife a brutal master to a laborious slave. Was a more horrid deed ever perpetrated, than that witnessed on the west coast by Byron, who saw a wretched mother pick up her bleeding dying infant-boy, whom her husband had mercilessly dashed on the stones for dropping a basket of sea-eggs! How little can the higher powers of the mind be brought into play: what is there for imagination to picture, for reason to compare, for judgment to decide upon? to knock a limpet from the rock does not require even cunning, that lowest power of the mind. Their skill in some respects may be compared to the instinct of animals; for it is not improved by experience: the canoe, their most ingenious work, poor as it is, has remained the same, as we know from Drake, for the last two hundred and fifty years.[7]

The differences Darwin observed between the three Fuegians on the *Beagle* and the Fuegians he met in their native habitat at the tip of South America had a profound effect on him. In his mind, FitzRoy had civilized them more in four years than what thousands of years of living in Tierra del Fuego had accomplished. In her definitive biography of Darwin, Janet Browne comments:

> Jemmy Button polished his shoes and made jokes. Fuegia Basket wore jewellery and an English bonnet. Their language, faith, and aspirations had altered, as he believed, to reflect an English education, short and basic as it was; and they themselves seemed to feel markedly different from their former compatriots.[8]

These three Fuegians had even been presented to the Queen of England and had conducted themselves quite appropriately. Browne states that without experiencing this amazing change in these Fuegians, Darwin "would never have had the breadth of vision" to include humans in the evolutionary process.[9] There is no question that, although it is an absurd comparison, Darwin saw the gradual evolution of humans from lower primates in the differences he observed between the Fuegians on the *Beagle* and the Fuegians in their homeland. He felt he was seeing evolution in action. As Browne puts it, "The story of the Fuegians metamorphosing into Europeans and back again showed it."[10]

In an obvious effort at damage control, to make Darwin appear less racist than he was, Browne writes, "Even so, Darwin did not slip into the contemporary solipsism of believing the aboriginal Fuegians really constituted another species separate from his own."[11] Browne's comment is a bit of a stretch. As is seen from the title of *The Origin*, Darwin used the terms "species" and "race" interchangeably, and the Fuegians were obviously of another race. Further, the term "species" was not as well defined as it is now. Browne does not have enough omniscience to know exactly how Darwin viewed the Fuegians with regard to the term "species." Nor do we.

However, there is another way of determining how Darwin viewed the Fuegians, and it reveals the true depth of his racism. It was his belief that the Fuegians were incapable of being evangelized. As we have seen, Darwin often compared the Indians of Tierra del Fuego to animals. Does that mean he considered them to be animals? Perhaps the best evidence of how lowly he viewed the Fuegians is seen in how he viewed them spiritually. Herein lies a fascinating tale.

The holy Scriptures make a clear and qualitative distinction between all humans and all animals. In Genesis 9, God gives humans the right to use any and all animals for food. Yet human life is protected as sacred because we are made in God's image. Anyone who kills a human being in what we call "Murder 1" must forfeit his own life.

Darwin's knowledge of Scripture was abysmal. Yet, having studied for the ministry at Cambridge, he had to be aware of the distinction that Scripture makes between humans and animals. This is shown by the fact that although he loved to hunt animals, he was strongly opposed to slavery. He seems to have had some respect for the gap between humans and animals. Although Darwin later denied human uniqueness, he was aware that the Bible taught that only humans were created in God's image and that Christ commanded his disciples to evangelize all humans.

There is a delightful story about St. Francis of Assisi preaching to the animals and birds. We have no report as to whether or not any of the animals responded to St. Francis's preaching. Other than that, the Christian church has never tried to evangelize animals. It has, however, gone to the ends of the earth to reach every people group. The Bible states regarding Christ, "You were slain, and with your blood you purchased men for God from every tribe and language and people and nation" (Rev. 5:9).

On board the *Beagle* was a missionary sent by the Anglican Church to evangelize the Indians of Tierra del Fuego. Later, seven missionaries died there of exposure and starvation because their supply ship was two months late. In 1844, the Patagonian Missionary Society was founded; its name was later changed to the South American Missionary Society. Early on, there was much hostility by the Fuegians, but eventually, there was a bountiful harvest of souls for the Lord Jesus Christ. By 1869, over four hundred Fuegians had believed and were baptized.

In the London *Daily News*, 24 April 1885, Admiral Sir James Sulivan, who as a lieutenant was a shipmate with Darwin on the *Beagle*, wrote:

> Mr. Darwin had often expressed to me his conviction that it was utterly useless to send Missionaries to such a set of savages as the Fuegians, probably the very

> lowest of the human race. I had always replied that I did not believe any human beings existed too low to comprehend the simple message of the Gospel of Christ.[12]

To Darwin's credit, he admitted he was wrong. In a letter to Sulivan, dated 30 June 1870, Darwin wrote, "I had never heard a word about the success of the T. del Fuego mission. It is most wonderful, and shames me, as I always prophesied utter failure." In another letter to Sulivan, dated 20 March 1881, Darwin wrote, "I certainly should have predicted that not all the Missionaries in the world could have done what has been done."[13]

Without question, Darwin lived in a racist society. Yet the fact that Darwin would have denied the Indians of Tierra del Fuego the gospel, whereas other Englishmen at great sacrifice did give those same Indians the gospel, suggests that his incipient ideas on evolution, even at that early date, caused Darwin to be even more racist than some of his peers. And the theory of evolution he developed is equally racist. Since Darwin was a child of his times, we can forgive him for his racism. It is more difficult to excuse his followers today for not recognizing evolution for what it is.

It was exactly forty years after Darwin left Plymouth on that fateful voyage that he published *The Descent of Man*. He had no objective evidence for human evolution because there is none. However, his brilliantly reasoned philosophical arguments captivated the world. But it was seeing the Fuegians in their native habitat as compared to the Fuegians on the *Beagle* that, apparently, first convinced him that humans had evolved.

CHAPTER 14

TASMANIAN DEVILS

> Before and after Darwin, this false evolutionary equation, that low-tech means "sub-human", was an easy justification for the sinful, racist, and incredibly brutal treatment of these people. The Tasmanians were regarded as "wild beasts whom it is lawful to extirpate" [eradicate].
>
> —CREATIONIST CARL WIELAND

> When civilised nations come into contact with barbarians the struggle is short.
>
> —DARWIN, ON THE EXTINCTION OF HUMAN RACES BY THE EVOLUTIONARY PROCESS OF NATURAL SELECTION

I ONCE THOUGHT it was disgusting to name an animal "Tasmanian devil." That is, until I saw one! Unfortunately, not every "Tasmanian devil" was black and had four legs. The genocide of the Tasmanian Aboriginals in the 1800s is one of the saddest chapters in human behavior. The problem was European chauvinism and the twisted evolutionary and racist conception of what constituted a true human being.

The Tasmanian Aboriginals were black, very black. They lived on their tiny island—south of Australia and about the size of Ireland—in isolation for thousands of years. In fact, Tasmania is said to hold the record for the longest-known isolation in human history. Jared Diamond, in a 1993 *Discover* article on Tasmania, wrote, "Australia, lying at the end of the inhabited world, was an outpost; Tasmania, an outpost of that outpost."[1] Although Tasmanians always knew that their island existed, it was "discovered" by whites in 1642. Tasmania is now one of the states of Australia.

When the Europeans arrived, the population of Tasmania was about five thousand, consisting of about nine tribes speaking five or more languages and dialects. These languages had no obvious relationship to the Aboriginal Australian languages or to any other languages in the world. Tasmanians lived at forty-five degrees south latitude, similar to Boston or Seattle in the northern hemisphere. Yet they had hair as woolly as that of Africans living at the equator. Their eyes were deep set, overhung by Neandertal-like browridges. They were nomadic hunter-gatherers, depending solely on foraging for their livelihood.

The early explorers stated that the Tasmanians readily understood their gestures. They described these people as intelligent, friendly, polite, cheerful, graceful, kind, having strong and loving family relationships, and living in simplicity with nature. The women were described as excellent mothers displaying marked maternal tenderness.[2]

Unfortunately, the Europeans' cordial relationships with the Tasmanians changed rapidly and drastically. James Bonwick, in a book published in 1870, noted that although the Tasmanians were at first regarded positively by the English colonists, in just a few years they were considered as beings "whose destruction would be a deed of merit, as well as an act of necessity."[3] In terms of brutality, the Tasmanian genocide equaled the treatment of the Jews by the Nazis during World War II. Hitler's "final solution" for the Jewish problem was actually carried out on the Tasmanians by the British.

The first major skirmish with the Tasmanians occurred in 1804, when a British officer ordered his men to fire on them, killing or mortally wounding at least fifty. The reason for firing on them was simply to see them flee, the officer being drunk from an "over-dose of rations' rum." This was the beginning of a deep-seated hatred of the whites by the Tasmanians that never died out.[4]

In seeking to understand this genocide, there are two separate issues to consider—the cause and the justification. First, the *cause*. The Tasmanian genocide is a textbook illustration of the clash of two mutually exclusive cultures. The Tasmanians were hunter-gatherers. Theirs was a free-ranging hunting and gathering lifestyle that involved no specific property ownership or settled living conditions. In a sense, the whole island was their hunting grounds. When the settlers came, their culture was to settle down in a specific area, own specific property, grow crops on it or fence it for cattle or sheep, build roads for access, and build towns and villages. The former hunting and gathering areas of the Tasmanians were now owned and occupied by invaders and were off-limits to the Tasmanians. A clash was inevitable. Since the invading settlers had guns and the Tasmanians didn't, the outcome was also inevitable. By 1842, about 40 years after the first settlers arrived, only 135 out of an original 5,000 Tasmanians remained.

In the Word of God, such genocide is unthinkable. Only God has the right to take human life or to give government the authority to execute those who murder. Yet Darwin, using the Tasmanian genocide to illustrate his point, informs us that extinction is to be expected in an evolutionary world. In his section "On the Extinction of the Races of Man" in *The Descent of Man*, he writes, "Extinction follows chiefly from the competition of tribe with tribe, and race with race."[5] He almost excuses the genocide as a result of the inevitable cultural clash by saying, "When civilised nations come into contact with barbarians the struggle is short. . . . We can see that the cultivation of the land will be fatal in many ways to savages, for they cannot, or will not, change their habits."[6]

Jared Diamond vividly describes the European attitudes that led to the atrocities committed on the Tasmanians.

> When British settlers poured into Tasmania in the 1820s . . . racial conflict intensified. Settlers regarded Tasmanians as little more than animals and treated them accordingly. Tactics for hunting down Tasmanians included riding out on horseback to shoot them, setting out steel traps to catch them, and putting out poison flour where they might find and eat it. Shepherds cut off the penis and testicles of aboriginal men, to watch the men run a few yards before dying. At a hill christened Mount Victory, settlers slaughtered 30 Tasmanians and threw their bodies over a cliff. One party of police killed 70 Tasmanians and dashed out the children's brains.
>
> In 1828 the governor of Tasmania declared martial law, permitting Europeans to shoot on sight any aborigine found in European-settled areas. That was followed by roving search-and-capture parties ("five convicts of good character led by a field police constable") and by a bounty established in 1830 of £5 per Tasmanian adult, £2 per child [captured alive, probably to be sold as slaves].[7]

Diamond also reports the practice of "tying Tasmanian women to logs and burning them with firebrands, or forcing a woman to wear the head of her freshly murdered husband on a string around her neck."[8]

We could assume that these atrocities were committed willfully against British or Tasmanian law. We may be right, but those laws were not enforced. An item in the April 1836 issue of the *Hobart Town Times* reads, "The Government, to its shame be it recorded, in no one instance, on no single occasion, ever punished, or threatened to punish, the acknowledged murders of the aboriginal inhabitants."[9]

What was the *justification* for such acts? People who commit such atrocities must have, for the sake of their consciences, some justification for doing so. They did. It was a belief in *evolution.* Only the most naive person thinks that the concept of evolution first entered the human thought-stream when Darwin published the *Origin* in 1859. Evolution, as a philosophical concept, was well-known long before Darwin published. Desmond King-Hele, author of a biography of Erasmus Darwin (Charles Darwin's grandfather, who was a famous philosophical evolutionist), writes, "After 1794 [sixty-five years before Darwin published the *Origin*], statements on the principle of natural selection and evolution came fairly thick and fast."[10]

During the nineteenth century, and long before Darwin published, the scientific community was vitally concerned about the Tasmanian genocide. Its concern, however, was not moral or ethical. It was far more interested in dead Tasmanians than in living ones. This was the golden age of "scientific racism," and the belief was widespread that the Tasmanians were the missing link between Stone Age men and fully evolved whites. Because evolution was still mainly a philosophical belief, the race was on to find scientific proof that would put it on more solid ground. This race intensified after Darwin published his famous work.

The scientific community wanted Tasmanian skulls and body parts for measurements to prove that the Tasmanians were "missing links," or evidence for evolution. Darwin's grandfather, Erasmus Darwin, was one of the first researchers to dig up an Aboriginal from the grave in order to stuff and exhibit the stolen body at the Royal College of Surgeons—the first of over ten thousand Tasmanian skeletons and skulls that would be in the Royal College collection. Some of the greatest names in British science were involved in this body-snatching business, including Sir Richard Owen and Sir Arthur Keith. Charles Darwin himself was implicated through letters written in the 1870s and found in a Hobart, Tasmania, archive in the 1970s.[11] This racism on the part of the scientific community was the direct result of evolutionary thinking.

Jared Diamond describes the case of the very last full-blooded Tasmanian:

> Before Truganini, the last woman, died in 1876, she was terrified of similar post mortem mutilation and asked in vain to be buried at sea. As she had feared, the Royal Society dug up her skeleton and put it on public display in the Tasmanian Museum, where it remained until 1947. In that year, the museum finally yielded to complaints of poor taste and transferred Truganini's skeleton to a room where only scientists could view it. . . . Finally in 1976—the centenary year of Truganini's death—her skeleton was cremated over the museum's objections, and her ashes were scattered at sea as she had requested.[12]

"Today the number of people of Tasmanian aboriginal descent is estimated to be around 4,000," Diamond writes, "though it could be considerably higher because racial prejudices make people reluctant to be identified as Tasmanian."[13] In spite of one hundred years of being considered "animals" and "missing links," the Tasmanians continually passed the one and only valid test proving that they were fully human—the test of interfertility with other humans. Those 4,000 descendants of the original Tasmanians are a mixture of Tasmanian and white blood. They constitute 4,000 items of hard, rigorous, factual evidence that the Tasmanian Aboriginals were fully human in every way.

Their full humanity was doubted because the evolutionary world applied the false test of culture to determine their humanity, and their culture was said to be very primitive. Diamond states that any anthropologist would describe the Tasmanians as "the most primitive people still alive in recent centuries."[14] Of all the people in the world, they were considered among the least technologically advanced. Hence they were considered less evolved than most other people.

From an evolutionary point of view, it was quite proper to think that less evolved people (Tasmanians) were living at the same time as modern humans. Many animals living today are considered, from an evolutionary point of view, "primitive." For instance, sharks have a skeleton of cartilage rather than of bone. Hence they are considered "primitive," because the fishes are said to have evolved further and have bony skeletons. Yet sharks, known as "wolves of the sea," are the most efficient hunters in the world. The problem is not with the sharks. The problem is that evolutionists have imposed a false template on the world God has created.

Based on evolutionary presuppositions that are inherently racist, a horrible mistake was made a century ago regarding the Tasmanians. The result was one of the most brutal episodes in human history. The full-blooded Tasmanians are now gone—gone forever. In light of those who committed the Tasmanian genocide, can anyone doubt that not all the devils in Tasmania were black and had four legs?

CHAPTER 15

The Elephant in the Living Room

THE LATE DR. A. E. WILDER-SMITH tells of an incident regarding Adolf Eichmann, known as "Hitler's Hangman." Eichmann was responsible, directly or indirectly, for the deaths of over five million Jews. After World War II, Eichmann escaped to Argentina and lived in hiding there for some years. The State of Israel eventually discovered Eichmann's whereabouts, sent in a commando party in May 1960 to apprehend him, and brought him back to Israel to be tried for war crimes. He was found guilty and was sentenced to be hanged.

Wilder-Smith continues:

> The British sent over a believing man of God, a chaplain, to talk to this poor fellow. Just before he was hanged, Eichmann had a last consultation with the chaplain. The chaplain said, "Herr Eichmann, before you see God tomorrow, wouldn't you like to get absolution? Wouldn't you like to confess?"
>
> Eichmann reared up and said: "Confess? What have I got to confess? I've done nothing wrong!"
>
> The chaplain replied, "You've done nothing wrong? Do I understand you?"
>
> "Yes," Eichmann replied. "I've done only right!"

> The chaplain said: "Would you please explain yourself."
>
> "Certainly I will," said Eichmann. "Both the churches in Germany, the Catholic and the Protestant, believe in Theistic Evolution. Both of them believe that God's method of creation was to wipe out the handicapped and to wipe out the less fitted. And as the Jews are less fitted than our people, I have only helped God in his methods. I have only catalyzed God's way of working. And when I meet God I shall tell Him so."[1]

Wilder-Smith, who had three earned doctorates, was one of the most prestigious creationists of the twentieth century. He taught extensively in Europe and then at the University of Illinois Medical Center in Chicago. His wife grew up in prewar Nazi Germany, where her father ran an institution to train deacons in the care of physically and mentally handicapped children who were cast off by the Nazis. These children were allegedly "unfit" and thus were using up valuable resources for no good purpose. The Nazi response was to eliminate them, and hundreds of them from that institution were eventually sent to the gas chambers.

The most astonishing aspect of the whole Nazi affair is the failure of the world to realize the motivation behind those gas chambers. It was evolution. Adolph Hitler was indeed a *bad* man, but he was not a *mad* man. He was an absolutely consistent evolutionist. He was wicked enough to drive evolution to its logical conclusion. He believed that the German people were genetically, thus inherently, superior and that the Jewish people were genetically, thus inherently, inferior. To have a superior race coexist side by side with an inferior race was unnatural from an evolutionary point of view. If that condition continued, the superior German genes would be compromised and polluted by the inferior Jewish genes. The German gene pool would not be able to maintain its superiority, let alone evolve even further. Hence "the ultimate solution to the Jewish problem" was put into play.

I do not wish to imply that evolution was the *cause* of the almost universal acceptance of racism (in the form of anti-Semitism) in Germany in that era. Racism has been practiced throughout all of human history. Its cause is the sin nature that all humans possess (Genesis 3). I am saddened to say that my German ancestors have a history of anti-Semitism going all the way back to the Middle Ages. No, evolution is not the *cause* of racism. The *sin* of evolution is that evolution has given humans an allegedly scientific *justification* for racism.

There is an elephant in the living room of evolution. It is racism. Everyone is stepping around the elephant, pretending it isn't there or pretending not to see it. In the era of "scientific racism," evolutionists enjoyed having the elephant in the living room and fed it well. But the Nazi Holocaust brought the evils of scientific racism to public consciousness as nothing else had done. Now that we know the evils of racism, evolutionists are frantically trying to make the elephant as invisible as possible.

Scientific racism was one of the fruits of evolution in the nineteenth and early twentieth centuries. It began even before Darwin published his famous work in 1859. As we saw in the account of the Tasmanian genocide, there was a frantic search for hard evidence for scientific racism when evolution was still just a philosophy. Scientific racism came to full flower after Darwin's publication.

Lamenting that awful era of scientific racism, the late Stephen Jay Gould recounted the story of a group of Dutch settlers in the 1800s who shot and ate an African Bushman. They believed that he represented such a low race of beings that he was no different from a Malay orangutan and that, since he was an animal, eating him was perfectly all right. Recognizing that the scientific (read "evolutionary") community was a driving force in promoting racism until recently, Gould continues:

> The tragedy of this is that the history of scientific views, for the most part and until quite recently on the subject of race, has supported this social and political notion that human groups are separated by profound and innate inequalities in intellectual abilities and moral behaviors.[2]

Since then, evolutionists, knowing that evolution is intrinsically racist, have tried desperately to distance the theory from any obvious racial overtones. This is why the racist issue erupted in evolutionary circles in the 1960s. It centered around a Boston-born and Harvard-educated anthropologist, Carleton S. Coon.

Carleton Coon's special interest was how the various human races had evolved, especially how they differed before they evolved into *Homo sapiens*. He tried to develop his theories on strictly "scientific" principles, without regard to dogma, emotion, or political agenda.

Coon held that there were five basic human races. He defined them as Caucasoids (a group that included most Europeans, North Africans, Near Easterners, and the peoples of India and Pakistan); Mongoloids (a group that included the peoples of East and Southeast Asia and most of Indonesia and the Polynesians, Micronesians, and American Indians); Australoids (a group that included the native peoples of Australia, New Guinea, and Melanesia; the Negroid dwarfs of Indonesia and South Asia; and certain aboriginal tribes of India); Capoids (a group that included Bushmen and Hottentots); and Congoids (a group that included African pygmies and Negroes). According to Coon, each of these five races had evolved separately into *Homo sapiens*, and each had evolved from a *Homo erectus* stock.

The Caucasoids, Coon believed, had evolved from what he classified as European *Homo erectus* individuals represented by fossils such as Swanscombe,

Steinheim, and Heidelberg Man (Mauer mandible), which were dated at that time from about 250,000 to 500,000 ya. The Mongoloids, he believed, had evolved from Asian *Homo erectus* individuals represented by fossils such as Java Man and Peking Man, which were dated at that time from about 400,000 to 500,000 ya. The Congoids, he felt, had evolved from African *Homo erectus* individuals represented by fossils such as the Saldanha skull and Rhodesian Man (also known as Broken Hill skull or Kabwe skull) that were dated at that time at about 40,000 ya.

Although Coon tried to build on strictly "scientific" evidence and to avoid racial overtones, the racial implications were quite obvious. If the Europeans and the Asians began to evolve from *Homo erectus* toward the *Homo sapiens* condition as early as 500,000 ya, and the Africans began evolving from *Homo erectus* toward that same *Homo sapiens* condition only 40,000 ya, the Africans were the last to cross the *Homo erectus–Homo sapiens* boundary and hence were less evolved or less civilized than were the other races.

I remember the situation well. Only three possibilities existed to quell the firestorm that erupted over these racial implications. The first was to discredit Coon as a scientist. The second was to redate the Rhodesian and Saldanha fossils. The third was to change (upgrade) the classification of the Rhodesian and Saldanha fossils. All three were predictable, and all three happened.

The redating of the Rhodesian and Saldanha fossils was highly questionable (see the following chapter on Rhodesian Man). On the other hand, the reclassification (upgrading) of the two fossils was certainly proper and should have gone even further, since they were both fully human individuals. Meanwhile, Carleton Coon died in 1981 in relative disgrace. Pat Shipman (Pennsylvania State University) comments that "Coon had wrestled with the painful subject of evolution of human races and had come away covered in burning welts that turned to lasting scars. He never regained his prominence in his field."[3]

Coon's problem was not that he proposed an evolutionary theory that had racist implications. His real problem was that every theory of human evolution has racist implications. Coon thought that by being strictly "scientific," he could avoid racism. He couldn't. There is no question that, according to evolutionary theory, if one race evolved from a *Homo erectus* condition 500,000 ya, it would be far more evolved than a race that evolved from that same condition just 40,000 ya.

Coon's theory is somewhat similar to the current theory of human evolution known as the Multiregional Continuity Model. The principal proponents of this view are Milford Wolpoff and Alan Thorne. Since this view is especially vulnerable to racist implications, it is not surprising that Wolpoff and his wife and fellow paleoanthropologist, Rachel Caspari, wrote *Race and Human Evolution* in 1997 purposely to distance their model from the charge of racism.

Because racism always lurks more obviously in the shadows of this model, however, the Multiregional Continuity Model is not the model of choice among most paleoanthropologists today.

The experience of Carleton Coon suggests that the best way for paleoanthropologists to escape the charge of racism is to come up with a model in which: (1) all living humans evolved from the very same human stock; (2) all evolved in the same time period; and (3) thus the racial distinctions among humans today have a very shallow rather than a very deep history. This is exactly what the Out of Africa Model proposes. There is no question that the Out of Africa Model, with its African Eve, arose more out of political correctness than out of scientific data. To support my charge that human evolution is intrinsically racist, I call on "Mr. Evolution" himself, Stephen Jay Gould.

In 1987, Gould delivered a lecture at The College of Wooster, Wooster, Ohio, on "Evolution and Human Equality." For fifteen years, little was mentioned of this lecture, until it was made available on videotape in 2002 by *Natural History* magazine. This magazine is an organ of the American Museum of Natural History in New York. Gould begins by stating his premise: "If anything, we are now beginning to realize that there is a profound equality among the different groups of us, based on the history of our evolution."[4] In other words, science has now shown that all humans are biologically equal, and we are all biologically equal because of our evolution. Evolution, after being used in various ways for one hundred years to prove that some humans were superior to others, is now being used to prove our equality.

Gould then confirms that evolution is intrinsically racist. However, we humans happen to be all equal because we are just plain lucky.

> Human equality is a contingent [accidental] fact of history. That is, human evolution might have unfolded or happened in a thousand other ways that would have been perfectly sensible, and each of those other ways might have left us with a variety of human groups profoundly unequal with respect to their intellectual capabilities and moral achievements. But it just didn't happen that way. It could have happened that way, it just plain didn't.

In an evolutionary scenario, human equality is a crap shoot. Mother Nature rolled the dice. We had only about one chance in a thousand of being biologically equal as humans, but somehow we were incredibly lucky. What I find so utterly amazing about this lecture is that in trying to prove that we are all equal because of evolution, Gould doesn't seem to mind that he is actually confirming that evolution is intrinsically racist.

This lecture was given just seven months after Cann, Stoneking, and Wilson made their dramatic announcement in *Nature* about all humans being descended from one woman, African Eve, about 200,000 ya. Of the several assumptions in the theory, Gould admits that the assumption about timing "is admittedly more tenuous." If that assumption holds, he says, then the "stunning conclusion" is that all humans are biologically equal.

When I received my copy of the video, I was shocked to realize that this was a lecture given fifteen years before. I immediately questioned why it was not released sooner. Although there may be other reasons, I suspect the main reason is that the "science" upon which African Eve is based has been continually challenged since 1987. One of the first problems was the discovery by one of the researchers that the timescale for "Eve" was determined by the order in which the data was entered into the computer. Entering the data in a different order gives a different timescale. The present status of African Eve is summed up by Jonathan Marks (Yale University): "And with each new genetic study that claims to validate 'Eve' conclusively, there comes an equal and opposite reaction, showing the study's weakness."[5]

That this video was released in 2002 in spite of the scientific weaknesses on which human equality is based reveals how important it is for evolutionists to disclaim racism by promoting the Out of Africa Model. Gould reveals how crucial it is that all of the human races have a very recent evolutionary origin. Citing past versions of human evolution in which the various human races evolved separately from a much older source, *Homo erectus*, Gould warns of the implications of our having evolved directly from older evolutionary ancestors.

> That point of origin [evolving directly from *Homo erectus*] was so long ago, that human races have been so separate for so long that substantial genetic differences translating to inequalities have evolved, and are now so fixed in the various races that for all it means in terms of the possibility of ever eradicating those differences, you might as well say that they are original.

As I stated earlier, to avoid any racist implications in the evolution of modern humans it is crucially important that we all come from the same *Homo sapiens* stock, that we all come from the very same person in that stock, and that all of the human races evolve in the same short time-span. This is the Out of Africa Model. The political correctness of this model far outweighs its scientific shortcomings. That's why it is the model of choice among evolutionists today.

It is important that I make a distinction between evolution as a racist theory and evolutionists as racists. I cannot emphasize strongly enough that none of the many evolutionists I have known personally were racists. Nor has any of the evolutionary literature written since the mid-twentieth century that I am aware of been racist. And that is my point. Evolutionists today are not racist, but their theory is racist to the core. In dealing with human evolution, evolutionists have sacrificed their theory on the altar of political correctness. They are either denying or purposely ignoring the basic implications of their theory. Whatever the reason, I am very thankful that evolutionists do not practice the implications of their theory.

I know of only one evolutionist who has had the courage—if that be the word for it—to openly recognize the inherent racism of evolution. He was one of the twentieth century's greatest evolutionary biologists, the late W. D. "Bill" Hamilton. He saw clearly that evolution demanded the elimination of the weak, not their preservation. He was a consistent evolutionist in that he despised efforts to keep children with genetic defects alive and abhorred the fact that cesarean sections allowed women with narrow hips to give birth. He foresaw "an imminent mutational meltdown—the crippling accumulation of deleterious mutations—as a consequence of modern medicine."[6] While I abhor his beliefs, I admire his intellectual honesty in taking evolution to its logical and ethical conclusions.

The Scriptures clearly teach three things. First, all humans were created by God in his image. "God said, 'Let us make man in our image, in our likeness'" (Gen. 1:26). Thus, we are all genetically equal, not by accident but by God's eternal plan and purpose. Second, all humans constitute one interrelated family. "From one man he made every nation of men, that they should inhabit the whole earth" (Acts 17:26). Third, since all humans are made in the image of God and have a spiritual component in their makeup, they are to be evangelized because they are all potential candidates for the salvation God has provided. Jesus said, "All authority in heaven and on earth has been given to me. Therefore go and make disciples of all nations" (Matt. 28:18–19). Only these facts give true dignity and equality to every individual, regardless of our superficial differences, weaknesses, or limitations. Nothing could be clearer in the Bible than the unity, solidarity, and equality of the human family.

A few paragraphs ago I stated that evolutionists have finally discovered, after 150 years, that in order to avoid biological or racial distinctions in the human family, it is crucially important that we all come from the same stock, that we all come from the very same person in that stock, and that human races have the same short time-span. It is no accident that this is exactly what the Bible teaches! The Bible teaches clearly that all humans came originally from a single created couple, Adam and Eve, just thousands of years ago. We didn't evolve equally by accident. We were created equally by design.

There is an elephant in the living room of evolution. It is racism. Everyone is stepping around the elephant, pretending it isn't there or pretending not to see it. Yet, behind the scenes, the two major versions of human evolution are frantically trying to make the elephant as invisible as possible. How the more popular model is doing it is the subject of a later chapter, "African Eve: A Woman of No Importance."

CHAPTER 16

Rhodesian Man

The Fastest-Aging Man in the World

THE CASE OF RHODESIAN MAN is a classic example of: (1) how fast a fossil can age after it is out of the ground; (2) how philosophical and political considerations can take precedence over scientific matters in the dating of a fossil; and (3) how the shape of the skull has no relationship to human evolution or to the age of a fossil human. The rapid aging of Rhodesian Man is a reaction to Carleton Coon's human fossil scenario in the 1960s and a desire to ensure that Rhodesian Man can never again be used in a racist interpretation of the human fossils.

The original report of the discovery of Rhodesian Man appeared in *Nature*, 17 November 1921, and was authored by Dr. Arthur Smith Woodward of the British Museum (Natural History).[1] Rhodesian Man was so named because the skull was found in what was then known as Northern Rhodesia, now

Zambia. The fossil is also called Broken Hill Man (after the mine in which it was found) or Kabwe Man (after the city near which it was found). The skull was found on 17 June 1921, at the farthest and deepest point of a cave, sixty feet below ground level. The cave was part of the Broken Hill Mine, which produced lead and zinc ore.

Because the browridges on this fossil skull are more severe than those found on any other human fossil, no human fossil appears to be more "primitive," "savage," or "apelike" than does Rhodesian Man. Woodward remarked that it is more apish than Neandertal Man. Yet Rhodesian Man's cranial capacity of 1280 cc is so large that he demands to be classified as *Homo sapiens.* We need to be reminded that there is nothing in the contours of the skull of an individual that gives any clue as to his degree of civilization, culture, or morality.

An indication that Rhodesian Man was not a totally unique individual is that the Saldanha, South Africa, skull (known also as the Hopefield or Elandsfontein skull) has almost the same "savage" morphology (shape). Although this fossil is considered to be a bit older than Rhodesian Man, there is no question that it also is from a fully human individual.

Other human remains found in the cave were a bone from the bottom of the spinal column (sacrum), a lower leg bone (tibia), two ends of the upper leg bone (femur), and an upper jaw fragment (maxilla) of another individual. Woodward reported that the limb bones, the tibia, and the femur are very modern in shape and do not differ from that of modern humans. Since Rhodesian Man is today classified by Klein as "early *Homo sapiens,*" the very modern fossils present could indicate the association of persons at the site whom evolutionists place in two different evolutionary categories.

For evolutionists, a major problem with all of the fossils found before the late 1950s, which was the advent of radiometric dating, is to establish a date for them. For instance, Rhodesian Man was found in the process of mining operations by miners who had no knowledge of, or interest in, the stratigraphic layers of the cave. The entire site was later destroyed by mining. Thus the possibility of establishing a sequence of events for "relative" dating was lost. The discovery was also before the advent of the radioactive dating methods, so there was no possibility of establishing an "absolute" date for the fossil. (The radioactive dating methods and their so-called "absolute" dates are science fiction, as will be demonstrated later.)

In the case of fossils found before about 1950, there are several ways of attempting to estimate a date for the site: (1) Working to reconstruct the sedimentary sequences of the sites after the fact; (2) Attempting to find other sites that can be dated and are similar enough for comparison with a site that can't be dated; (3) Imposing the template of cultural evolution with its time scenario and date on the cultural artifacts found at a site; (4) Interpreting the

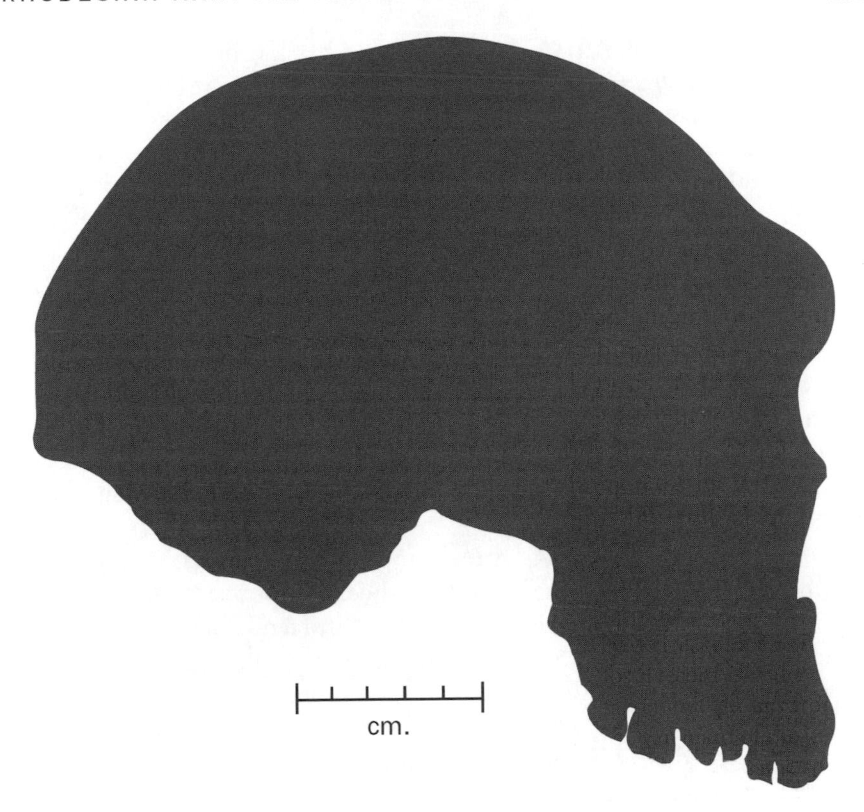

Rhodesian Man Skull, Broken Hill Mine, Zambia
archaic Homo sapiens

plant and animal fossils at the site in an evolutionary framework (it is believed that some indication of age involves the number of fossils of contemporary plants and animals at the site as compared to the number of extinct plants and animals at the site); (5) Determining the degree of mineralization of the fossil bones at the site to shed light on its age. It is obvious, however, that all of these methods are highly subjective because a number of conditions other than the passage of time could give a false indication of age.

Cultural artifacts associated with Rhodesian Man include Middle Stone Age quartz and chert stone tools, several bola stones, and a few bone tools.[2] The bone tools are significant because in Europe it is claimed that bone tools are not found in the Middle Stone Age (with the Neandertals) but in the Upper Stone Age (with the modern Cro-Magnon people). Bola stones are used even today in South America in hunting animals on horseback.

According to Michael Day, the animal fossils found in association with Rhodesian Man include the following: large extinct baboon, mongoose, lion, leopard, extinct serval, African elephant, zebra, black rhinoceros, and extinct long-horned buffalo.[3] Carleton Coon claimed that there were only two extinct animals in the assemblage at the site, the serval and a white rhinoceros.[4] Richard Klein states that there was a minimum of twenty-three large mammal species in the assemblage, including at least five extinct species—the baboon, a saber-toothed cat, a small warthog, a short-necked giraffe, and the long-horned buffalo.[5]

This confusion regarding the number of extinct species at the site is best explained by Coon.[6] The cave was filled with fossilized and mineralized (lead and zinc) bones. These bones were discovered by the miners in 1907 and were smelted. Only in 1921 did the miners notice the human bones. Obviously, the miners would not have bothered to identify the animal bones or to keep records. Only when the human bones were found would people be involved who were interested in recording the relative location and description of both the human and the animal fossils.

For the evolutionist, there is virtually nothing upon which to base a date for Rhodesian Man. "Absolute" dating methods had not yet been developed, and the possibilities for "relative," faunal, and artifactual dating had all been compromised by the disturbance of and the eventual destruction of the site without adequate records. In this situation, the original 1921 report in *Nature* by Dr. Arthur Smith Woodward of the British Museum (Natural History) would likely be the most reliable, with later dating estimates influenced by evolutionary or political considerations. In light of the fervor over Coon's "racist" interpretation of fossil humans, it is not surprising how rapidly Rhodesian Man aged after the fossil was out of the ground.

The original 1921 report describing Rhodesian Man says, "Its large and heavy face is even more simian [apelike] in appearance than that of Neanderthal man." Yet the report goes on to say, "The skull is in a remarkably fresh state of preservation, the bone having merely lost its animal matter and not having been in the least mineralised." Regarding the animal bones, the report states that those that have been identified "belong to species still living in Rhodesia or to others only slightly different from these." An estimate of the age of the fossil is also given: "The occupation of the cave, therefore, seems to have been at no distant date—it may not even have been so remote as the Pleistocene period."[7]

Writers of that era were hesitant to ascribe specific dates to fossils and geologic periods, but today the end of the Pleistocene Period (Ice Age) is considered to be about 11,000 years ago. Woodward compares the fossil to some Neandertals, to Heidelberg Man, and to Java Man, concluding that Rhodesian Man is more modern than the Neandertals, even though it is more apish in appearance.

Four conclusions appear rather obvious from the known details of the fossil site: (1) The lack of fossilization or mineralization of the skull, even though it was in a heavily mineralizing environment, suggests that the skull is of very recent age. (2) Further, bone tools are considered by evolutionists to be one of the cultural marks of modern humans, even more advanced than the Neandertals. (3) Although there is some question as to which of the human fossils are early (archaic) and which are modern, all authorities seem to agree that there is a mixture of early (archaic) and modern human fossils in the assemblage.[8] Chemical analysis of these bones indicates that they are all of about the same age.[9] The implication is that the evolutionist's distinction between modern *Homo sapiens* and archaic or early *Homo sapiens* is a false one, since both types may have been living at the same time in the same place. (4) The presence of humans in a cave rich in lead and zinc ore suggests that they may have been mining these minerals. If this is the case, it indicates a high degree of culture. The book of Genesis clearly confirms the advanced culture and technology of the ancients, specifically mentioning metallurgy in Genesis 4:22 and music in Genesis 4:21.

Found under other circumstances and given an earlier age, Rhodesian Man, with its low cranial doming and very heavy browridges, would have served evolutionists as an excellent illustration of an evolutionary transitional form between apes and humans. We now know how wrong that concept is. In spite of the obvious lessons to be learned from the Rhodesian and Saldanha skulls, evolutionists continue to base much of their evidence for human evolution on the alleged primitive-to-advanced contours of fossil skulls. This is probably one of the reasons for Rhodesian Man's rapid aging after he came out of the ground: His supposedly "savage" morphology does not fit with a young age. Creationists, on the other hand, maintain that in light of the evidence of the wide genetic diversity in the human family, skull contour is an inadequate basis for determining relationships or age.

The Rapid Aging of Rhodesian Man

11,000 years of age (?). Arthur Smith Woodward, in 1921. Possibly post-Pleistocene (Ice Age) based on: (1) The skull being in a remarkably fresh state of preservation—not fossilized or mineralized; (2) The skull having more modern features than the Neandertals; (3) Some associated human limb bones that were very modern in shape; (4) The associated animal bones that were modern or only slightly different; and (5) Bone tools that were more modern than the Neandertals.

40,000 years of age. Carleton Coon, in 1962. Based on early radiocarbon dates of the Stone Ages in Africa and comparisons with the Saldanha site.

125,000+ years of age. Richard G. Klein, in 1973.[10] Based on: (1) Five animal species found in association with Rhodesian Man that are extinct forms; (2) A revision of the particular African tool industry found in association with the skull, placing it at the end of the Early Stone Age; (3) The similarity of the Saldanha skull to Rhodesian Man, with Saldanha considered to be a bit older; and (4) Klein's belief that Coon's dates are wrong.

300,000–400,000 years of age. Ian Tattersall, in 1999.[11] No specific reasons are given for the older dates. However, there is no doubt that Tattersall's commitment to the Out of Africa view is involved. He must allow time for the Rhodesian Man type, with heavy browridges, to evolve into more modern *Homo sapiens* so that these modern humans can then leave Africa and move out into the rest of the world by about 200,000 years ago. Tattersall is often guilty of dating fossil humans by their shape, to fit into his evolutionary scenario, rather than by objective evidence.

Thus Rhodesian Man aged about 350,000 years in just 82 years, even though there was not a shred of rigorous evidence indicating that he was other than a few thousand years old. Here we see the power of philosophy over fact.

SECTION IV

THE EMPIRE'S NEW CLOTHES

The once-popular fresco showing a single file of marching hominids becoming ever more vertical, tall, and hairless now appears to be a fiction.

—J. J. HUBLIN, *NATURE* 403 (27 JANUARY 2000): 363

INTRODUCTION TO SECTION IV

THE FAKE PARADE

No clever arrangement of rotten eggs can make a good omelet.

—C. S. LEWIS

EVERYONE LOVES A PARADE. There are very famous parades such as the Saint Patrick's Day Parade in New York City and the New Year's Day Rose Bowl Parade in Pasadena. Hundreds of thousands of people line the streets to witness these parades, and millions more view them on television.

However, there is an even more famous "parade" that has been seen by billions of people worldwide. It is a different kind of parade, but it is a parade nonetheless. It is a parade of fifteen figures, artists' drawings, used to demonstrate the alleged evolution of humans, starting with protoapes and showing their evolution all the way up to modern humans. This parade first appeared in the Time-Life Nature Library series *Early Man* by F. Clark Howell, originally published in 1965, with the parade on pages 41 to 45.

This parade has become an almost universal icon for human evolution. I have held a number of creation seminars in foreign countries. The parade was well-known in Japan. The Australians and Tasmanians were very familiar with the parade. In 1995, my wife and I were involved in a humanitarian mission to the Baltic country of Lithuania. The Lithuanians knew about the parade. The Canadian schoolchildren in British Columbia were knowledgeable of the parade. The students at the University of the West Indies, Barbados, had seen the parade. To illustrate how this parade has infected the human thought-stream, a few years ago my daughter sent me a birthday card. On the front of it was one of the more apish individuals taken from the original Time-Life parade. Inside, it said, "You've come a long way, Baby!"

I mentioned this parade in chapter 2 as an illustration of outrageous and deceptive evolutionist artistry. I now want to explain why this parade, the most brilliant propaganda tool evolutionists have ever devised, suddenly became politically incorrect.

Evolutionists have known for many years that this parade was fiction. The first edition of this book revealed it to be fiction back in 1992. Yet the first "official" admission of that fact that I am aware of was not until the year 2000 when J. J. Hublin (Max Planck Institute for Evolutionary Anthropology, Leipzig) wrote, "The once-popular fresco showing a single file of marching hominids becoming ever more vertical, tall, and hairless now appears to be a fiction" (*Nature* 403 [27 January 2000]: 363). To add insult to injury, Hublin seems to suggest that this condition was a *recent* discovery by evolutionists.

The parade was scientific fraud, but for human evolution it was brilliant advertising. However, we no longer see the parade in reputable science books and journals. It's not because the evolutionist community finally became conscience-stricken regarding the parade. It was of such incalculable propaganda value that I believe we would still see it if it were not for a deeper reason—the parade was incredibly and explosively racist. And racism is no longer politically correct.

The last six figures in that famous parade were all labeled *Homo sapiens.* The six figures were:

1. Early *Homo sapiens,* based on the Swanscombe, Steinheim, and Montmaurin fossils and called "the earliest examples of man's modern species."
2. Solo Man, based upon the Solo fossils from Java and called "an extinct race."
3. Rhodesian Man, based upon the Rhodesian skull and called "another extinct race of *Homo sapiens.*"
4. Neandertal Man, which at that time was considered a subspecies of modern man.
5. Cro-Magnon Man, called "only a cultural step away from modern man."
6. Modern Man

The evolutionist empire needed new clothes. The parade had been milked for all it was worth, but a whole new scenario for human evolution was needed. First, the origin of humans must take a backseat, temporarily, and the origin of *modern* humans must become the main issue.

Second, the problem addressed by Stephen Jay Gould (chapter 15) must be solved. If different human groups had different fossil lines of descent with different time spans, racist charges would be inevitable. Racism in the human family hundreds of thousands of years ago would concern no one. But with five fossil members of *Homo sapiens* in the parade already (plus modern man) and several more members of the genus *Homo* discovered since, a nonracist origin of *modern* humans was desperately needed.

Third, all possible lines of descent for modern humans, other than African Eve, needed to be erased or pushed back. Cro-Magnon was so similar to modern man that these two groups could be combined. Rhodesian Man, now given a respectable evolutionary age (chapter 16), was made the evolutionary ancestor of African Eve and the newly discovered Herto fossils (chapter 19). The Swanscombe, Steinheim, and Montmaurin fossils have been assigned to the Neandertals (chapter 20), and all of the Neandertals are now considered extinct and have been moved out of the species of modern man completely (section V). Solo Man could be something of a problem. These fossils were recently redated as possibly only 27,000 years old on the evolutionist's time scale (see Carl Swisher, Garniss Curtis, and Roger Lewin, *Java Man* [New York: Scribner, 2000]: 223). The best response was to challenge the new date and to consider these fossils to be late surviving *Homo erectus* population that also became extinct without issue.

Unfortunately, the fossil evidence for African Eve was skimpy. Chemistry came to the rescue. This section tells the story.

CHAPTER 17

African Eve

A Woman of No Importance

THE EVOLUTION EMPIRE'S NEW CLOTHES—its new, fashionable, nonracist look—began with African Eve in 1987. The African Eve view, also known as the "Out of Africa" or "Mitochondrial Eve" theory, originated in a chemistry test tube at the University of California, Berkeley. This African Eve has no relationship to the biblical Eve of Genesis 2 and 3.

Particulars of the African Eve Model are:

1. Archaic humans and/or *Homo erectus* originally evolved in Africa from some form of *Homo habilis* or australopithecine stock. Some of those people had moved out of Africa into Asia and possibly into Europe about 2 million years ago. This earlier migration out of Africa represents the populations that were eventually replaced by the African Eve people. Some workers suggest the possibility of many migrations out of Africa

by many primitive *Homo* populations before the migration of the African Eve people.

2. Modern humans evolved in Africa, stemming from an "Eve" who lived in Africa about 200,000 years ago. Unlike the biblical Eve, this imaginary Eve was not the first woman living. She was one of a population of about 10,000 living at that time, having evolved from an earlier *Homo erectus* population. All other lines of mtDNA descent eventually disappeared without issue. Only hers survived to repopulate the world and be the "mother of us all."
3. These modern humans migrated out of Africa about 100,000 to 150,000 years ago.
4. These modern humans then migrated across Europe and into Asia.
5. They eliminated all other humans with little or no interbreeding, eventually replacing the Neandertals in Europe and replacing all other "primitive" humans in the world, including *Homo erectus.*
6. Africa thus became the birthplace of all original humans (*Homo erectus*) as well as the origin of all modern humans.
7. The validity of this view must be established by fossil evidence.

Several factors have been interpreted as evidence for this replacement of the Neandertals by modern humans. Anatomically modern human fossils were found at a site known as Qafzeh in Israel. Originally, the date for this site was thought to be about 30,000 ya, the same as some Neandertal sites in the Near East. This would be about the time that the Neandertals disappeared and anatomically modern humans appeared in Europe. However, thermoluminescence dating of the Qafzeh site gave the amazing date of 90,000 ya.[1] This date not only implies the very early appearance of modern humans in the Near East but also implies a 60,000 year period when Neandertals and modern humans allegedly lived together without any genetic exchange or hybridization.

From the Kebara Cave in Israel came the first complete pelvic bone of a Neandertal. It appears that the structure and orientation of the sockets into which the thigh bones fit were a bit different from those of modern humans. Yoel Rak and Baruch Arensburg (both of Tel-Aviv University) interpret this as differences in "locomotion and posture-related biomechanics" between the Neandertals and modern humans.[2] While the true significance of this difference remains unclear, it is interpreted by some evolutionists as further evidence that there was no close evolutionary relationship between the Neandertals and anatomically modern humans.

Based on the belief that Neandertals and anatomically modern humans were reproductively isolated for about 60,000 years (though they may have been neighbors geographically), some evolutionists now suggest that the Neandertals were more distinct from modern humans than has been realized. Stephen Jay Gould and Chris Stringer (Natural History Museum, London) even suggest that the Neandertals be removed from our species (*sapiens*) and once again be given their earlier designation of *Homo neanderthalensis.*[3,4] But Stringer does suggest that Neandertals and modern humans "were probably sufficiently closely related to allow hybridization."[5] (This comment reveals why the scientific word *species* and the Genesis word *kind* are not the same and should never be used as synonyms. *Kind* implies a genetic barrier that cannot be crossed.)

The African Eve theory exploded on the paleontological world in 1987, when three Berkeley biochemists, Rebecca Cann, Mark Stoneking, and Allan Wilson, published a paper in *Nature.*[6] They explored a new way of tracing human origins using tracer DNA, called mitochondria (mtDNA), from inside the cell. Each of our cells contains many mitochondria, the cell's powerhouses that crank out a cascade of energy-rich molecules. Unlike our regular chromosomes, the chromosomes of mitochondria are allegedly passed unchanged from a mother to her offspring. The father's mtDNA ends up "on the cutting room floor." The assumption is that, were it not for occasional mutations, everyone in the world would have identical mitochondrial chromosomes. But mutations do occur, and each mutation establishes a new mitochondrial type.

The inheritance of mitochondria is very much like the inheritance of surnames, except that mitochondria pass down the female rather than the male line. Surnames are a good analogy, however, and I will use them to illustrate the process the Berkeley biochemists used when they sought to determine the evolutionary origin of modern humans.

We start with a small group of people who have the surname Smith. In each generation a random "mutation" that changes one letter of the old name occurs in each of the family lines. In the first generation a Smith could become Sbith, and in the next generation it could become Qbith. In another line Smith might become Smjth, and in a third line it could become Smifh. In some lines the name might be lost completely if the males carrying it do not reproduce or if they have only female children. Over many generations, a large population would be traceable back to a single ancestral name—a name that may no longer be in existence but that could be reconstructed from the modified names.

The African Eve theory seemed to be rather brilliantly conceived. The Berkeley biochemists made several reasonable, although unprovable, assumptions. They first assumed no mixing from generation to generation, the chromosomes of mitochondria being passed unchanged from a mother to her offspring. (This assumption has recently experienced a serious challenge and may be false.[7]) They then assumed that all changes in the mtDNA were the result of mutations over time. They further assumed that these mutations occurred at a constant rate. On the basis of these assumptions, the researchers believed they had access to a "molecular clock." Since mtDNA is thought to mutate faster than other DNA, it was favored because it would lend itself to a more fine-grained index of time.

The original 1987 study used mtDNA from 136 women from many parts of the world who had various racial backgrounds. The analysis seemed to lead back to a single ancestral mtDNA molecule from a woman living in sub-Saharan Africa about 200,000 years ago. A subsequent and more rigorous 1991 study seemed to confirm and secure the theory.

Unfortunately, there was a serpent stalking this "Eve" as well as the first Eve. The researchers used a computer program designed to reveal a maximum parsimony phylogeny. This is the family tree with the smallest number of mutational changes, based on the assumption that evolution would have taken the most direct and efficient path—a rather strange assumption, considering the presumed random and haphazard nature of evolutionary change. The computer program was, however, far more complicated than the biochemists realized. They did not know that the result of their single computer run was biased by the order in which the data were entered. Others have determined that with thousands of computer runs and with the data entered in different random orders, an African origin for modern humans is not preferred. Some have also suggested that in the original study the biochemists were influenced in their interpretation of the computer data by their awareness of other evidence that seemed to favor an African origin.

Henry Gee, from the editorial staff of *Nature*, describes the results of the mtDNA study as "garbage." He states that, considering the number of items involved (136 mtDNA sequences), the number of maximally parsimonious trees exceeds one billion.[8] Geneticist Alan Templeton (Washington University, St. Louis) suggests that low-level mixing among early human populations may have scrambled the DNA sequences sufficiently so that the question of the origin of modern humans and a date for Eve can never be settled by mtDNA.[9] In a letter to *Science*, Mark Stoneking (one of the original researchers, now at the Max Planck Institute for Evolutionary Anthropology, Leipzig) acknowledges that African Eve has been invalidated.[10] There is general recognition that Africans have greater genetic diversity than other people, but the significance of that fact remains unclear.

The African Eve theory represented the second major attempt by biochemists to contribute to the question of human origins. Earlier, Berkeley biochemist Vincent Sarich had estimated, based on molecular studies, that the chimpanzee-human separation took place between five and seven million years ago. Although that date was much later than had been estimated from fossils, Sarich's date is now almost universally accepted.

In an article written before but published after the recent challenge to African Eve, Wilson (who died in 1991) and Cann (now at the University of Hawaii,

Manoa) laud the virtues of molecular biology in addressing human origins. They say that "living genes must have ancestors, whereas dead fossils may not have descendants." The molecular approach, they claim, "concerns itself with a set of characteristics that is complete and objective." In contrast, the fossil record is spotty. "Fossils cannot, in principle, be interpreted objectively."[11] They conclude that the method of the paleoanthropologists tends toward circular reasoning. They are right. Creationists have asserted that fact for many years.

However, Wilson and Cann were not able to see the logical fallacy in their molecular biology when it addressed phylogeny. This approach, known as molecular taxonomy, molecular genetics, or the newer related field of molecular archaeology, also traffics in circular reasoning. Molecular genetics, hiding behind the respect we all have for the science of genetics and the objectivity of that science, is highly infused with subjective evolutionary assumptions. In this field, the commitment to evolution is so complete that Wilson and Cann understand "objective evidence" as "evidence that has not been defined, at the outset, by any particular *evolutionary* model" (emphasis added).[12]

The mtDNA study of African Eve, as well as other aspects of molecular genetics, is based on mutations in the DNA nucleotides. Perhaps we could be forgiven for asking this question: When an evolutionist looks at human DNA nucleotides, how does he know which ones are the results of mutations and which ones have remained unchanged? Obviously, to answer that question he must know what the original or ancient sequences were. Since only God is omniscient, how does the evolutionist get the information about those sequences that he believes existed millions of years ago? He uses as his guide the DNA of the chimpanzee.[13] In other words, the studies that seek to prove that human DNA evolved from chimp DNA start with the assumption that chimp DNA represents the original condition (or close to it) from which human DNA diverged. That is circularity with a vengeance.

Evolutionists also must determine the rate of mutational changes in the DNA if these mutational changes are to be used as a "molecular clock." Since there is nothing in the nuclear DNA or the mtDNA molecules to indicate how often they mutate, we might also ask how the evolutionist calibrates his "molecular clock." Sarich, one of the pioneers of the molecular clock concept, began by calculating the mutation rates of various species "whose divergence [evolution] could be reliably dated from fossils."[14] He then applied that calibration to the chimpanzee-human split, dating that split at from five to seven million years ago. Wilson and Cann applied Sarich's mutation calibrations to their mtDNA studies, comparing "the ratio of mitochondrial DNA divergence among humans to that between humans and chimpanzees."[15] By this method they arrived at a date of approximately 200,000 years ago for African Eve. Hence an evolutionary time scale obtained from an evolutionary interpretation of fossils was superimposed on the DNA molecules. Once again, the circularity is obvious. The alleged evidence for evolution from the DNA molecules is not an independent confirmation of evolution but is instead based on an evolutionary interpretation of fossils as its starting point.

We humans are enamored with our ability to develop sophisticated experiments and to process massive amounts of data. Our problem is that our ability to process data has outstripped our ability to evaluate the quality of that data. Computers are not able to independently generate "truth," nor can they cleanse and purify data. With the recognition that mtDNA studies are incapable of determining the origin of modern humans, biochemists are now turning to nuclear DNA to help them solve the problem. More and more, molecular genetics and sophisticated computer programs are being enlisted in the service of evolution. The results are advertised as independent confirmations of evolution when in reality they are not. Molecular studies are the wave of the future for evolutionary studies. This approach is very convincing because it appears to be so scientific to those who do not recognize its evolutionary presuppositions.

Ever since the African Eve theory was first invented, there have been attempts to sanitize and refine it because it is so desperately needed to protect human evolution against racism. At this time, the African fossil evidence for African Eve is both sparse and questionable. The present status of African Eve is summed up by Jonathan Marks (Yale University): "And with each new genetic study that claims to validate 'Eve' conclusively, there comes an equal and opposite reaction, showing the study's weakness."[16]

The Multiregional Continuity Model is the competing theory of the origin of modern humans. Its leading advocates, Milford Wolpoff and Alan Thorne, argue that the ancient ancestors of various human groups in Asia, Africa, and Europe lived more or less where those groups are found today.

Particulars of the Multiregional Continuity Model are:

1. Some archaic humans and/or *Homo erectus* started leaving Africa by 2 million years ago.
2. They were not replaced wholesale by more recent migrations out of Africa.
3. In Africa, Asia, and Europe, these pre-*sapiens* all gradually evolved into *Homo sapiens* in their own areas (parallel evolution).
4. There was sufficient "cross-pollination" between geographic areas to maintain the genetic unity of the human family.
5. *Homo erectus* and *Homo sapiens* should be classified as one species having geographic and genetic variation.

The need for cross-pollination in order for groups in all geographic areas to evolve into the same species, *Homo sapiens,* seems to be a tacit admission that evolution is inherently racist and that without sufficient cross-pollination, evolution would produce groups that are inherently unequal.

Matt Cartmill (Duke University) verifies my charge that behind the Empire's new clothes—the African Eve Model—lurks the concern of racism.

> The upshot of the controversy over [Carleton] Coon's book was that some anthropologists began to view the whole regional-continuity model of human origins as a theory tainted by racism.
>
> For this reason, some fossil experts hail every piece of evidence for the out-of-Africa theory as further proof of modern human equality. (If we're all descended from an African "Eve" who lived just 100,000 years ago, we can't be very different from one another, can we?)
>
> This debate may be more vehement than it needs to be because partisans on each side see themselves as defending the unity and equality of the Family of Man against attacks by the opposing camp.[17]

Wolpoff and Thorne list five predictions of the African Eve theory that the fossil evidence should support.

> The first and major premise is that modern humans from Africa must have completely replaced all other human groups. Second, implicit within this idea is that the earliest modern humans appeared in Africa. Third, it also follows that the earliest modern humans in other areas should have African features. Fourth, modern humans and the people that they replaced should never have mixed or interbred. Fifth, outside of Africa an anatomical discontinuity should be evident between the human fossils before and after the replacement.[18]

The major flaw in the Multiregional Continuity Model is that it is an *evolutionary* model involving the vast time scale of evolution. In one area, it is correct: It states that *Homo erectus* and *Homo sapiens* should be classified as one species having geographic and genetic variation.

In contrast, all of the premises of the African Eve Model are falsified by the fossil record except the second one, which is falsified by the Genesis record. Humans did not first appear in Africa. After the Genesis flood, the first humans appeared in Turkey or Armenia.

The third premise is that the earliest modern humans in other areas should have African features. Wolpoff makes the point that the African Eve view demands "that people came from Africa, replaced the natives around the world, but then came to look just like the natives." The fourth premise, that modern humans and the people that they replaced should never have mixed or interbred, is doubted by an increasing number of African Eve believers themselves. Erik Trinkaus declares regarding the two opposing models, "Continuity [Multiregional Continuity Model] versus replacement [African Eve Model] is dead." The debate is now over "trivial amounts of admixture versus major amounts of admixture."[19]

The first and the fifth premises of the African Eve Model are completely falsified by the fossil record in Asia. The Asian fossils do not show replacement or discontinuity. In fact, there is an amazing continuity of *Homo erectus*-like fossils in Asia beginning at 2 million years ago all the way to 6,000 years ago and possibly to just a few hundred years ago (the Cossack skull).

Richard G. Klein, who advocates the African Eve view, is courageous enough to state what could falsify the African Eve Model. He writes:

> The occupation of Australia by modern humans at or before 60,000 years ago would argue not only against a radical behavioral shift between 50,000 and 40,000 years ago, it would also require modifications to the more fundamental hypothesis that modern humans originated in Africa. The early Australian dates are revolutionary if they are correct, but they have encountered serious skepticism.[20]

Klein is referring to a redating of Mungo Man 3 at Lake Mungo, New South Wales, Australia. Alan Thorne and his associates had just dated Mungo Man at 62,000 years old. The problem is not just the age of Mungo Man, although that in itself is a problem. The real problem is that Mungo Man is anatomically modern in shape, whereas some Australian fossils dated at about 10,000 years ago are robust, more *erectus*-like. That is the opposite of what the African Eve theory would predict. A firestorm has broken out because the African Eve (Out of Africa) Model is seriously threatened, and with it the protection that human evolution has against the charge of racism. It was predictable that Mungo Man would be redated. This time the date for Mungo Man was 40,000 years old. However, there is evidence that humans were present in that area up to 50,000 years ago.[21] African Eve is definitely in danger. Stay tuned.

Why is the dating of Mungo Man so emotionally charged? Emma Young, writing in *NewScientist*, comments regarding that 40,000 year date, "That date will reassure many palaeontologists, because if it were significantly older, Mungo Man would have challenged the *dearly held* 'Out of Africa' hypothesis" (emphasis added).[22] Question! Why is the Out of Africa Model "dearly held"? Obviously, we are dealing here with more than just "science"; we are dealing with politics.

It is commonly claimed that the Neandertals in Europe were replaced, yet the fossil record does not confirm that claim. Gradations from Neandertals to modern humans are seen in the fossil record. We are not referring to an evolutionary transition from earlier Neandertals to later modern humans. We are referring to morphological gradations between Neandertals and modern humans, both having the same dates, living at the same time as contemporaries, and representing a single human population. Whereas evolutionists have chosen to divide these Europeans into two categories—Neandertals and anatomically modern *Homo sapiens*—individual fossils are not always that easy to categorize. There is wide variation among modern humans, and there is variation within the Neandertal category as well. A number of fossils in each group are very close to that subjective line and could be categorized either way. These fossils constitute a gradation between Neandertals and modern humans, demonstrating that the distinction made by evolutionists is an artificial one.

Among the fossils usually classified as Neandertal are at least twenty-six individuals from six different sites who are clearly close to that subjective dividing line. These fossils constitute part of that continuum or gradation from Neandertals to modern humans found in the fossil record. Evolutionists recognize these fossils as departing from the classic Neandertal morphology and describe them as "progressive" or "advanced" Neandertals. Their shape is sometimes explained as the result of gene flow (hybridization) with more modern populations. This would refute the interpretation of mtDNA and other evidence that the Neandertals and modern humans are not the same species—since reproduction is on the species level. Those sites having "advanced" Neandertals are:

Vindija Cave remains, Croatia, twelve individuals[23]
Hahnöfersand frontal bone, Germany, one individual[24,25]
Starosel'e remains, Ukraine, CIS, two individuals[26]
Stetten 3 humerus, cave deposits, Germany, one individual[27]
Ehringsdorf (Weimar) remains, Germany, nine individuals[28]
Krapina 1 (formerly Krapina A) skull, Croatia, one individual[29]

Completing that continuum or gradation from Neandertals to modern humans are at least 107 individuals from 5 sites who are usually grouped with fossils of anatomically modern humans. Since they are close to the line that divides them from the Neandertals, however, they are often described as "archaic moderns" or stated to have "Neandertal affinities" or "Neandertal features." These five sites are:

Oberkassel remains, Germany, two individuals[30]
Mladec (Lautsch) cave remains, Czech Republic, a minimum of ninety-eight individuals[31,32,33]

Velika Pecina Cave skull fragments, Croatia, one individual[34,35]
Bacho Kiro Cave mandibles, Bulgaria, two individuals[36]
Pontnewydd Cave remains, Wales, four individuals[37]

Creationists have long maintained that the differences found in the human fossil material are the result of geography, not evolution. Notice that of the 133 fossil individuals that are "close to the line" between Neandertal and modern European morphology, all but four of them are from eastern or central Europe. If the differences between Neandertal and modern Europeans are entirely genetic (other possibilities will be suggested later), perhaps eastern Europe is where the hybridization or the change in gene frequencies began.

G. A. Clark summarizes the evidence that the Neandertals are the ancestors of at least some modern humans: "Those who would argue that Neandertals became extinct without issue should show how it could have occurred without leaving traces of disjunction in the archaeological record and in the fossils themselves."[38]

Australian archaeologist Robert Bednarik, describing his discovery in the Moroccan city of Tan-Tan of what appears to be a 400,000-year-old figurine carved out of quartzite rock, says, "The only way to maintain the Eve hypothesis is by drawing a thick line between moderns and totally different archaic people. That's not what we see."[39]

It's too bad. African Eve seems like the kind of person you would love to join for a cup of Starbucks. It's a shame she's just the product of a test tube.

A Really Clever Trick

I am continually amazed that the theory of human evolution, so totally racist, has escaped public outcry. However, a novel theory proposed by Ian Tattersall allows human evolution to escape the racist charge. The proposal is both bold and absurd. Tattersall suggests that human evolution has stopped! Evolution, after allegedly going on for three to four billion years, has stopped—at least the human aspect of it. How on earth would anyone know? Tattersall does not state that he made the suggestion to escape the racism inherent in human evolution, but the idea does succeed in doing just that. It is clearly politically correct. It allows racism in the past so that humans could evolve to their present condition but removes human evolution in the present from the racist stigma.

Tattersall bases his claim entirely on evolutionary theory. He states, "If any meaningful innovations are to become fixed in a population or if a population is to become established as a new species, it is essential that the population be small. Large populations simply have too much genetic inertia."[40] With a human population of six billion, with humans very mobile, with mixed marriages common, and with the boundaries of various ethnic groups becoming more blurred, Tattersall feels that the possibility for significant evolutionary change in humans has disappeared.

Tattersall's statement is an admission of the inherent racism of evolution. He states that for a population to be established as a new species, that population must be small. He can't be referring to a "new species" of humans that would be genetically inferior to those living today. According to evolutionary theory, inferior species are quickly eliminated. Nor can he be referring to a "new species" of humans equal to those living today. Evolution doesn't do that sort of thing. He could only be talking about a "new species" of humans that would be genetically superior to those living today. And that is racism.

So, while Tattersall has advanced an idea that could possibly take human evolution "off the hook" for being racist in the present time, he is the most ardent of racists when it comes to the interpretation of the fossils of our ancestors.

I trust that my friends of African descent will not be offended by what I have said about African Eve. I have not dishonored you. Your true honor is that you and I have both been created in the image of the living God. Our further honor is that by trusting in our risen savior, Jesus Christ, we can live with him as his dear children forever. There is no real honor in being descended from a woman who never existed.

CHAPTER 18

SPLITTERS VERSUS LUMPERS

"SPLITTERS" AND "LUMPERS" ARE terms used to describe two different philosophies in interpreting the fossil record. Regarding the human fossil record, the King of Splitters is Ian Tattersall (American Museum of Natural History, New York). The Crown Prince of Lumpers is Milford Wolpoff (University of Michigan). Let me spin a story, simplistic though it is, to explain what these terms mean.

In the distant future, a fleet of fifty silver spacecraft from the planet Zox, in the galaxy Sigma, gently set down at various places on planet Earth. They uniformly report that Earth is a wasteland with no apparent evidence of life. Yet, long before these Zoxians had intergalactic travel ability, their Very Long Array radio telescopes had confirmed that there was intelligent life on planet Earth. Their ultra-sensitive receivers had listened to "Earth chatter" coming from a very high life-form with intelligence, communication skills, and technical ability. Cryptologists had deciphered enough of "Earth chatter" to know that the scientific term for these Intelligent Beings was *Homo*. They also sensed that there were two types of these *Homo* beings, which in "Earth chatter" had been referred to as XX (Earth females) and XY (Earth males). Unfortunately, the Zoxians had not been able to discern what the distinctions were between the XXs and the XYs.

They start to dig. At many of the fifty sites, they uncover fossils of numerous life-forms (animals and plants). These they take to the Earth Central Field Laboratory that they have established. Overseeing the entire project are two of their most eminent scientists, Dr. Tattersplit, from the Zoxian Association for the Advancement of Science, and Dr. Wollump, representing the Zoxian Academy of Science. Tattersplit and Wollump are studying these animal life-forms to determine which one might be that "Being of Highest Intelligence."

Then, at one of the sites (China), they uncover a fossil life-form that apparently had walked upright, was about five feet tall, and had a very large brain. Comparing its brain size to its body size, they decide that this must be the "Intelligent Being" of planet Earth that was the object of their search. They call it *Homo magnum* (Being of Greatness) and catalogue the fossil skeleton as Earth Homo-1, EH-1.

Thus, EH-1 becomes the type specimen for the other *Homo* fossils they hope to discover. A *type specimen* is the first significant fossil of a new kind that is capable of being described for purposes of comparison. It becomes a sort of yardstick for evaluating similar fossils as to whether or not they belong in the same category. (I have always felt that the type specimen idea was a bit strange, because a type specimen is not necessarily representative of its group. In terms of the total variation within a group, the type specimen could be nearer to the edge of the group's variation than the middle. I suppose that from the standpoint of probability, a random sample is more likely to be close to the median than to the edge, but we have no way of knowing. Sometimes, statistics has a way of betraying us. In the minds of many paleontologists, the type specimen tends to be thought of as being the median in variation, whether or not it actually is.)

Later, a second *Homo* fossil is found at another site. This one is similar to the first one but is only about four feet tall (possibly a Mbuti pygmy from Zaire). It is catalogued as EH-2. Both Tattersplit and Wollump agree that EH-1 and EH-2 probably belonged to the same group of Earth *Homo*. Some time later, at two other sites, two more *Homo* fossils are found—one about six feet tall and the other about seven feet tall (possibly the remains of an NBA basketball player). They are catalogued as EH-3 and EH-4, respectively. At this point, controversy breaks out.

Dr. Wollump believes that there are enough similarities among the four fossils that all of them can be classified as being members of the same group of *Homo.* Although neither he nor Dr. Tattersplit know anything about sexual dimorphism (size differences between males and females), Wollump also speculates about the possibility of a size difference between the *Homo* XXs and the *Homo* XYs, with EH-2 being a small XX and EH-4 being a large XY, or the reverse.

While not denying the similarities of the *Homo* fossils, Tattersplit feels that the degree of size difference in the *Homo* fossils goes far beyond that which one would find in a single Earth species. He cites the studies they have conducted on other Earth animals and claims that the size variation among these four *Homo* fossils exceeds that of any of the other animal groups. Further, he emphasizes that the four *Homo* fossils were found at distant places from each other on planet Earth—places separated by what had been vast oceans. The animal fossils tended to reveal species differences in different geographic areas of Earth. Since these *Homo* beings were just a part of the animal life of planet Earth, they ought to reflect that diversity.

Tattersplit is especially impressed with the results of one of their digs on a large island off a large land mass (Madagascar, off the eastern coast of Africa). Here they have discovered fossils of lots of little animals, pre-pre-pre-pre-pre-*Homo*-like beings (lemurs) that have a lot of variation, suggesting different species. He feels that the *Homo* group should reflect that kind of diversity. He submits the following classification:

Earth *Homo* Fossil EH-2	*Homo minor*
Earth *Homo* Fossil EH-1	*Homo minor magnum*
Earth *Homo* Fossil EH-3	*Homo major*
Earth *Homo* Fossil EH-4	*Homo major magnum*

Wollump is adamantly opposed to such an extreme "splitter" classification. Not only does the total variation in the *Homo* fossils not warrant such splitting, but, he points out, Fossil EH-2, the shortest fossil *Homo* of the group at four feet, has the largest brain size compared to body size. Therefore, it presumably was the most intelligent. In Tattersplit's classification, it should be the one classified as *Homo major magnum.*

Unfortunately, a financial crisis in scientific research funding on planet Zox forces an end to the project. The spaceship fleet is recalled before more *Homo* fossils could be discovered. However, the fleet takes their trove of fossils with them back to Zox for further study. We understand that the sharp discord between Tattersplit and Wollump continues unabated.

Milford Wolpoff: So Near, Yet So Far

Milford Wolpoff's view of the human fossil record is a reflection of his view on the evolution of modern humans—the Multiregional Continuity Model. He has long held that the differences between *Homo erectus* and *Homo sapiens* are within the range of genetic variation and that the primary difference is one of time. Time alone, he feels, is not a sufficient basis for a species distinction.

Wolpoff has been one of the most vocal evolutionists calling for the "sinking" of the taxon *Homo erectus* into *Homo sapiens.* He writes in conjunction with Wu Xin Zhi (Institute of Paleoanthropology, Beijing) and Alan G. Thorne (Australian National University), "In our view, there are two alternatives. We should either admit that the *Homo erectus/Homo sapiens* boundary is arbitrary and use nonmorphological (i.e., temporal) criteria for determining it, or *Homo erectus* should be sunk [into *Homo sapiens*]."[1]

To sink, that is, combine, *Homo erectus* with *Homo sapiens* means that all of those "species" that allegedly lie between *Homo erectus* and *Homo sapiens* on the evolutionary scale should be considered *Homo sapiens* as well. These would include early *Homo sapiens, Homo heidelbergensis, Homo antecessor, Homo ergaster,* and the Neandertals. Wolpoff is right in stating that all of these should be combined, since they are all fully human. He is wrong in his belief that *Homo erectus* and the others evolved into anatomically modern humans and that there is a large time distinction separating them.

Ian Tattersall: Diversity Gone Crazy

In the world of science fiction, if you were seated at the *Star Wars* bar, you would find yourself surrounded by some very strange individuals from various places in the galaxy. In a similar manner, according to Ian Tattersall, if you could visit an African "watering hole" in the past, at 100,000-year intervals, you might meet some "human" species rather different from yourself. Tattersall claims these could be your very early ancestors.

Depending on the time of your visit, Tattersall feels that you could meet one or more of up to nineteen different "human" species of several genera (plural of genus; see Technical Section below).[2] Exactly how many different "human" species you might meet would depend on whether the evolutionist classifying them is a "Splitter" or a "Lumper." Tattersall is a world-class "Splitter." A "Splitter" tends to classify fossil populations that differ only slightly from each other into different species. A "Lumper" demands much larger differences between fossil populations before making species distinctions.

In fact, Tattersall is convinced that many more evolutionary "cousins" remain to be discovered in the fossil record. He is leading the charge to "manufacture" species in the human fossil record so that the human fossil record resembles the alleged fossil record of other mammals. One suspects that if the entire human race were to suddenly become fossilized, some future "Splitter" would classify the Pygmy and Watusi tribes of central Africa as two separate species.

It is always important to know the evidence paleoanthropologists use for their pronouncements on human evolution. Before Ian Tattersall entered the field of human evolution, he worked on the island of Madagascar studying lemurs. Lemurs are called "prosimians" and are lower primates from which the apes are thought to have evolved. Bernard Wood (George Washington University, Washington, D.C.) is perhaps the leading authority on the classification of human fossils. In reviewing one of Tattersall's books, Wood gives us an insight into Tattersall's reasoning:

> Tattersall came relatively late to human evolution. The study of lemurs was his introduction to evolutionary biology. . . . Tattersall looks at the [human fossil] evidence, compares it with what he saw in the lemurs, and interprets the human fossil record as evidence for discontinuity. Thus, he recognises many more human taxonomic groups from the record. This same prejudice explains why he is in the vanguard of those who interpret the Neanderthals as a species in their own right [not the same species as modern humans].[3]

Wood has not given us the whole story. When evolutionists say that "humans are just a part of nature," it sounds rather benign. However, they do not mean that we eat the same food and breathe the same air as the animals. They mean that we are the product of Nature (with a capital "N") with no divine intervention whatsoever. God was not involved in our evolution in any way. Mainstream evolutionists cannot tolerate the concept of creation because creation smacks of God. However, they have an even deeper intolerance—if that is possible. It is the biblical teaching that humans were created *Imago dei*—in the image of God. There is no way to describe how offensive the idea is to mainstream evolutionists that we humans are "special" in the eyes of God—separate and distinct from all animals.

Tattersall studied the lemurs of Madagascar. He noticed that there were just a few living species but many extinct (fossil) species. Since he believes that "humans are just a part of nature," he reasoned that although there is just one species of humans living today, there must have been many human species that have become extinct. These would be what he and other evolutionists call "Nature's failed experiments." Not only does this concept fit well with evolution, but it also removes the idea that we humans are "special." Any god having so many failed experiments is no smarter than the average village tinker. And the more extinct species Tattersall can invent, the more it looks like the work of imperfect Nature, not the work of an all-wise God.

The fossil record is now being reinterpreted by Tattersall and others (especially by those who hold the Out of Africa Model) to bring human origins more in line with the rest of nature. Evolutionary trees are out. Evolutionary bushes are in. *Homo habilis* has been split into at least two separate species, *Homo habilis* and *Homo rudolfensis*. *Homo erectus* has been split into two separate species, *Homo erectus* and *Homo ergaster*. The Neandertals are just one of at least five twigs on the human evolutionary bush. Evolutionists do not know, and say that they may never know, from which of the twigs modern humans evolved. However, the Neandertal twig, based on the evolutionists' interpretation of this mtDNA recovery, has been eliminated from modern human ancestry. Since 1964, the Neandertals have been considered a subspecies of modern humans. They have now been moved out of our species, at least by the splitters, and back into a separate species, *Homo neanderthalensis*.

To split a species, such as *Homo erectus*, into an Asian species and an African species implies a thorough knowledge and study by the splitters of the fossils involved. However, Susan C. Antón (Rutgers University), who has probably had as much exposure to the Javanese fossils as anyone, writes regarding a division between the Asian and the African *Homo erectus* fossils:

> The earliest Southeast Asian fossils have not been fully described nor directly compared with their Western Asian or African contemporaries, particularly with regard to their dental anatomy. Thus, relationships among these early dispersers remain unclear.[4]

The lack of a full description of the Southeast Asian (Javanese) *Homo erectus* fossils is probably due to the lack of access to those Javanese fossils that we described in chapter 1. However, this supports our claim that the splitting of the human fossils into many different species is not based upon the genuine differences in the fossils themselves but upon the philosophical belief that the human fossils should conform to the multiple species in the animal world.

A problem, however, is human equality. The Bible has always taught that all humans are equal, but it is a relatively new concept for evolution. In fact, it is incompatible with evolution. Hence there is the need for smoke and mirrors and the wild-eyed enthusiasm for African Eve. Most of the human species Tattersall and others hold to can be relegated to the deep past—hundreds of thousands of years ago. Few will mind, and virtually none will notice the racism involved. The Neandertals are the problem. They are the subject of a later section.

Humans Are Unique

A remarkable feature of the human family is that, worldwide, we are all one species, or to use the biblical word, "kind." We are all interfertile. In this regard, humans are unique. Scripture clearly teaches this unity: "From one man he made every nation of men, that they should inhabit the whole earth" (Acts 17: 26). In the evolutionary classification, almost all other forms of life—living or fossil—are represented by a cluster of similar species, such as the dog and its related species, the wolf, the coyote, the jackal, and the hyena. Since humans are just a part of the evolutionary process, and since there are no other species of humans living today, evolutionists claim that there must have been many species of humans or semihumans living in the past. We are now alone. The others have all become extinct.

Although some have considered them to be synonyms, the words *species* and *kind* are not exactly the same. The basis for both concepts is reproductivity. However, the scientific word *species* deals with reproducing populations in the world of the present. Further, it is not a static condition. If an interbreeding population becomes divided and each section becomes permanently isolated from the other, they would be reclassified scientifically as two separate species.

In contrast, the biblical word *kind* deals with organisms as they were at the beginning—at the time of their creation. Over time, various populations of organisms have become geographically isolated. Due to mutations, which are always harmful or at best neutral, they are no longer interfertile. Hence scientifically they would be classified as separate species. Biblically, however, they would still be members of their original kind.

To illustrate, there are two separate populations of squirrels on the north and south rims of the Grand Canyon. If put together, these two populations would be interfertile. However, the Grand Canyon presents a permanent geographic barrier. Although they would still be considered a part of the original "squirrel kind" biblically, they are considered two separate species scientifically. The Bible teaches that the original genetic barrier between kinds that was established at creation cannot be crossed. This can be demonstrated scientifically. The Bible teaches the fixity of kinds. It does not teach the fixity of species. It is not unusual for those who do not understand the difference between the two terms to think that when they have proven that species are not fixed, they have automatically proven the Bible to be in error.

Technical Section: Tattersall's Cast of Characters

A splitter's classification and description of the hominid fossils

Members of the Genus Homo

Homo sapiens sapiens represents modern humans worldwide, the lone "survivors" in the evolutionary lottery. We allegedly arose in Africa about 150,000 years ago.

Homo neanderthalensis is the Neandertals of Europe and Western Asia, the closest extinct "cousins" of modern humans. Tattersall and some others no longer consider the Neandertals to be in the same species as modern humans. They lived from about 200,000 to about 30,000 years ago.

Homo heidelbergensis, formerly called Archaic *Homo sapiens,* is found throughout Europe. These fossils date from 600,000 to about 200,000 years ago. This taxon may comprise more than one species and may have given rise to the Neandertals.

Homo antecessor is found in Spain and lived about 800,000 years ago.

Homo erectus is found in China and Java and lived from about 1.8 million years ago to as late as 40,000 years ago.

Homo ergaster is the East African version of *Homo erectus* and may have invented the hand ax. The best specimen is the "Turkana Boy," KNM-WT 15,000, whose skeleton is similar to modern humans. *Homo ergaster* lived from 1.9 million years ago to 1.6 million years ago.

Homo rudolfensis is found in eastern Africa, based primarily on skull KNM-ER 1470, which is relatively large-brained, and dated at about 1.8 million years ago. These fossils were originally included in *Homo habilis* but now are considered distinct from that group.

Homo habilis is found in Sub-Saharan Africa. His name means "handy man" because he was originally thought to be a toolmaker. That now may be questioned. The collection consists of an odd assortment of fossils from 2.5 million years ago to 1.8 million years ago that probably involves several species.

Members of the Genus Australopithecus

Australopithecus garhi is found in Ethiopia and dated at 2.5 million years ago.

Australopithecus africanus is found in South Africa and dated at 3.0 million years ago to 2.0 million years ago.

A new, unnamed, and undescribed South African fossil species is dated at 3.3 million years ago.

Australopithecus bahrelghazali is based on a jawbone found in Chad and dated at 3.5 million years ago to 3.0 million years ago.

Australopithecus afarensis is found in Ethiopia and Tanzania, includes the famous "Lucy," and is dated at 3.8 million years ago to 3.0 million years ago. This taxon may be a mixture of several species.

Australopithecus anamensis is found in Kenya and is very much like *A. afarensis.* It is dated at 4.2 million years ago.

Members of the Genus Paranthropus

Paranthropus boisei, found in eastern Africa, has massive jaws and teeth and was probably a vegetarian. Dated at 2.0 million years ago to 1.4 million years ago, it is not considered to be in the direct human line. It includes Zinjanthropus, the "Nutcracker Man" found by Louis and Mary Leakey that they claimed at first was a toolmaker. That claim was later rescinded.

Paranthropus robustus is found in South Africa and is dated at 1.6 million years ago. It is not considered to be in the direct human line.

Paranthropus crassidens is found in South Africa and is dated at 1.6 million years ago. It is very similar to *P. robustus* and is not considered to be in the direct human line.

Paranthropus aethiopicus is found in Eastern Africa. It is based primarily on the "Black Skull," KNM-WT 17,000, and dated at 2.5 million years ago. It is not considered to be in the direct human line.

Member of the Genus Ardipithecus

Ardipithecus ramidus is found in Ethiopia and dated at 4.4 million years ago. It is quite apelike and is the oldest known potential human evolutionary ancestor.

CHAPTER 19

"We Are All African"

A Study in Political Correctness

IN LATE 1997, three remarkable early *Homo sapiens* fossil skulls were discovered at Herto, Middle Awash, Ethiopia, about 140 miles south of Addis Ababa. It took the discoverers, Tim White (University of California, Berkeley) and his team, more than three years to clean, prepare, assemble, and analyze the fossils for the public announcements and reports. I find these reports a fascinating study in political correctness for three reasons: (1) the high degree of excitement concerning this discovery; (2) the admission that up to this time the African Eve Model had very little fossil evidence; and (3) the claim that the majority view—the African Eve Model—has been proven. It is all a desperate attempt to give legitimacy to African Eve.

The primary reports are by Tim White and J. Desmond Clark and their associates.[1,2] They are found in the 12 June 2003 issue of *Nature*. In that same issue of *Nature* was a commentary on the discovery by Chris Stringer.[3] Other articles are found in *Science*, in *NewScientist*, in *Time*, and in the *San Diego Union-Tribune*.[4,5,6,7,8,9] These articles represent an interesting mix of professional and popular reporting on this scientific discovery.

The discovery consists of two adult crania, one child partial cranium, and bits and pieces of skulls and teeth of seven other individuals. The skulls are thick with big brains, robust features, rather high doming, and moderate prognathism (jutting forward of the upper and lower jaws). One skull has a cranial capacity of about 1450 cc, a bit above today's average. There are no postcranial (body) remains. Argon radioisotope dating gives an age of about 160,000 years. These fossils are the most complete and best-dated fossils from this time period found in Africa thus far.

Stone tools found in association with the fossils consist of 640 items of Acheulean and Middle Stone Age technologies, including hand axes, flake tools, cores, flakes, and blades. There is evidence of heavy butchering of hippopotamus and the presence of bovid, horse, crocodile, and catfish. Cut marks and other indications suggest a deliberate mortuary practice. Cannibalism is ruled out because there is no meat on a person's skull at the places of the cut marks. The child's skull has both cut marks and a kind of polish on it as if it had been handled repeatedly and lovingly, resembling that seen on skulls handled in rituals in New Guinea in recent times.

Tim White *et al.* gave a detailed description and classification of the Herto fossils. It is one of the most confusing descriptions I have ever read. It sounds more like a justification for the naming of a new species or, in this case, a new subspecies. In paleoanthropology, you receive more fame if you find something new than if you find more of what others have already found. I will spare you the details of the description and give you the authors' conclusions: "The Herto hominids, although clarifying evolutionary questions, raise taxonomic issues." Because the Herto fossils, shapewise, are just beyond the range of variation seen in anatomically modern *Homo sapiens*, "and because they differ from other known fossil hominids, we recognize them here as *Homo sapiens idaltu*, a new palaeosubspecies of *Homo sapiens*." *Idaltu* is the word for "elder" in the local Afar language. They suggest that *Homo* evolution in Africa went from *Homo rhodesiensis* (Bodo and Kabwe) to *Homo sapiens idaltu* (Herto fossils) and then to fully modern people.

Studying the details, I felt the way Rick Gore (senior assistant editor of *National Geographic*) must have felt when Antonio Ascenzi (University of Rome) was pointing out to him some of the nuances of the Ceprano fossil skull. Gore was working on a series of articles for *National Geographic* entitled "The Dawn of Humans." In preparation, Gore traveled the world to view fossils and talk to paleoanthropologists. In Rome, with the Ceprano fossil skull on a desk in front of him, Gore wrote, "Tracing his finger across the skull, Ascenzi points out features that seem subtle, if not arcane [hidden, secret], to me but that indicate to specialists that Ceprano man was a *Homo erectus*, but with some distinctive features."[10]

Some "specialists" see things that others can't see and that may not even be there. Tim White is one of those splitter specialists. Chris Stringer, also known for his splitting ways, makes a comment regarding the most complete of the Herto crania: "It shows an interesting combination of features from archaic, early modern and recent humans." Later, he observes "that human populations in this era showed a great deal of anatomical variation."[11]

The solution is not difficult. The Middle Pleistocene early *Homo sapiens* fossils in Africa show the same high degree of variation that the Middle Pleistocene fossils in Europe show—the ones formerly called archaic *Homo sapiens*. These European fossils were a mystery to evolutionists until the Spanish Sima de los Huesos fossils revealed that all of those European fossils could fit under the same umbrella (see chapter 20). These diverse fossils are now all called Neandertals or pre-Neandertals by a growing number of paleoanthropologists. Unfortunately, there is not yet a Sima de los Huesos–like site in Africa to enlighten Tim White and his fellow workers. They are still trying to fabricate an evolutionary progression out of scattered fossil sites.

There is something about the reports of this fossil discovery, however, that I find very fascinating. Under the surface there seems to be an element of excitement—if one could ever call a scientific paper "exciting." From the emphasis given to it as well as the emotional language used, it is obvious that the true significance of this discovery in the minds of the discoverers is confirmation of the African Eve (Out of Africa) Model of modern human origins. One can almost feel a sigh of relief because they have finally found that for which they have been looking.

Tim White *et al.* say of the Herto fossils, "Their anatomy and antiquity constitute strong evidence of modern-human emergence in Africa." Later they claim, "The Herto crania fail to confirm such 'multiregional' speculation and conform more closely to most molecular predictions [of the Out of Africa Model]."[12]

Chris Stringer observes, "The fact that the geological age of these fossils is close to some estimates obtained by genetic analyses for the origin of modern human variation only heightens their importance."[13] James Randerson quotes Stringer as calling the Herto fossils "landmark finds in unravelling [*sic*] our origins." Randerson goes on to state, "Now, in a coup for the Out of Africa theory, the Herto fossils place modern-looking humans in the right place at the right time."[14]

Ann Gibbons emphasizes, "Now for the first time, fossils fit the genetic data." She then quotes Tim White as saying, "Now we have a great sequence of fossils showing our species evolved in Africa, not all over the globe."[15] Science writer Roger Lewin comments that the Out of Africa believers claim that the Herto fossils settle the question of where, when, and how modern humans evolved. Lewin concludes, "Although it is always risky in science to say 'case closed', that's how it looks."[16]

Time magazine quotes paleoanthropologist G. Philip Rightmire (State University of New York, Binghamton) regarding the certainty that the Herto people were the ancestors of modern humans: "It's as near as we're going to get." *Time* continues, " 'It's not just another nail in the coffin for the multiregional view,' says Rightmire. 'It lowers the coffin into the ground.' "[17]

According to *New York Times* writer John Noble Wilford, after declaring that the Herto people are our ancestors, not the Neandertals, the research team said, "In this sense, we are all African."[18] I cannot recall hearing scientists and science writers make such colorful and excited statements about any other fossil discovery.

There is another aspect of these Herto fossil reports that fascinates me. Let me give you a bit of background. It is the continual practice in evolutionary literature to present almost every aspect of evolution in a very positive, authoritative manner—especially in articles written for the general public. When a new discovery is made or a new methodology is developed, only then do they honestly tell of the weaknesses of the situation before this new development. I will recount this situation again regarding ostrich eggshell dating when we deal with the methods used to date the human fossils. Here, with the Herto fossils, we have another striking example of this deceptive technique.

In former accounts of the Out of Africa theory, the major evidence has been the alleged genetic evidence. But there was always an attempt to put a smiley face on the fossil evidence as well. Obviously, the paleoanthropologists were well aware of the serious lack of good fossil evidence for the Out of Africa theory. But that lack wasn't stressed. Now all the reports of the Herto fossil discovery talk about the former lack of fossil evidence as if it were common knowledge—even though the Out of Africa theory has been the theory of choice for over fifteen years.

Tim White *et al.* state that the origin of modern humans and the fate of the Neandertals have been fundamental questions for evolutionists for over a century. "A key barrier to the resolution of these questions has been the lack of substantial and accurately dated African hominid fossils from between 100,000 and 300,000 years ago." Concerning the genetic evidence for the Out of Africa Model, they write, "Fossil confirmation of these predictions has been lacking."[19] J. Desmond Clark *et al.* also refer to the lack of quality African fossils and difficulties in dating them when they speak of "the lack of chronological control over the widely scattered hominid fossils that are often singular and fragmentary."[20]

Chris Stringer likewise complained that "hard evidence for the inferred African origin of modern humans has remained somewhat elusive."[21] Ann Gibbons claims that the genetic evidence pointed "firmly" to Africa as our origin, but there were few fossils to support it.[22] Randerson speaks of the absence of "convincing African human fossils" in the right time period.[23] Lewin refers to "fragmentary" fossils with few good dates.[24] *Time* records that "until now, paleontologists could only speculate about the existence of such people [the Herto]."[25] And Wilford speaks of "the gap in fossil evidence between 100,000 and 300,000 years ago."[26]

I must smile at these quotations. The evidence was never "firm," nor is it now. The Out of Africa Model was seldom viewed by its believers as just a "prediction" or an "inference." Evolutionists did more than just "speculate" about the existence of people like the Herto. And one seldom heard about the "gap" of 100,000 to 300,000 years in the African fossil record. Politics is not the only place where we have revisionist history.

I am intrigued by these articles on the Herto fossil discovery because of the overblown claim of proof for the Out of Africa theory and the wide acceptance of that theory in the absence of fossil proof for it by, of all people, paleoanthropologists. This is especially intriguing because the genetic evidence for this theory has always been controversial. Remember, the statements regarding the popularity of the Out of Africa theory refer to the situation before this Herto fossil "proof" was discovered. Common sense tells me that when scientists are enraptured with a "scientific theory" that lacks adequate proof, there is a hidden reason that goes well beyond science.

It all goes back to the Carleton Coon fiasco back in the 1960s. Although most people have forgotten about Coon, the paleoanthropological community has not. Human evolution is as racist as ever, but the evolutionists are determined that such a scandal will never blot their science again.

I believe that the real reason paleoanthropologists have lovingly embraced the Out of Africa Model in spite of inadequate evidence and the reason for the evangelistic enthusiasm for this new Herto fossil discovery—very atypical of objective scientists—is that this theory represents to them the final deliverance from any possibility of the concept of human evolution being accused of being racist. Of course, the Out of Africa Model does not remove racism from the concept of human evolution. Racism is intrinsic to all of evolution. But it doesn't matter now, for few will notice. In the Out of Africa Model, all humans have evolved from the very same ancestors in so short a time span that there has not been time for significant racial differences to develop. That's all that matters. *Viva la* "Out of Africa."

In April and May 2003, PBS aired a series of programs entitled *Race: The Power of an Illusion*, produced by Larry Adelman. The series was reviewed in the journal *Science* by Mildred Cho and Maren Grainger-Monsen (both at the Stanford Center for Biomedical Ethics).[27] The reviewing authors state that the programs show the similarities between how science, in the past, used "ethnographic and anthropomorphic studies as 'proof' of race and the current use of genetics to do the same." The comparison is "particularly striking" because of the many nineteenth-century scientific studies showing racial distinctions that are now "completely discredited." The researchers in the film series "then illustrate how science is selectively invoked to validate and further particular political agendas, as well as how the personal opinions and agendas of individual scientists can greatly influence broad social policy." The reviewers suggest that the series should challenge scientists "to examine whether pervasive societal concepts have caused unconscious bias in their own research."

In the past, evolution was a "natural" to prove racial inequalities. Now evolution and genetics, via the Out of Africa Model, are being used to "prove" human equality. I am offended by the fact that evolution, the most racist theory ever to gain acceptance by the scientific community in the last 150 years, is being given credit for "proving" human equality. Human equality does not need to be proved. It needs to be accepted. It is an intrinsic part of God's creation. To credit evolution for it is an insult to the Creator.

The discoverers of the Herto fossils claim that the Herto people were our direct ancestors. They said, "In this sense, we are all African."[28] One senses a large dose of political correctness in that statement. When the first Neandertal fossils were found, they were thought by many to represent our oldest direct ancestor. Yet no one said, "In this sense, we are all German." When Java Man was discovered, he was thought to be our oldest direct ancestor. Yet no one said, "In this sense, we are all Javanese." Later, the Peking Man fossils were discovered. They were thought by many to also represent our oldest direct ancestors. Yet no one said, "In this sense, we are all Chinese." Piltdown Man was a forgery. For many years, however, it was believed that he was the transitional form between apes and humans. Hence he was thought to be our oldest ancestor. Yet no one said, "In this sense, we are all Englishmen." Now, with the discovery of the Herto fossils, we are told that, "In this sense, we are all African." I have absolutely no problem with my being African if it were true. But it is not.

The Bible states that after the flood, the ark came to rest on the mountains of Ararat (Gen. 8:4). Mount Ararat is a 17,000–foot-high mountain in extreme eastern Turkey. *Ararat* is also the old name for Armenia, one of the Confederation of Independent States (the old USSR), about 20 miles east of Mount Ararat. So, if one insists on going back that far, we are *not* all Africans. We are *all* either Turks or Armenians.

Yesterday my wife and I went into a small tailor shop, a mom-and-pop operation run by a Vietnamese couple. We communicated with the man behind the counter using gestures because he spoke virtually no English. Looking around, I noticed three pictures of Jesus on the wall—pictures more ornate than those I usually see. Perhaps Catholic? I sensed that this man wanted to tell his customers something. Pointing to one of the pictures, I said, "I love Jesus too." His face lit up like a flaming birthday cake. Although we could hardly communicate, tears filled my eyes as I realized that this man, in the very deepest sense, was my brother. We were both members of the family of God, and we would sit together forever at the feet of Jesus. Only God has made us all equal as humans, and only God, through our trust in the death of Jesus Christ for forgiveness of our sins, can make us truly brothers.

CHAPTER 20

A Cave in Spain Makes It All Very Plain

THERE IS A GROUP of fossils from Europe, Africa, and Asia that I have not described thus far. It is the "unofficial" group formerly known as "archaic *Homo sapiens.*" This group included fossil individuals who did not fit into either the Neandertal or the *Homo erectus* categories as defined by evolutionists. The reasons are that: (1) They have a somewhat different skull morphology from the "classic" Neandertals; (2) Many of this group are dated much earlier than the "classic" Neandertals, although more than half of them are contemporaries of "classic" Neandertals; and (3) They have a cranial capacity that is too large for them to be classified as *Homo erectus.*

Members of this group had been given various names, such as anti-Neandertals, pre-Neandertals, Neandertaloids, or African Neandertals. Some workers suggested that the entire category be given the scientific name *Homo sapiens rhodesiensis,* after the best-known fossil in this category, Rhodesian Man. The original purpose of this category was to distinguish these fossils from the Neandertals, contrasting the Neandertals and these archaic *Homo sapiens* fossils.

In the most general sense, the archaic *Homo sapiens* category could be described as follows: (1) They have a low, sloping skull with a cranial capacity of approximately 1100 to 1300 cc; (2) They have very heavy ridges over the eyes (especially true of Rhodesian Man); (3) The rear of their skulls are more rounded and lack the Neandertal "bun;" (4) They have large, long faces with jaws that jut forward; and (5) Their postcranial bones are essentially indistinguishable from modern humans or very different from modern humans, depending upon which fossil is being described and which paleoanthropologist does the describing.

In reality, the taxon "archaic *Homo sapiens*" was an artificial storage bin of convenience. This category served as a dumping ground for a heterogeneous assortment of fossils that did not fit elsewhere. By making a taxon of this assortment, evolutionists attempted to give it the appearance of a transitional group between *Homo erectus* and the Neandertals or between *Homo erectus* and anatomically modern *Homo sapiens.* However, the dating of these fossils by evolutionists revealed that this was not a valid picture. These archaic *Homo sapiens* fossils date from the most recent ones at about 5,000 years ago (some African and Asian individuals) all the way back to the oldest ones at 700,000 years ago on the evolution time scale. This chronological spread reveals that these fossil individuals were not part of an evolutionary sequence but were instead contemporaries of *Homo erectus,* Neandertal, and anatomically modern *Homo sapiens.* The intrinsic racism of evolution worked to create distinctions that were not valid. All these groups are fully human in their own right and are part of the human family created by God. Creationists felt that the dating and morphology of the archaic *Homo sapiens* category helped to falsify the concept of human evolution and served to demonstrate the wide degree of skeletal and cranial diversity that is found in the human family.

Ian Tattersall testified that these fossils presented a problem to the evolutionist who tried to classify them. With delicate understatement he wrote, "The hominid fossil record of the past 300–400 kyr (thousand years) offers a remarkable degree of morphological variety. Yet (late-persisting *Homo erectus* aside), conventional wisdom assigns all these fossils to *Homo sapiens,* albeit of 'archaic' varieties."[1]

William Howells (Harvard University) commented on the lack of formal definition for this fossil category. He wrote:

> Others have agreed that *H. sapiens* should indeed include a number of specimens or populations that were not anatomically modern but that for various reasons (brain size, loss of *erectus* occipital or temporal traits) could hardly be included in *H. erectus.* They are usually given the slightly evasive appellation of "archaic" *H. sapiens,* and in fact seem to be accepted as such simply on the basis of not being *H. erectus.*[2]

Richard Klein, in his 1989 work, didn't even attempt to give a description of the archaic *Homo sapiens* fossils as a group. As he described each of the individual fossils, he made the following statement: "Readers who are less concerned with the nature of the early *Homo sapiens* sample than with the conclusions that can be drawn from it may wish only to skim the next three sections, attending mostly to the accompanying illustrations."[3]

Obviously, Klein wished to spare readers exposure to one of the major anomalies in the concept of human evolution by having them skim over the evidence and proceed to the comfort of the "party line." In the conclusion, readers are assured that the concept of human evolution is secure.

The European fossils that were included in this "archaic *Homo sapiens*" category, mostly dated prior to the Neandertals, were the Le Lazaret Cave skull fragment, France; the Montmaurin Cave remains, France; the Casal dé Pazzi skull fragment, Italy; the Reilingen partial skull, Germany; the Steinheim skull, Germany; the Swanscombe skull, United Kingdom; the Bilzingsleben skull fragments, Germany; the Castel di Guido skull fragments, Italy; the Sima de los Huesos Cave remains, Spain; the Altamura Cave skeleton, Italy; the Arago Cave remains, France; the Vértesszöllös skull fragments, Hungary; the Karain (Black Cave) cranium, Anatolia, Turkey; the Boxgrove lower leg bone, United Kingdom; the Petralona skull, Greece; and the Mauer mandible (Heidelberg Man), Germany. Recently, a more respectful category was devised for these European fossils: *Homo heidelbergensis*, named after Heidelberg Man. However, this did nothing to solve the problem of their relationship to *Homo erectus* or to the Neandertals.

No matter what these European fossil individuals were called, evolutionists, bent upon finding transitional forms, found them a strange assortment. Some had very large faces and rather small braincases. Others had big braincases but small faces. Some were more *erectus*-like, others were more Neandertal-like, and still others were more like modern humans. Sometimes, what they looked like depended on whether you viewed them from the front or the back. Most evolutionists felt strongly that these Europeans were not Neandertals, and all evolutionists felt that they were far too old to be considered modern humans. Michael Day, looking for that transition from *Homo erectus* to modern humans, referred to these fossils as "the muddle in the middle." Now, thanks to a cave in Spain, we have the answer. And for the evolutionist, it's not a pretty picture.

The Cave in Spain

In 1992, a team led by Juan Luis Arsuaga (Complutense University, Madrid) finally reached undisturbed fossil deposits in the Sima de los Huesos Cave, "the pit of the bones." This cave is just one cave in a complex of caves in the Sierra de Atapuerca limestone hills of northern Spain. Hundreds of fossil fragments had been taken out of this cave by amateurs since the 1970s. Because the layers had been so disturbed in the process, however, it was impossible to properly interpret those previous fossil finds.

"The pit of the bones" is no ordinary cave. It is very deep and very narrow, and the bottom is not visible from the entrance. Hence no ancient humans ever lived in this cave. Instead, this cave was used as a burial ground or a cemetery by the ancient inhabitants of that area. Arsuaga writes:

> There is only one known case that suggests any funerary practice before the time of the [classic] Neanderthals and modern humans and that is at Sima de los Huesos. Although bodies were amassed one atop the other in a place chosen for that purpose at the Sima, they were not *buried,* a process defined as the digging of a grave and the placing of a body within it.[4]

Arsuaga explains their mortuary practice:

> When a member of the group died in the caves at Sierra de Atapuerca or nearby, the body was carried to this hidden niche and deposited there. That is why we came to call this place Sima de los Huesos, which in English means bone pit or depository, in other words, an ossuary [a receptacle to preserve the bones of the dead].[5]

In the undisturbed deposits, Arsuaga and his associates found 3 beautifully preserved fossil skulls. Since then, the remains of a minimum of 33 individuals have been recovered and dated at about 400,000 years ago. The results are nothing short of stunning. These remains have so much variation within one contemporaneous population that it demonstrates that all that "muddle in the middle," all those European fossil individuals that appeared to be so different, belong to one population—the Neandertals. For instance, one of the Sima de los Huesos adult skulls is one of the smallest ever recovered from that time period, while another one is one of the largest. The physical variation found in this one assemblage of fossils encompasses all the other European archaic *Homo sapiens* fossils.

Chris Stringer is one of the foremost advocates of the Out of Africa Model and a believer that the European fossils involved many species. After seeing the tremendous variation and the Neandertal affinities of the Sima de los Huesos fossils, he states, "In spite of all the variation they display, they get sucked in with the Neanderthals. Once that happens, it becomes very difficult to prevent the rest of the European material from getting sucked in as well."[6]

In a *Nature* article, Stringer comes about as close as we can expect a scholar to come in admitting he was mistaken about these European fossils comprising many species. Stringer gives a list of fifteen cranial characteristics. He compares the Sima de los Huesos fossils to fossils of *Homo erectus*, Neandertal, and *Homo sapiens*. He finds that they have seven similarities with *Homo erectus*, seven similarities with *Homo sapiens*, and ten similarities with the Neandertals. He then gives four possible interpretations of the Sima de los Huesos fossils. His preference is to call them all Neandertals, including all of the European archaic *Homo sapiens* fossils.[7] Some workers prefer to call this group "pre-Neandertals " in a desperate effort to give it some sort of evolutionary spin.

Unfortunately, Stringer does make the Neandertals a separate species from modern *Homo sapiens*. Richard Klein's new 1999 view is the same.[8] Ian Tattersall, the king of splitters, reconstructed one of the fossil skulls from Sima de los Huesos. He states, "It's a relative of Neanderthal, with lots of Neanderthal features, but not a Neanderthal."[9] This is the statement of a true splitter. All three of these workers, and many others, believe that the Neandertals became extinct without issue and have no direct relationship to modern humans.

However, many workers now consider all of the European archaic *Homo sapiens* fossils to be Neandertals or pre-Neandertals. Because the Neandertals are a unique European development, the African and Asian archaic *Homo sapiens* fossils are not included with them. These African and Asian fossils are now called "early *Homo sapiens*." They are still considered to be a distinct evolutionary group transitional between *Homo erectus* and modern humans.

The Sima de los Huesos fossil assemblage has powerful and profound implications for creationists. Because of the relative isolation of the various areas of the ancient world and the slow means of transportation, this extreme variation within populations, such as what is seen at Sima de los Huesos, is exactly what one would expect.

Further, thanks to the extreme variation seen in the Sima de los Huesos fossil collection, the distinctions made by evolutionists between *Homo erectus*, early *Homo sapiens*, Neandertal, and anatomically modern *Homo sapiens* now fade into insignificance. It is a remarkable affirmation of the biblical statement from Acts 17:26: "From one man he made every nation of men, that they should inhabit the whole earth."

Again, it is obvious that the extreme variation seen in the Sima de los Huesos fossil collection was not caused by evolution. Since they were all a part of the same population living at approximately the same time, evolution cannot be the explanation. Hence it strongly suggests that the differences between *Homo erectus*, early *Homo sapiens*, Neandertal, and anatomically modern *Homo sapiens* have a nonevolutionary cause.

The Sima de los Huesos fossil assemblage reveals the absurdity of attempting to determine species distinctions in fossil humans. The only legitimate species criteria is the fertility test. Since this test obviously cannot be applied to fossil humans, invalid criteria have been used to establish an evolutionary sequence. The confusion regarding the European fossils is a classic example.

Between 1968 and 1976 it was decided that there were no *Homo erectus* fossils in Europe. In 1968, F. Clark Howell (University of California, Berkeley) spoke of a *Homo erectus* presence in Europe.[10] In 1976, Howell, speaking at a symposium in Ontario, announced that *Homo erectus* did not exist in Europe.[11] The probable reason for this change in thinking was that the *Homo erectus* fossils in Asia and Africa were thought to be much older than any fossils discovered in Europe. Hence, it was assumed that humans hadn't reached Europe until they had evolved beyond the *Homo erectus* condition.

However, before that "decision" was made, most of the "archaic" fossils that had been discovered by that time in Europe were classified as *Homo erectus* by one or more competent paleoanthropologists. These fossils include Montmaurin, Bilzingsleben, Castel di Guido, Arago, Vértesszöllös, Mauer, and Petralona. These and other fossils were then reclassified as "archaic *Homo sapiens*" in order to distance them from both *Homo erectus* and the Neandertals. But now, as a result of a discovery at just *one* fossil site, all these fossils are classified as Neandertals. I can't think of any scenario that more clearly demonstrates the meaninglessness of these categories. Nor can I think of a scenario that more clearly demonstrates the tenuousness of human evolution when a discovery at just *one* fossil site causes such an upheaval in the interpretation of so many fossils. It involves a minimum of sixty-eight fossil individuals at sixteen sites.

Perhaps the most unique differences in the human fossil record are found in the "classic" Neandertals, such as the Neandertal from La Chapelle-aux-Saints (see the illustration in chapter 6). These distinctions could well have accumulated gradually because of their isolation during glacial conditions. The continental ice sheets in the north and east and the Mediterranean Sea in the south would have severely limited gene exchange with populations outside of Europe. This could have allowed the genetic expression of the morphology that characterizes the "classic" Neandertals. The end of the Ice Age would see: (1) The increase in direct sunshine; (2) A greater variety in diet; (3) The lessening of the use of the jaws as a tool; and (4) Most important of all, the genetic exchange with other populations moving into Europe. The Neandertal morphology would thus be reduced and eventually be swamped. There are a number of European fossils that demonstrate this change in morphology (see chapter 17).

Evolutionists claim that the changes from Neandertals to modern humans could take up to 500,000 years. That might be true in the world of evolutionary make-believe, because alleged evolutionary changes result from mutations in genes. The original genes are permanently changed, which is why the process is said to be nonreversible. The process is also very slow because sheer chance determines when and where the mutations take place.

In real life, however, nonevolutionary morphological changes can take place rather quickly due to genetic isolation or the ending of genetic isolation and the resulting genetic recombination. This would be especially true if a larger population interacts with a much smaller population. And it is believed that the Neandertals were a rather small population.

Evolutionary estimates of the time involved for such morphological change are subjective and rather elastic, as the following illustration demonstrates.

Evolutionists had always assumed that the earliest Americans would resemble modern Native Americans. However, based upon the 9,300-year-old skeleton of Kennewick Man found in Washington state, a 10,500-year-old skeleton found in Brazil, and other skeletons found in South America and Baja California, Mexico, it has been discovered that the earliest Americans had longer, narrower skulls and lower faces than modern Native Americans.

Instead of looking like modern Native Americans, Walter Neves (University of São Paulo, Brazil) states that these earliest Americans looked like modern Australians and Africans. They did not look like the people of northern Asia, thought to be their closest relatives. Neves feels that the skull changes after 8,000 years ago are too sudden for evolution. However, Joseph Powell (University of New Mexico) believes that they could have evolved into modern Native Americans in the time allowed through adaptation and microevolution (read genetic recombination).[12] It seems that when it suits their purposes, evolutionists can devise a way for cranial morphology to change in just a few thousand years. Why can't it apply to the Neandertals also?

Kennewick Man, found in 1996, has been a source of bitter conflict between anthropologists and northwest Indian tribes. The anthropologists want to study him. The Indian tribes, claiming that he is one of their own, want to bury him according to their cultural and religious practices. In seeking to decide if his non-Native morphology and his very early age could still qualify Kennewick Man legally as a "Native American," Federal Judge Ronald Gould said, "That's a metaphysical question that's outside my pay scale."[13]

Whether it involves the Neandertals or Native Americans, it is extremely difficult to determine genetic relationships after thousands of years have passed. A cave in Spain tells us that trying to determine those relationships by morphology is impossible.

The Evolution Empire's new clothes are not clothes at all. The Empire is naked, and human evolution is a myth. A cave in Spain makes it all very plain.

SECTION V

TRASHING NEANDERTALS

As recently as 35,000 years ago western Europe was still occupied by Neanderthals, primitive beings for whom art and progress scarcely existed. Then there was an abrupt change. Anatomically modern people appeared in Europe, and suddenly so did sculpture, musical instruments, lamps, trade, and innovation. Within a few thousand years the Neanderthals were gone.

—JARED DIAMOND

In essence this boils down to stating that the Neanderthals were so different from ourselves that a firm line can be drawn between them and us, a view that is by no means universally held. To shore up this approach, all the growing body of evidence for "art" before 40,000 years ago is simply dismissed and ignored.

—PREHISTORIAN PAUL BAHN
(*NATURE* 369, 16 JUNE 1994, P. 531)

INTRODUCTION TO SECTION V

The "Neandertal Storm"

PHILIP LIEBERMAN (BROWN UNIVERSITY) called it the "Neandertal Storm." Although Lieberman was referring particularly to the issue of Neandertal speech, the term "storm" could well refer to almost every area of Neandertal research: their humanity, their brain power, their manual dexterity, their locomotion, and their culture—to name just a few. To refer to "Neandertal Discussions" on any of these subjects would be too mild a phrase, considering the emotions that these ancient people—worthy ancestors of ours—evoke. After studying the Neandertals for forty years, I still find myself shocked at the prejudice that exists against them in the scientific literature.

At the heart of the "Neandertal Storm" is the question, "Who were these people who are so little understood by evolutionists?" The question, itself, is surprising because: (1) We have known about the Neandertals since 1856; (2) We have more fossils of them than we have of any other human category; and (3) They are the most recent of all the "extinct" humans and hence should be the easiest to understand and study.

The charge against the Neandertals is that they were not fully human. This charge is based on their alleged mental, physical, and cultural deficiencies as well as mtDNA evidence. All of these charges will be answered in this section of the book or elsewhere. If Neandertals were living today, these tough ancestors of ours could dispatch their evolutionist critics in mere moments. Since they are not here, I will serve as their defense attorney. We Americans enjoy defending the underdog.

After detailing and answering the charges against the Neandertals, the ultimate question is, "What does 'human' mean?" This is a problem for evolutionists, but the Word of God gives ultimate answers. The key to being human is our being created in the "image of God" and our being able to reproduce with *all* humans and *only* with humans. Since fossils cannot reproduce, this reproductive test cannot be applied to fossil humans. Hence evolutionists have applied an inappropriate test: culture. The Neandertals have been charged with being culture thin and thus subhuman. If this same cultural test were applied to some remote tribes in recent history, logic would demand that they also be considered subhuman. Remember, just over 150 years ago, the Tasmanians and the Tierra del Fuegians were considered by many evolutionists to be subhuman because of their alleged lack of culture.

Evolution is inherently racist. Although evolutionists are not racist regarding the people in today's world, their evolutionary racism comes through loud and clear when they deal with our fossil ancestors, especially the Neandertals. Unfortunately, this type of racism is not as obvious and not as offensive.

Based upon artifacts and other evidence found in association with the Neandertals, there is no question that the Neandertals were full members of the human family and probably part of the post-flood/Ice Age European population. Evolutionists, blinded by their philosophy, cannot see the Neandertals as they really were. Instead, they impose upon the Neandertals an evolutionary template that does not really fit.

CHAPTER 21

"OTHERIZING" THE NEANDERTALS

Neandertals were "humans, and yet not really human."

—JARED DIAMOND

THE UNIVERSITY OF TEXAS, Austin, is not exactly a right-wing think tank. One of the departments there has coined the term "to otherize." It means "to place a person or persons into a group other than one's own, to divide, to separate oneself from others," and so "to otherize someone." This is what many paleoanthropologists are doing to the Neandertals and to other true fossil humans. The reason is to foist on the human fossils a particular evolutionary scenario.

Previously, we recorded the difficult struggle the Neandertals had to achieve respectability and full human status. It wasn't until the 1960s that they were finally classified as *Homo sapiens neanderthalensis*, a subspecies of, and equal to, modern humans. However, their victory was short-lived. The battle soon began to reduce them to lesser human status and to again give them the evolutionary classification *Homo neanderthalensis*—human, but lower than *Homo sapiens*. In other words, they are being "otherized." It is necessary. In order for the present

Homo sapiens population to have evolved from African Eve free of racism, all racism in human evolution must be relegated to the past. Our extinct ancestors were "inherently inferior," starting with the Neandertals. It must be shown that those extinct ancestors have no direct ancestral ties to us.

Of the two basic models regarding the evolution of modern humans, the Multiregional Continuity Model holds that the Neandertals, who lived in Europe and the Near East, evolved into the modern humans who now live in those areas. Although this model involves the errors of all evolutionary models, it does recognize that the Neandertals were the ancestors of modern Europeans.

The other model, the African Eve Model, is more exotic, more widely known, and more widely held. There are many variations of it. Basically, it holds that humans evolved modern physical bodies first in Africa. However, they still did not exhibit modern behavior. As they migrated out of Africa over a period of thousands of years, they eventually evolved into humans who had modern behavior as we have. Moving throughout the world, they eliminated, either actively or passively, all lesser evolved humans who were living in those areas. In Europe, they replaced the Neandertals with little, if any, genetic mixing.

My favorite *Far Side* cartoon shows a Neandertal couple sitting on a tree limb. Below them another couple passes by, obviously looking more modern. Mrs. Neandertal says to Mr. Neandertal, "There go the Smiths. It looks like everyone is evolving but us." In the African Eve Model, the Neandertals didn't evolve into us. They became extinct. They were replaced or eliminated by our own direct ancestors who came out of Africa.

A major popularizer of the Out of Africa Model is Jared Diamond. Although others have written more extensively on this model, I use his 1989 article in *Discover* magazine for reasons that will be obvious in the next chapter.

Diamond maintains that for several million years human evolution went forward at a snail's pace and that during most of that time humans were not much more than glorified baboons. Then came what he and others call the "Great Leap Forward."

> As recently as 35,000 years ago western Europe was still occupied by Neanderthals, primitive beings for whom art and progress scarcely existed. Then there was an abrupt change. Anatomically modern people appeared in Europe, and suddenly so did sculpture, musical instruments, lamps, trade, and innovation. Within a few thousand years the Neanderthals were gone.[1]

Diamond claims that the Neandertals were rather ineffective hunters because their tools and weapons were rather primitive. Although the average brain size of the "classic" Neandertals was about ten percent larger than ours, they obviously were not as smart as we are. Their brains were not "wired" as well as ours are. For many years, evolutionists claimed that smaller brains (the size of *Homo erectus* humans) meant inferior conceptual ability. We now know that is not true. It is rather humorous to see evolutionists now argue that the larger brains of the classic Neandertals also indicate inferior conceptual ability.

The African Eve Model and the Great Leap Forward idea tend to go together, although there are probably as many variations of each one as there are paleoanthropologists who hold them. African Eve explains the origin of the European invaders. These people, represented by the Cro-Magnon (pronounced "Cro Man-yon") fossils found in Europe, possessed innovation. The Neandertals, Diamond and others say, didn't. In a few thousand years the Neandertals were gone. The Great Leap Forward had taken place, and the foundation for all of the culture and technology we know today was laid. While recognizing the physical prowess of the Neandertals, Diamond suggests that in various ways the Cro-Magnon people caused the extinction of the Neandertals because, in the long run, brains always win over brawn.

In the evolution of modern humans, the most important unsolved problem is, What were the evolutionary events or the genetic mutations in those anatomically modern humans that made them truly modern in their behavior, like us? Diamond believes that the key to our modern behavior is language, which involved evolutionary changes in "the structure of the larynx, tongue, and associated muscles that gives us fine control over spoken sounds." The Neandertals could not speak well. We can. It has made all the difference.

At this point Diamond makes a truly unique contribution to the problem. He claims that the Neandertals lacked that "precious human quality: innovation." We moderns have it. Diamond proves that he is truly one of us—he has innovation. He reaches into the world of make-believe for an explanation of the evolutionary event that marked the transition toward modern humans. He calls it a "magic twist." In his *Discover* article, he uses the term eight times. From the third time on, he capitalizes the term, working his own special magic to suggest that by capitalizing the term, it becomes an objective, scientific term conveying useful information. I quote several sentences in which he uses "Magic Twist" in order to demonstrate his sleight of hand.

> By perhaps 60,000 years ago, some magic twist of behavior had been added to the modern anatomy. That twist . . . produced innovative, fully modern people who proceeded to spread westward into Europe, quickly supplanting the Neanderthals.[2]

> The Magic Twist that produced the Great Leap Forward doesn't show up in fossil skeletons.[3]

> The Magic Twist may have been a change in only 0.1 percent of our genes.[4]

> Thus, the Magic Twist may have been some modifications of the protohuman vocal tract to give us finer control and permit formation of a much greater variety of sounds.[5]

> But if the Magic Twist did consist of changes in our vocal tract that permitted fine control of sounds, then the capacity for innovation that constitutes the Great Leap Forward would follow eventually. It was the spoken word that made us free.[6]

To substitute "Magic Twist" for a solid scientific explanation, to do it with a straight face, and then to be able to sell it is quite a bit of magic in itself.

For Diamond, the key that made that Great Leap Forward possible was language; Neandertal did not have adequate language skills, and Cro-Magnon did. However, the fossil evidence now favors Neandertal language capability. At the very time Diamond's article suggesting that the Neandertals lacked sophisticated speech appeared in *Discover*, the British journal *Nature* published a report regarding a Neandertal skeleton discovered at Kebara Cave, Mount Carmel, Israel. The report concerned the hyoid bone of the Neandertal individual known as Kebara 2. The hyoid is a small bone lying at the base of the tongue and connected to the larynx by eleven small muscles important in speech. Since the hyoid bone of Kebara 2 is almost identical in size and shape to that of modern humans, the inference is that this part of human anatomy has shown great stability over time. The report continues:

> A related inference would be that the associated larynx beneath the hyoid has scarcely changed in position, form, relationships or size during the past 60,000 years of human evolution. If indeed this inference is warranted, the morphological basis for human speech capability appears to have been fully developed during the Middle Paleolithic [Neandertal times], contrary to the views of some researchers.[7]

In spite of this strong evidence, Neandertal speech capability is still hotly debated among evolutionists. Since Adam and Eve could speak, and speak well, there is no question about what creationists believe on this subject.

Other authorities who accept the Great Leap Forward tend to agree with Diamond that the Neandertals were "humans, and yet not really human." Their sparse cultural inventory reveals it. According to Diamond, the Neandertals had no glue or adhesives for hafting tools; no unequivocal art objects; no boats, canoes, or ships; no bows and arrows; no cave paintings; no domesticated animals or plants; no hooks, nets, or spears for fishing; no lamps; no metallurgy; no mortars and pestles; no musical instruments; no needles or awls for sewing; no ropes for carrying things; no sculpture; and no long-distance overland trade.

Diamond claims that for the Neandertals, progress scarcely existed. Their clothing and their dwellings were crude. Describing their tools, he sums up his case against the Neandertals: "In short, Neanderthal tools had no variation in time or space to suggest that most human of characteristics, *innovation.*"[8] He adds, "To us innovation is utterly natural. To Neanderthals it was evidently unthinkable."[9] He refers to them as "strange creatures of the Ice Age—humans, and yet not really human."[10]

Diamond's racist view of the Neandertals is based entirely on culture, not on reproductive ability, which is the only true test. Archaeologist Paul Mellars (Cambridge University), who himself holds to the Great Leap Forward view, cautions against such judgments regarding the mental abilities of the Neandertals. In his 1996 book *The Neanderthal Legacy*, he writes, "There is an obvious danger of making simplistic equations between 'simplicity of behaviour' and 'simplicity of mind' which somehow short-circuit scientific analysis, and effectively assume what one should be attempting to find out."[11]

Archaeology is a science that seeks to extract a maximum of information out of a minimum of evidence. In their rush to form conclusions, archaeologists and others tend to forget that "absence of evidence is not evidence of absence." Just because an item has not yet been discovered does not necessarily prove that the Neandertals didn't possess it. Some very recent discoveries demonstrate that fact. The pitfalls of basing extensive conclusions on archaeological evidence alone are pointed out by another advocate of the Great Leap Forward view, Richard G. Klein (Stanford University), who, in reviewing Mellar's book, writes that there are several other possible conclusions from the evidence, "including the conclusion that there can be no conclusion because archaeological interpretation is inevitably a product of personal judgment applied to circumstantial evidence."[12]

The truth is that the evidence for the full humanity of the Neandertals has already been discovered, and it is compelling. It will be examined in a later chapter. Because of political considerations, however, this evidence is being ignored. We will demonstrate that the Neandertals were card-carrying members of the human family, descendants of Adam, and probably a part of the post-flood population. God does not "otherize" humans he created in his image. God is not a racist.

CHAPTER 22

NEANDERTALS, TASMANIANS, AND THE WILD AND WOOLLY WEST

THE ASSOCIATION OF THE Tasmanians with the Neandertals was not uncommon and was probably not intended as a compliment to either group. As early as 1926, Harris Wilder wrote that the Australians were "strongly reminiscent of the species Neandertalensis."[1] In 1990, Ann Shepherd wrote about the importance of Tasmanian skeletons for scientific study: "The discovery of the remains of Neanderthal man paralleled the discovery of the Tasmanians, societies that were almost equally primitive."[2] Others observed that the Tasmanian Old Stone Age tools seemed to be even less complex than those of the Neandertals. In fact, early on, the English attitude toward the Tasmanian Aboriginals was exactly the same as Diamond's attitude toward the Neandertals—"human, but not fully human."

In a previous chapter, we emphasized that the interfertility test could not be applied to the Neandertals. In his 1989 article in *Discover*,[3] Jared Diamond thus imposed the false test of culture and found that *the Neandertals were not fully human* because they had no glue or adhesives for hafting tools; no unequivocal art objects; no boats, canoes, or ships; no bows and arrows; no cave paintings; no domesticated animals or plants; no hooks, nets, or spears for fishing; no lamps; no metallurgy; no mortars and pestles; no musical instruments; no needles or awls for sewing; no ropes for carrying things; no sculpture; and no long-distance overland trade.

Yet in a 1993 article in *Discover*, this same Jared Diamond recognizes that *the Tasmanians were fully human* even though they had no glue or adhesives for hafting tools; no unequivocal art objects; canoes that quickly sank; no bows and arrows; no cave paintings; no domesticated animals or plants; no hooks, nets, or spears for fishing; no lamps; no metallurgy; no mortars and pestles; no musical instruments; no needles or awls for sewing; no sculpture; and (being on a rather small island) no long-distance overwater trade.[4]

Why would Diamond consider the Neandertals to be "subhuman" solely on the basis of their alleged limited cultural inventory, when he considers the Tasmanians, *having the very same limited cultural inventory*, to be fully human? This is one of the most glaring lacks of logic I have ever read in the scientific literature. In a more perfect world, evolutionists would be required to take a course in logic. I cannot explain such an obvious inconsistency on the part of this scientist; I can only report the problem. I know that evolution blinds the soul. I have reason to believe that it also blinds the mind.

The cultural inventory of the Tasmanians, listed above, is what they had when they were discovered in 1642. Diamond then proceeds to detonate an archaeological bombshell—revealing something that is "taboo" in an evolutionary scenario. The evidence is strong that the Tasmanians actually *abandoned* several cultural practices that they had many years before. This evidence is so vital both in our defense of the Neandertals and in our refutation of human evolution that I must give you some background.

Evolutionists insist that the evolutionary process is nondirectional. This is because mutations, the raw material of evolutionary change, are random. However, if you look at an evolutionary chart of the history of life, that chart is clearly directional. It goes from nonlife all the way up the organizational ladder to humans. The human brain is the most complex aggregate of matter known in the universe. To go from utter simplicity to incredible complexity is directional, not nondirectional.

This same directionality is found in the evolutionary charts of individual animals and plants, in worldwide rock correlation, and in archaeology. Family trees of all animals and plants are constructed by evolutionists on the assumption that these entities increased in complexity over time. Worldwide rock correlation is based on the assumption that the animals and plants they contain (as fossils) go from simple (or primitive) to complex (or more modern). Rocks containing the fossils of what are assumed to be "primitive" humans are considered to be older than rocks containing the fossils of more modern humans. "Primitive" stone tools are normally thought to represent an older and more primitive culture in contrast to tools that are more advanced, specialized, or artistically refined. The alleged limited cultural inventory of the Neandertals is why they are considered to be more "primitive" and not quite fully human.

This directionality in the history of life in the construction of family trees, in rock correlation, and in archaeology has been evolutionary dogma. Any evidence of reversals or regression in any of these areas has been either ignored or denied. The reason is obvious. If evidence for reversals in any of these areas was found to be extensive, it could undermine the entire concept of evolution. To get your attention, let me state that again. *If evidence for reversals in any of these areas was found to be extensive, it could undermine the entire concept of evolution.*

There is a personal side to this "reversals" story. In the late 1970s, I attended a lecture given by University of Michigan paleontologist Gerald R. Smith. He described his discovery of reversals in the record of the Lake Idaho fossil fishes.[5] I was so impressed with how this concept could undermine evolution that I undertook a research project on the possibility of reversals in the fossil record. In the paleontological literature I discovered a number of well-documented cases of reversals in the fossil record of insects, worms, ammonites, fishes, mammals, and humans. These were cases where organisms had seemingly gone from a specialized to a more generalized condition. Because the paleontological literature is so vast, I suspect that the results of my research were just the tip of the iceberg. I have no doubt that further research would reveal many more well-documented cases of reversals.

In the paper I wrote on the subject, I pointed out the tremendous implications reversals pose for worldwide rock correlation (where ammonite fossils are used extensively) and for evolutionary relationships. If reversals should prove to be extensive, this could call into question many, if not all, evolutionary relationships as they are presently understood.

I submitted my paper for publication to Dr. Richard H. Bube (Stanford University). He was one of the most influential theistic evolutionists in the United States and was then editor of the *Journal of the American Scientific Affiliation* (now known as *Perspectives on Science and Christian Faith*), published by an organization of which I have been a member for more than thirty years. The American Scientific Affiliation is composed largely of Christians who work in the sciences. Its orientation now is overwhelmingly one of theistic evolution, although that was not the case when it was first founded. As a young earth creationist, I am not an extinct species in that organization, but certainly I am an endangered species.

My paper was rejected. Although it was very well-documented, Bube said that I must be mistaken. He did not explain. It was not clear if he recognized the serious implications reversals have for evolution and preferred to ignore them, if he thought that the issue was irrelevant, or if he thought that I was an idiot who could not understand the English language. The paper was later published in the *Creation Research Society Quarterly*.[6]

That was almost thirty years ago. Today reversals are openly talked about in the evolutionist literature. Regarding one of the contemporary views of human evolution, Richard Klein (Stanford University), citing McHenry, states that to go from *Homo habilis* to *Homo erectus* to *Homo sapiens* involves reversals in both cranial thickness and in browridge development.[7] In spite of the references in today's literature to reversals, however, seldom is there any hint that evolutionists understand the serious implications these reversals could have for their theory.

Jared Diamond confirms that reversals or retrogression is as abhorrent in archaeology as it is in paleontology. He writes:

> The Tasmanians actually *abandoned* some practices that they shared with Australia 10,000 years ago. This idea violates cherished views of human nature, since we tend to assume that history is a long record of continual progress. Nevertheless, it is now clear that Tasmanians did abandon at least two important practices.[8]

The first practice Diamond mentions was the production of bone tools. Bone tools are important in a culture because with bone one can make items, such as needles, that cannot be made with stone or wood. We know that 7,000 years ago Tasmanians made bone tools that resembled Australia's awls, reamers, and needles, but by 3,500 years ago that practice had stopped.

The second cultural item that the Tasmanians abandoned was the practice of eating fish. Although most Tasmanians lived on the coast, European explorers were astonished that the Tasmanians did not eat fish, whereas the Tasmanians were astonished that the Europeans did. Yet remains at archaeological sites show that early on the Tasmanians did eat fish—the same species that are found in Tasmanian waters today.

How does all of this relate to the Neandertals? Most authorities agree that the Neandertals were big-game hunters. If that was their lifestyle, then the items Diamond criticizes them for not having could be the very things they gave up as unnecessary hindrances to their hunting success. If their lifestyle was a minimalistic one in which they were continually on the move following herds of big game, hunting them, and living on them, why would they need art objects; boats; canoes; ships; cave paintings; domesticated animals or plants; hooks, nets, or spears for fishing; metallurgy; mortars and pestles; musical instruments; sculpture; and long-distance overland trade? Criticizing the Neandertals for not having these items is like criticizing a cat for not having fins. Yet, as we shall see in a later chapter, the Neandertals actually did have some of these cultural items.

To show the absurdity of this position regarding the Neandertals, let's do a "thought experiment." Using the same logic that is used regarding the Neandertals, we will show that the Plains Indians of North America were not fully evolved humans in the 1600s and that the Great Plains experienced two episodes of the Great Leap Forward in just 250 years. (My apologies to my Native American friends.) We will use the criteria, cultural deficiencies, that Jared Diamond, Richard Klein, and many other evolutionists use. "Thought experiments" should be quite acceptable to the scientific community, since Albert Einstein used them to demonstrate many of his most important concepts.

First, a bit of background. When I was a child in North Dakota, we played Cowboys and Indians. That game is not politically correct today, but I don't recall that we cared whether we were cowboys or Indians. Kit Carson and Buffalo Bill were household names to us. But so were Sitting Bull, Red Cloud, and Geronimo. To us, they were all brave, courageous men. After all, we thought the northern Great Plains was where the *real* Indians were—the Sioux, the Mandan, and the Dakota. We were reliving a colorful part of the wild and woolly West.

Further, my grandparents were early pioneers in North Dakota. Their names are emblazoned on a plaque at the entrance of a restored Indian fort and historic site, Fort Abercrombie, on the banks of the Red River of the North. It's even possible that my grandparents had contact with some of those very Indians. However, if they or my parents knew that our nation's dealings with the Indians were not always honorable, they never told us. For us, it was a time of relative innocence.

As a child, my knowledge of the exploits of the Plains Indians came mainly from western movies and novels. The works of Louis L'Amour, a fellow North Dakotan, had not yet been written. But earlier western writers like Zane Grey were idols. We thought of the Plains Indians on horseback, sweeping across the prairie after buffalo or engaging in tribal warfare. What we didn't realize was that our concept of the Plains Indians was largely a product of European contact. For most of the history of the American Indians, life was not like that at all.

Before AD 1700, many of the tribes associated with the Great Plains were farming people. Only a few tribes, such as the Comanche, Kiowa, and Shoshone, lived a nomadic life on the Great Plains. Bison were difficult to hunt, kill, and transport on foot because of the vast distances involved. The Plains tribes were materially impoverished and militarily weak.

The introduction of the horse by the Spanish resulted in a major cultural revolution. By 1700, horses were available in great numbers to the tribes of the Great Plains. This precipitated a cultural revolution among the Plains Indians, the first "Great Leap Forward." Horses enlarged their economy and radically altered their lifestyle. On horseback an Indian, using bow and arrow, could find and kill enough bison in a few months to supply his family with food for a whole year. With a horse he could transport meat and other items long distances. The Indians' tipis became larger, and clothing and material goods became more abundant. Art became more common and decoration more elaborate. Now distant tribes could gather by the thousands in large camps for a season. In other words, the horse quickly elevated the Plains Indians to a new prosperity.[9]

It was *these* Plains Indians, having horses, that met the people involved in the westward migration on the Oregon and California trails. This was the largest migration in the history of the world up to that time, involving about 500,000 people between 1840 and 1860. Here we have a second "Great Leap Forward" going from the Advance Stone Age Indians to the Iron Age. Within 250 years the Great Plains experienced two dramatic cultural revolutions, both involving entirely modern humans, *Homo sapiens sapiens.*

The account of the Plains Indians might serve to illustrate the transition in Europe from the Neandertals to what are called "modern" humans and how evolutionists have erred in thinking that this cultural transition, this "Great Leap Forward," involved a change from lesser evolved humans to modern *Homo sapiens.*

In our thought experiment, we are evolutionary archaeologists. We are living 50,000 years in the future, doing archaeological digs on the Great Plains of North America. In our thought experiment, we must empty our minds of any knowledge of historical records of what happened there in the past. Our *only* knowledge of the inhabitants of the Great Plains must come from our digs—the way archaeologists today must study the Neandertals. Again, in our thought experiment, we must *close our minds* to all that we know of the history of the Great Plains and depend solely on what our archaeological digs tell us happened there. Although it is not my purpose to disparage archaeology, I want to show that archaeology is a field seeking to derive maximum information from minimum evidence.

The lowest level of our digs on the Great Plains, dated at about AD 1600, suggests a vast continent relatively devoid of humans. Among the large mammal fossils found are *Bison,* a bovid, probably the North American bison, later to become almost extinct. The skeletal remains of *Homo* reveal that the inhabitants were anatomically modern. However, their culture reveals that they had not yet reached a fully modern condition mentally or conceptually. They had virtually no art, no metallurgy, no ships, and no long-distance overland trade, except perhaps trading in obsidian for arrows. The Advance Stone Age best describes their culture. Since their stone arrows were perhaps the most distinctive feature of their culture, we have called these inhabitants "The Arrow People."

Their seeming cultural poverty surprises us because at this time Europe was well into the Iron Age, and the wheel had been invented over 5,000 years earlier. Yet there was no evidence in any of our digs of the use of iron. Nor was there evidence that the inhabitants of the Great Plains had invented or possessed the wheel. These inhabitants obviously had not evolved into a fully human condition. At one site, near a dried-up river channel, we found what may have been a hollowed-out log, possibly used in river navigation. But its condition was such that it was impossible to determine if the log had been worked on by humans or if what looked like a hollowing out was simply the result of decay.

The next higher level reveals a striking cultural change. It has been dated at about AD 1750. Among the large mammal fossils found are *Bison* and *Equus* (horse). The skeletal remains of *Homo* reveal that these inhabitants were also anatomically modern, and their tool assemblage still represented an Advance Stone Age culture. Although that culture revealed that they also had not yet reached the fully modern condition mentally or conceptually, there is more evidence of art, especially ornaments of stone, obsidian, and bone. We also found evidence of what might have been some sort of watercraft made of birch bark, but other researchers questioned if these people were capable of making such sophisticated craft. For the first time, there is evidence of long-distance overland trade as well as of warfare with distant communities.

The striking cultural changes in such an amazingly short time make it virtually impossible for these changes to be the result of in situ evolution. It is more likely that the Arrow People were replaced by an invasion of humans from elsewhere with more highly evolved conceptual ability, resulting in this "Great Leap Forward." The evidences of warfare at many sites would tend to support that interpretation. We have called these invaders "The Advanced Arrow People." But the source of these Advanced Arrow People is as yet unknown.

The next higher level reveals the most dramatic cultural change of all. Excavations done primarily at a site known as Fort Laramie, Wyoming, dated at about AD 1850, reveal a revolutionary shift from an Advance Stone Age culture to an Iron Age culture. In a geologic formation known as the Arikaree Sandstone are channels or gouges so deep that they must have been made by hundreds, if not thousands, of animal-drawn vehicles with iron-rimmed wooden wheels. Some remains of these vehicles and their iron-rimmed wheels have been found. The width of the grooves in the rock matches the width of the wheels. There is no question that this "Great Leap Forward" represents an invasion from the East of fully evolved *Homo sapiens*—both anatomically and conceptually. We have called these invaders "The East People." Among the large mammal fossils found are *Bison* and *Equus*, but also *Bos* (ox). The association of oxen with the vehicles suggests that the majority of the vehicles were pulled by oxen.

The nature of this invasion from the East is mystifying. Strangely lacking are the military items that were used by invading armies at that time. Remains of the firearms uncovered seem to be more of a hunting type rather than a military type, and as a whole they would be classified as nondescript. The animal-drawn vehicle routes seem to be basically east-west, and grave sites along these routes include many graves of women and children as well as of men. If this was an invading army, it seems to have been a rather ragtag operation. The evidence seems to suggest that this was more a migration than a military invasion, with the animal-drawn vehicles going through the Great Plains to destinations farther west. It does not seem to be an invasion or migration into the Great Plains. Nonetheless, the culture of the East People was indelibly stamped on the Great Plains, never to be eradicated.

In this interim report of our work on the archaeology of the Great Plains, we document two dramatic cultural revolutions, two "Great Leaps Forward," in just 250 years, the second one being even more profound than the first. These cultural changes were so sudden that they could not have occurred through in-situ evolution of the indigenous populations. Both revolutions had to be the result of invasions of more highly evolved *Homo* populations from outside the area who then imposed their cultures upon the local inhabitants. The second revolution obviously heralded the arrival of fully modern *Homo sapiens sapiens.* While we hope to continue our work on the Great Plains, we first plan to shift our focus to the area of the East People to determine which factors brought about the evolution of fully modern *Homo sapiens sapiens* from lesser evolved people in that area.

Back to reality. The next chapters demonstrate that the Great Leap Forward does not describe the transition from the Neandertals to anatomically modern *Homo sapiens.* Instead, the Great Leap Forward exists only in the mind of the evolutionist.

CHAPTER 23

Technical Section

mtDNA Neandertal Park— A Catch-22

> The fact that they managed to find [Neandertal] DNA from a region of prime importance is proof that there is a God who likes paleoanthropology.
>
> —Dan Lieberman

THE RECOVERY OF MITOCHONDRIAL DNA (mtDNA) from the right arm bone (humerus) of the original Neandertal fossil (discovered in 1856 in Feldhofer Cave in the Neander Valley, near Dusseldorf, Germany) has been hailed as a stunning feat of modern biochemistry. Christopher Stringer exclaimed, "For human evolution, this is as exciting as the Mars landing."[1] The achievement was announced in the 11 July 1997 issue of the journal *Cell*.[2] Since then, recoveries have been made from Neandertals in Mezmaiskaya Cave, Georgia, CIS, and in Croatia; from two Cro-Magnon fossils in Italy; and from ten fossils of ancient Australians.[3,4] This chapter will primarily address the original Feldhofer Cave recovery, which has received by far the most coverage and publicity.

There is no question that the accomplishment was both conceptually and experimentally brilliant. The brilliance of the methodology, however, does not guarantee the accuracy of the interpretation that evolutionists have placed on the data. That interpretation includes (1) overenthusiastic and cavalier claims of confirmation of the Out of Africa Model of human evolution and (2) denials that the Neandertals were fully human.

Evolutionists interpret the differences between the Neandertal mtDNA and modern human mtDNA to mean that the Neandertal line diverged from the line leading to modern humans about 550,000 to 690,000 years ago and that the Neandertals became extinct without contributing mtDNA to modern humans. This implies that the Neandertals did not evolve into fully modern humans, that they were a different species from modern humans, and that they were just one of many protohuman types that were failed evolutionary experiments. We alone evolved to full humanity.

Details of mtDNA Recovery

DNA is the incredibly complex molecule involved in the genetics of life. There is substantial breakdown of DNA within a few hours after the death of an organism because the DNA is deprived of the active repair mechanisms found in living cells. Causes of DNA degradation include water, oxygen, heat, pressure, time, exposure to transition metals (such as zinc), microbe attack, and background radiation. This degradation involves the breakage of the cross-linking of the DNA molecules, modification of sugars, alteration of bases, and the breakage of long strands of DNA into strands that eventually become so short that no information can be retrieved from them.

It is uncertain how long DNA will last. To last at all, DNA must be removed from degrading factors soon after biological death and preserved. Evolutionists estimate that it might last "tens of thousands of years."[5] Even under ideal conditions, background radiation will eventually erase all genetic information. Sensational reports about the recovery of DNA millions of years old are now discounted because researchers have not been able to repeat the results. *Jurassic Park* will always be fiction. Even amber is not the foolproof preservative it was once thought to be.[6]

In the past genetic material for experimentation was scarce. It was largely inaccessible because it was always embedded in a living system. Kary B. Mullis writes, "The truth is that in practice it is difficult to get a well-defined molecule of natural DNA from any organism except extremely simple viruses."[7] One of the most remarkable breakthroughs in modern biotechnology was the development in the 1980s of the polymerase chain reaction (PCR). Kary Mullis shared the 1993 Nobel Prize in chemistry for his "invention." The PCR technique can make unlimited copies of a specific DNA sequence independent of the organism from which it came. "With PCR, tiny bits of embedded, often hidden, genetic information can be amplified into large quantities of accessible, identifiable, and analyzable material."[8]

In dealing with the Feldhofer Cave Neandertal specimen, the scientific team, led by Svante Pääbo (Max Planck Institute for Evolutionary Anthropology, Leipzig), decided to search for mitochondrial DNA rather than nuclear DNA. Whereas there are only two copies of DNA in the nucleus of each cell, there are 500 to 1,000 copies of mtDNA in each cell. Hence the possibility was far greater that some of the ancient mtDNA might be preserved. It has been assumed that mtDNA passes unchanged from a mother to her offspring, unlike nuclear DNA. The father's mtDNA was assumed to end up "on the cutting room floor." Since changes in mtDNA are from mutations rather than from genetic recombination, evolutionists believe that mtDNA is a more accurate reflection of evolutionary history. Further, because mtDNA has no repair enzymes, mtDNA accumulates mutations at about ten times the rate of nuclear DNA, making it, evolutionists believe, a more fine-grained index of time.

The most serious problem in analyzing ancient DNA is the possibility of contamination from modern DNA. This contamination could come from anyone who has ever handled the fossil since its discovery, from laboratory personnel, from laboratory equipment, and even from the heating and cooling system in the laboratory. Even a single cell of modern human contamination would have its DNA amplified blindly and preferentially by the PCR because of its superior state of preservation over the older material. The PCR technique is "notoriously contamination-sensitive."[9] The problem is so serious that some contamination from modern DNA is unavoidable. Ann Gibbons and Patricia Kahn express the problem:

> It's tough to distinguish DNA intrinsic to an ancient sample from the modern DNA that unavoidably contaminates it—the source of many false claims in the past. Ancient human samples are especially tricky, because their sequences might not differ much from that of contaminating modern human DNA, so it's hard to get a believable result.[10]

Whether or not genuine Neandertal mtDNA has been retrieved is impossible for an outside observer to say. Knowing the unstable nature of the DNA molecule, if DNA was retrieved from that Neandertal fossil, it is evidence that the fossil is not nearly as old as evolutionists claim (30,000 to 40,000 years).

The chronologies of Genesis, as historical documents, do not allow an age of much more than 10,000 years for the age of the universe.

As far as the recovery of Neandertal mtDNA is concerned, it is possible that the mtDNA is genuine. However, the evolutionary interpretation of those mtDNA sequences—that the Neandertals are a separate species and are not closely related to modern humans—is not scientifically or biblically justified. The reason is that the use of mtDNA as a dating system and as a species-distinguishing system is loaded with unproven and unprovable assumptions. Its use as "proof" is nothing short of a scientific scandal and illustrates the fact that it is favored because it supports a political agenda even though the method falls far short of scientific objectivity.

Assumptions and Flaws in the Neandertal mtDNA Interpretation

Unfounded Assumption #1: *That mtDNA is passed on only by the mother.*

Basic to the concept of mtDNA being an evolutionary dating and relationship method is the belief that, unlike nuclear DNA, mtDNA is passed unchanged from a mother to her offspring. It is not mixed with the father's mtDNA. Early on, experiments seemed to support this belief. Attempts to find evidence that paternal mtDNA was involved failed because the tests were not sensitive enough. The advent of PCR, with its incredible sensitivity, changed that. Evidence began to accumulate that some of the father's mtDNA was also passed on. This fact ought to have been enough to disqualify the use of mtDNA as a dating and relationship tool. But by this time a huge literature had been built up, dealing with evolutionary relationships, that gave evolutionists the alleged "proof" that they had been seeking for 150 years. To ask them to give this up is like asking a heroin addict to give up his habit. The new data has been ignored.

The late British biologist John Maynard Smith (University of Sussex), considered one of the greatest thinkers of the twentieth century, assembled evidence of maternal and paternal mixing of mtDNA. The journal *NewScientist* explains:

> In 1999, he and his colleagues published evidence that mitochondrial DNA undergoes recombination—mitochondria from your father and mother can swap genetic material. It may sound obscure, but if they are right it casts doubt on all the research that uses mtDNA as a molecular clock to unravel evolutionary history, including the dating of our oldest common ancestor—"mitochondrial Eve".

> Maynard Smith is frustrated but not surprised that the establishment chooses to ignore these findings.[11]

The most devastating evidence regarding this issue is in a paper Smith and his associates wrote in *Science* that gave evidence of the mixing of maternal and paternal mtDNA both in humans and chimpanzees.[12] In a commentary on this paper in the same issue of *Science*, Richard Hudson (University of Chicago) observes that this study "is pretty compelling and I can't think of good alternative explanations." Anthropologist Henry Harpending (University of Utah) adds, "There is a cottage industry of making gene trees in anthropology and then interpreting them. This paper will invalidate most of that."[13]

The particular ways in which the Out of Africa Model and Mitochondrial Eve would be invalidated by paternal contributions to mtDNA are very interesting. First, Harpending informs us that with recombination of mtDNA, there would be no single ancestral tree for modern humans because each part of the molecule would have its own unique history. Adam Eyre-Walker (University of Sussex), a coauthor of the paper with Smith, says, "there's not one woman to whom we can trace our mitochondria."[14] With modern humans originating from different ancestral trees and with different women as Eve, the potential for racism in evolution rears its ugly head.

Further, the new evidence would change the date for Eve. The present date for Eve is about 200,000 years ago. Eyre-Walker claims that recombination makes past events appear to be more recent than they really are and that Eve would actually be older than the mtDNA evidence now suggests.[15] That's bad because that could allow time for racial distinctions to evolve.

Other evidence also suggests that Smith and his associates are right. Writing in *Science*, Eleftherios Zouros *et al.* cite their studies showing "extensive paternal mitochondrial DNA inheritance in the marine mussel *Mytilus.*" They also cite evidence for paternal mtDNA inheritance in fruit flies, mice, and, by inference, in anchovies. They conclude, "This diverse collection of animals suggests that the phenomenon might be widespread, with obvious implications for the use of the mtDNA molecule for population and phylogenetic studies."[16] "A mere trickle of Adam's genes" is serious because such "leakage distorts the DNA clock."[17]

Thus far, the scientific community has had two responses to this new data that promises to undermine Out of Africa, its model of choice. The first response, as Maynard Smith stated, has been to ignore the new data. The second response is by Richard Hudson, who says that the new result "changes our view dramatically enough that we have to continue to think of other ways to explain it."[18]

Unfounded Assumption #2: *That mtDNA mutations are regular and serve as a molecular "clock."*

The most amazing development regarding the molecular "clock" is the possibility that mtDNA may mutate much faster than has been estimated. A recent article in *Science* states that the "clock" may be in error by as much as twentyfold. Neil Howard (University of Texas, Galveston) says, "We've been treating this like a stopwatch, and I'm concerned that it's as precise as a sun dial."[19] If the new rates hold up after further research, the results for evolutionary time estimates, such as for Mitochondrial Eve, could be startling. "Using the new clock, she would be a mere 6000 years old."[20]

Unfounded Assumption #3: *That mtDNA can be used to determine human and primate relationships.*

The basis of the interpretation that modern humans and Neandertals are separate species is the acceptance by evolutionists of the concept of the molecular "clock." Yet the authors of the mtDNA Neandertal study in *Cell* admit that "these dates rely on the calibration point of the chimpanzee-human divergence and have errors of unknown magnitude associated with them."[21] Their mtDNA interpretation assumes the legitimacy of the molecular "clock" as a means of determining the relationship of modern humans to chimpanzees and to Neandertals. Yet G. A. Clark writes:

> Molecular clock models are full of problematic assumptions. Leaving aside differences of opinion about the rate of base pair substitutions, how to calibrate a molecular clock, and whether or not mtDNA mutations are neutral, the fact that the Neandertal sequence . . . differs from those of modern humans does not resolve the question of whether or not "moderns" and "Neandertals" were different species.[22]

Jonathan Marks (Yale University) emphasizes the subjectivity involved in using mtDNA to determine relationships. He comments, "Most analyses of mitochondrial DNA are so equivocal as to render a clear solution impossible, the preferred phylogeny relying critically on the choice of outgroup and clustering technique."[23]

Unfounded Assumption #4: *That mtDNA can determine species distance and distinguish between species.*

Referring to the Out of Africa Model, the authors of the *Cell* article write, "The Neandertal mtDNA sequence thus supports a scenario in which modern humans arose recently in Africa as a distinct species and replaced Neandertals with little or no interbreeding."[24] This is one of the most irresponsible statements ever made in the name of science. The reason is that no one knows how different the human genome must be to represent a different species of humans. That statement is a philosophical statement, not a scientific statement.

Maryellen Ruvolo (Harvard University) points out that the genetic variation between the modern and Neandertal sequences is within the range of other single species of primates. She goes on to say, "There isn't a yardstick for genetic difference upon which you can define a species."[25] Geneticist Simon Easteal (Australian National University), noting that chimpanzees, gorillas, and other primates have much more within-species mtDNA diversity than modern humans do, states, "The amount of diversity between Neanderthals and living humans is not exceptional."[26] In other words, we do not know how many mtDNA substitutions it would take in the human family to make interfertility impossible. Species distinctions are based on interfertility, not on the number of mtDNA differences.

Perhaps the best way of indicating the worthlessness of determining human species relationship by mtDNA is to note the results of recoveries thus far:

Two anatomically modern Cro-Magnon recoveries from Italy—mtDNA very similar to modern humans living today.[27]

One anatomically modern Australian recovery, Mungo Man 3—mtDNA very different from modern humans living today.[28]

Three Neandertal (robust morphology) recoveries—mtDNA very different from modern humans living today.[29]

Ten robust, *erectus*-like Australian recoveries—mtDNA very similar to modern humans living today.[30]

Assuming that these were all legitimate mtDNA recoveries, the obvious conclusion is that whatever the cause of the extreme variation in human mtDNA, it has nothing to do with species distinctions.

Interpretive Flaw: ***A misinterpretation of the relationship of Neandertals to chimpanzees.***

Biochemist Dr. John P. Marcus (University of Queensland, Australia) makes a significant observation about a graph in the *Cell* article. He writes:

> This graph might lead one to think that Neandertal sequences are somewhere between modern human and chimp sequences. This could then give the impression that Neandertal is a link between chimps and humans. On closer examination,

however, this is not the case. As labeled, the graph shows the number of differences between human-human, human-Neandertal, and human-chimp pairs. Significantly, the authors do not show the distribution of Neandertal-chimp differences. The reason they do not show this last of four possible comparisons between the populations is not clear to me. What is clear, however, from the DNA distance comparisons that I performed, *is that the Neandertal sequence is actually further away from either of the two chimpanzee sequences than the modern human sequences are. My calculations show that every one of the human isolates that I used was "closer" to chimp than was the Neandertal.* The fact that Neandertal and modern human sequences are approximately equidistant from the chimpanzee outgroup seems to be a good indication that Neandertal and modern humans comprise one species. Clearly, the Neandertal is no more related to chimps than any of the humans. If anything, Neandertal is less related to chimps.[31]

Possible PCR Copying Errors

PCR copying errors on oxygen-damaged residues in the Neandertal mtDNA could result in the Neandertal mtDNA appearing to be more distant from that of modern humans than it actually is. John Marcus sees evidence of this in his own study of the *Cell* report. He observes possible PCR-induced systematic errors due to a uniform oxidation of particular residues in particular sequence contexts. He explains:

> When the nature of the differences between the modern human reference sequence and the Neandertal sequence was compared, it was noted that there were 27 differences. Twenty-four of these were transitions (G to A, and C to T) changes. Apparently it is easier for DNA polymerase to make this kind of substitution as it copies the template DNA. Since PCR also makes use of DNA polymerase to amplify the original template DNA, it is possible that the differences seen with the mtDNA from the Neandertal is actually a result of PCR induced errors. Some phenomena in the ancient DNA could actually cause a consistent misamplification of the DNA template present in the Neandertal bone. A possible example of this in the *Cell* paper can be seen in Figure 4 of the paper. At positions 107 and 108 as well as 111 and 112 there were a number of consistent variations that could be the result of bad copying by the DNA polymerase used. Tomas Lindahl, who writes a mini review at the beginning of the *Cell* volume, comments on this. Is it not then possible that a somewhat uniform oxidative process might damage the DNA in such a way that the original information present in the Neandertal mtDNA would be reproducibly "copied" wrongly?[32]

The *Cell* authors seem to hint at this type of possibility when they write, "Such changes that are present in only a few clones [I counted 14 out of 30] are likely to be due to misincorporations by the DNA polymerase during PCR, possibly compounded by damage [degradation] in the template DNA." Later, they write that "misincorporations in early cycles of the amplification might be misinterpreted as sequence differences." Further, they write, "Misincorporations are a fairly frequent phenomenon in the Neandertal extracts."[33] The article implies, however, that these problems have been taken into consideration and corrections made.

Possible Prejudice in the Amplification Procedure

There is an intrinsic catch-22 in the methodology involved in recovering mtDNA from Neandertals. It concerns the fact that: (1) Neandertal mtDNA is similar to modern human mtDNA; and (2) PCR is so extremely sensitive that it is virtually impossible to eliminate all modern human mtDNA contamination in the process. Thus, the closer the target mtDNA is to modern human mtDNA, the more difficult the problem of discrimination becomes. It is in the area of seeking to discriminate between Neandertal mtDNA and modern human mtDNA contamination that I see a serious problem in the methodology that could involve prejudice by the authors of the *Cell* article. I am certainly not implying deceit on the part of the researchers, nor am I suggesting that their scientific expertise is anything but top-drawer.

I am, however, suggesting something very human. Given the obvious bias of the researchers toward the Out of Africa Model, and the fact that one of the leading advocates of that model, Chris Stringer, has been involved, at least indirectly, in this and many other searches for Neandertal DNA, it would be quite normal for the researchers to expect any Neandertal mtDNA they find to be quite different from modern human mtDNA. This is what the Out of Africa Model predicts. Hence their tendency might be to not question their findings as rigorously. They obviously repeated the chemistry many times, but it is the approach to their methodology that I find questionable.

In testing new methods for determining if there is recoverable DNA in ancient specimens, the *Cell* authors write, "We excluded human remains because of the inherent difficulty of recognizing contamination from contemporary humans."[34] In other words, it is much easier to recognize modern human DNA contamination in ancient nonhuman specimens than in ancient human specimens. It is obvious that much of the contaminating DNA would come from the modern humans who are doing the research and who have handled fossils containing the ancient DNA. The closer ancient human DNA sequences are to modern ones, the harder it is to tell if they are truly ancient or if they are the result of modern human contamination.

The fossil evidence shows that the Neandertals were closely related to anatomically modern humans. Since the mtDNA evidence is being used to challenge that relationship, biochemist John P. Marcus forecasts a problem. "Knowing the bias of evolutionists, it would not be surprising if . . . true Neandertal mtDNA sequences were rejected on account of their being too close to modern human ones and therefore suspected of arising from modern human mtDNA contamination."[35]

This, in essence, is what I am suggesting. The process tended to suppress the mtDNA sequences that were similar to modern human mtDNA while amplifying those sequences that were distinctive. This gives the appearance that the Neandertal mtDNA was more distant from modern humans than it actually was.

A study of the *Cell* article reveals that in most of the processes the amplification product was composed of two classes of sequences, a small class of sequences that were very similar to modern humans and a much larger class that showed differences. To the researchers, the obvious explanation consistently was, "The former class of molecules probably reflects contamination of the specimen."[36] Knowing the extreme sensitivity of PCR, this explanation is certainly a legitimate one. However, it is not a conclusive one. There is simply no way of knowing if it is a matter of contamination or if that smaller class actually reflects some more modern sequences in the Neandertal mtDNA itself.

Believing that modern human mtDNA in the amplifications always represents contamination, they state, "Amplifications were performed using primers that are specific for a 104 bp product of the putative Neandertal sequence and that do not amplify contemporary human sequences." They then make the following claim: "Thus, the retrieval of the putative Neandertal sequence is not dependent on the primers used. Furthermore, most primer combinations yield a large excess of clones representing the putative Neandertal sequence over clones similar to contemporary human mtDNA."[37]

Nothing is more important in science than the ability to repeat an experiment for verification. Referring to a duplication of their procedure at Pennsylvania State University, the *Cell* authors comment, "Thus, while this third independent extract contains a larger amount of contemporary human DNA, probably stemming from laboratory contamination, it confirms that the putative Neandertal sequence is present in the fossil specimen."[38] However, Australian scientists contest this with good reason. They write:

> In the initial Neandertal report, independent sequence results were not achieved. Only contaminant sequences were obtained in the second laboratory until

> primers, based on the Neandertal sequence from the first laboratory, were used to amplify a small portion (about 10%) of the mtDNA segment studied. *This is not an independent replication* (emphasis added).[39]

In an effort to confirm their results, the researchers constructed primers that would amplify Neandertal mtDNA but not modern human mtDNA. They tested these primers on fifteen Africans, six Europeans, and two Asians. They state that "no amplification products were obtained, indicating that this sequence [the unique Neandertal mtDNA sequence] is not present in the genome of modern humans."[40] However, this is not the point. The absence of the unique Neandertal mtDNA in modern humans can be explained genetically. The real question is the possible presence of modern mtDNA in the Neandertals.

I mentioned earlier that there is an intrinsic catch-22 in the methodology involved in recovering mtDNA from Neandertals. It concerns the fact that: (1) Neandertal mtDNA is similar to modern human mtDNA; and (2) PCR is so extremely sensitive that it is virtually impossible to eliminate all modern human mtDNA contamination in the process. Further, a study of the *Cell* article reveals that in most of the processes the amplification product was composed of two classes of sequences, a small class of sequences that were very similar to modern humans and a much larger class that showed differences. Was that small class of sequences contamination? Or was it a part of the Neandertal mtDNA that, if recognized as such, would reveal that the Neandertal mtDNA was not as different from modern human mtDNA as the researchers claim? They will never know, and we will never know.

Unfortunately, even the concept of probability doesn't help us here. In fact, the probabilities are philosophical (read, "theological"). The evolutionist, imbedded in his naturalism, would claim that the much higher probability is that the smaller class of sequences is contamination and that the results give added confirmation for the Out of Africa Model of human evolution. On the other hand, we creationists, imbedded in our theism, accept the historical statement of Genesis 3:20: "Adam named his wife Eve, because she would become the mother of all the living [humans]." Thus the Neandertals are a part of the human family, and the higher probability is that the smaller class of sequences is a part of the Neandertal mtDNA, indicating that the Neandertals were our ancestors.

The fact that the Neandertal mtDNA is somewhat different from modern mtDNA is used as "proof" that it is genuine Neandertal mtDNA and not contamination from modern humans. Thus we have one of two possibilities: (1) This recovery of Neandertal mtDNA is a magnificent illustration of high-tech biochemistry; or (2) This recovery is a classic case of the logical fallacy known as "begging the question." First the researchers do all they can to remove all traces of modern mtDNA from the Neandertal sequences. Then they claim that the "proof" that they have recovered genuine Neandertal mtDNA is because it is so different from that of modern humans.

Of particular interest at this point are the alleged recoveries of Cro-Magnon mtDNA from Paglicci Cave, a paleolithic site near the Adriatic coast in southern Italy, and from Australia. (These recoveries were mentioned earlier in this chapter under "Unfounded Assumption #4.") The Cro-Magnon people are believed to be the earliest modern humans to have reached Europe and are considered to be the ancestors of modern Europeans. The two Italian fossils are 24,000 years old and the Australian fossil is at least 40,000 years old—the same age as the original Feldhofer Neandertal from Germany.

The Italian scientists who made the mtDNA recovery were Giorgio Bertorelle and Guido Barbujani (University of Ferrara). *NewScientist* reports that the Cro-Magnon mtDNA was very different from the Neandertal mtDNA. "But the Cro-Magnons' sequences were indistinguisable [sic] from modern humans'. '[The Cro-Magnons] had sequences that living individuals still have,' says Bertorelle, 'they are [sic] nothing to do with the Neanderthal sequences.'"[41] The Italian scientists claim that their study adds more weight to the Out of Africa Model.

However, the reaction to this announcement is not what one would expect. Svante Pääbo, who led the study on the original Feldhofer Neandertal from Germany, claims that "Cro-Magnon DNA is so similar to modern human DNA that there is no way to say whether what has been seen is real."[42] The problems associated with this technique, especially with PCR, are so fundamental that some experts are now doubtful that any conclusions can be drawn about the evolution of modern humans using this method. Pääbo and others are also questioning the value of studies such as Bertorelle's, because it is almost impossible to avoid contamination using PCR.

Obviously stung by the reaction of their fellow scientists, Bertorelle and Barbujani reply in a letter to *Nature* that they were very careful to use all of the methodology guidelines that had been worked out by one of their critics (Alan Cooper of Oxford) and then spell out the catch-22 that is associated with this methodology. "You report views that the DNA of Cro-Magnons can be studied only if they were different from us. If they had the bad luck to be like us, their sequences must remain unknown forever."[43]

The question I ask is this: In sequencing the mtDNA from the original Feldhofer Neandertal from Germany, were Pääbo and his associates correct in calling those modern mtDNA sequences "contamination," or did they, perhaps unconsciously, remove them or suppress them because they knew that the Neandertal mtDNA had to be different from modern human mtDNA in order to get a believable result? Unfortunately, we will probably never know the answer to that question.

PHILOSOPHICAL BIASES

Little attempt was made by Pääbo and his associates to hide their philosophical biases. These are: (1) A bias toward molecules over fossils. This is hardly surprising, since these researchers are biochemists and know virtually nothing about human fossils. (2) A bias toward the more politically correct Out of Africa Model of modern human origins, which demands a separation of the Neandertals from anatomically modern humans. This is not terribly surprising either, since one of their informal advisors was Chris Stringer, one of the leading advocates of the Out of Africa Model.

The Bias toward Molecules over Fossils

Ever since the advent of molecular taxonomy, paleontologists have been divided over whether molecules or fossils are the better interpreters of evolutionary history. Molecules seem so neat and tidy, so precise and objective, even though their use is based on the unproven assumption that every organism's evolutionary history is encoded in its genes. Fossils, on the other hand, seem so dirty and messy. Their interpretation is anything but objective. Paleontologists have felt the sting of the charge that their discipline is "non-experimental" and "resistant to falsification."[44] The newer fossil discoveries have not fulfilled their promise to clarify the picture of human origins. Instead, they have brought more confusion. Chris Stringer explains the problem with the fossils: "The study of human origins seems to be a field in which each discovery raises the debate to a more sophisticated level of uncertainty. Such appears to be the effect of the Kenyan, Tanzanian, and Ethiopian [fossil] finds."[45]

More and more, paleoanthropologists are using the new genetic evidence in the interpretation of fossils, even if the genetic data contradicts the fossil evidence. Kenneth A. R. Kennedy (Cornell University) comments, "This practice of forcing the paleontological and archaeological data to conform to the evolutionary and genetic models continues in reinterpretations of dates based upon the molecular clock of mitochondrial DNA as well as radiometric samples."[46]

The search for objectivity is an elusive one. Although the molecular data appear to be very objective and precise, John Marcus states that the interpretation of the molecular data is just as subjective as is the interpretation of the fossils. The molecular evidence is unfalsifiable, but "the scientist must always choose which piece(s) of DNA he is going to use to do his comparisons. Very often a particular piece of DNA will not give the 'right' answer and so it is dismissed as a poor indicator of the evolutionary process."[47]

The priority given to the mtDNA data over the fossil evidence is seen in the work of Pääbo and his associates. We have shown in chapter 17 that the nonevolutionary transition from Neandertals to modern Europeans clearly marks them as our ancestors and contradicts the interpretation of the Neandertal mtDNA.

The Bias toward the More Politically Correct Out of Africa Model of Modern Human Origins

The popularity of the Out of Africa Model is due, in part, to its being so politically correct, for the following reasons: (1) Modern humans are said to have originated in Africa, a source of satisfaction to non-Western people who may feel that they have been exploited by Westerners. (2) The model emphasizes the unity of all humans despite differences in external appearance. (3) For many people it is an advantage to have the Neandertals removed from their ancestry. After all, who wants to be related to a Neandertal? (4) A woman, Mitochondrial Eve, is the hero of the plot. We all owe our existence to her. (5) The sudden replacement of the Neandertals by modern humans favors the newer and more popular punctuated equilibrium evolution model.

We seem to be witnessing a classic struggle in paleoanthropology between the molecules and the fossils. Some paleoanthropologists are bewildered at how rapidly their fellow paleoanthropologists have forsaken the fossils for the molecules. It is all the more surprising because the human fossil evidence clearly contradicts the Out of Africa Model that the molecules promote. The Chinese fossil evidence is strongly against it, as Xinzhi Wu and Frank E. Poirier demonstrate.[48] The European, African, Javanese, and Australian fossils also witness against it, as this book shows.

The words of anthropologist Robert Foley (University of Cambridge) written about a book by geneticist Luigi Luca Cavalli-Sforza (Stanford University) sum up the work of Svante Pääbo and his team, who, in spite of brilliant biochemistry, "shows plainly the futility of trying to interpret genes without knowing so much more—about selection and drift, about processes of cultural transmission, about history and geography, about fossils, about anthropology, about statistics."[49]

After 140 years, the Neandertals still have to fight for their reputation.

CHAPTER 24

The Neandertal Next Door (Maybe)

TO CLAIM THAT THE Neandertals were culture thin involves an amazing lack of scholarship, considering the evidence now available. Perhaps I should say that it is amazing how scholarship can be influenced by prejudice and political correctness. Remember, the Neandertals were alleged to be not fully human because they had no glue or adhesives for hafting tools; no unequivocal art objects; no boats, canoes, or ships; no bows and arrows; no cave paintings; no domesticated animals or plants; no hooks, nets, or spears for fishing; no lamps; no metallurgy; no mortars and pestles; no musical instruments; no needles or awls for sewing; no ropes for carrying things; no sculpture; and no long-distance overland trade.

In one sense, the Neandertals can't win. In his 1999 work, Klein gives a number of reported examples of Neandertal culture and then proceeds to debunk each one of them for the flimsiest of reasons. If at a certain site a cultural item is found that the Neandertals were not supposed to have, Klein and many others change the charge from "the Neandertals didn't have them" to the charge "the Neandertals had only very few of them." And the prejudice continues.

Neandertals, it is claimed by the African Eve evolutionists, had no art. Yet of all the charges against the Neandertals, this one is the hardest to prove. It's hard to prove a negative anyway, because "the absence of evidence is not evidence of absence." But I am talking more about the problem of defining art. Just what is it? Art is so personal, so subjective, so emotional. Art may be a reflection of our culture, but it is also the victim of our culture. Contrast, for instance, Chinese art and Western art. Art knows no definition. Some call graffiti art. Some call pornography art. Some call a crucifix in a bottle of urine art. There is hardly anything made by humans that someone could not call art. If art ever had an objective definition in the past, it is obvious that the definition of art is much more elusive today.

Some years ago, I took my granddaughter Chelsea, age six, to the Museum of Modern Art in La Jolla, an upscale area of San Diego. As we entered the museum, she screamed, "Oh, look at the scribbling!" It was at that moment that I realized what an astute art critic she was. I am also aware, to my shame, that I am announcing my lack of appreciation for modern art when I say that the only two items that conveyed any message to me were the two "pictures" on the doors of the men's and women's restrooms. Yet I am not entirely devoid of appreciation for art. I do enjoy some of the French impressionists, and my wife and I have a large copy of a Renoir hanging in our bedroom. But in spite of my lack of appreciation for some art, I do have two friends who will testify that I am a fully evolved *Homo sapiens sapiens.* The Neandertals don't even seem to have that.

I question if it is possible to actually define "art." That is why it is impossible for evolutionists to prove that the Neandertals didn't have any. In truth, Neandertals did have art. By not considering the nature and the unique culture of the Neandertals, however, the African Eve advocates have defined art in such a way as to make it impossible for the Neandertals to qualify. Evolutionists then used the alleged lack of art by the Neandertals as one reason to disqualify them from full membership in the human family.

Regarding the Great Leap Forward, prehistorian Paul Bahn writes:

> In essence this boils down to stating that the Neanderthals were so different from ourselves that a firm line can be drawn between them and us, a view that is by no means universally held. To shore up this approach, all the growing body of evidence for "art" before 40,000 years ago is simply dismissed and ignored.[1]

Juan Luis Arsuaga states that making a stone tool is actually a work of art or sculpture. He writes, "Purposeful chipping at a stone is like sculpture in that it requires carefully chosen target points, very accurately aimed blows, a correctly calculated angle of impact, and well-regulated force."[2]

Thankfully, the cultural "redemption" of the Neandertals may be taking place. Recent discoveries include items of personal ornamentation and a flute used by Neandertals.[3,4] Archaeologist Randall White (New York University) says of the Neandertals, "The more this kind of evidence accumulates, the more they look like us."[5] It can now be said that virtually every type of evidence that we can reasonably expect from the fossil and archaeological record showing that the Neandertals were fully human has been discovered. But some authorities in this field fail to recognize it.

It has been claimed that the Neandertals had no boats, canoes, or ships. Why an Ice Age, big game–hunting people would need these things is unclear. However, it has not been proven that they did not have such capabilities. Both creationists and evolutionists believe that the European Neandertals originated elsewhere. Evolutionists believe that they probably evolved in Africa from *Homo erectus*. Long before African Eve had evolved, these pre-Neandertals left Africa and invaded Europe by land through Israel, Syria, and Turkey. Creationists believe that postflood Ice Age people would have spread out from the Near East to the rest of the world. Evolutionists are especially negative about an invasion of Europe from North Africa by sea, believing that humans at that time did not have such travel capabilities. Creationists feel that since their ancestors had been on the ark during the flood, postflood humans would have been aware of the tremendous possibilities of ocean travel.

With so much water locked up in continental ice sheets during the Ice Age, it is believed that ocean levels were about four hundred feet lower than they are today. A crossing from Morocco to Gibraltar and Spain would have involved minimal deep-water risk. A second possibility would be a crossing from Tunisia to Malta and Sicily, which probably were connected, and then a very short crossing to the toe of southern Italy. A third possibility would be a crossing from Tunisia to Sardinia and Corsica, which probably were connected, and then a very short crossing into northern Italy. The least likely, but still real, possibility would be a crossing from Libya to Crete and then a crossing to Greece.

The aversion of evolutionists to such possibilities is to be expected, considering their attitude toward the limited abilities of ancient humans. We saw in chapter 12, however, that there is evidence that *Homo erectus* had seaworthy craft 900,000 years ago. The settlement of Europe from North Africa by sea is not an impossibility. It is not without significance that the oldest fossils in Europe are those at the Gran Dolina site, Spain, and the calvarium (skull without the lower jaw) from Ceprano, Italy. Both sites are about 800,000 years old. Further, early stone tools have been found on Sicily, and the Ceprano pre-Neandertal calvarium fits very well with the *Homo erectus* mandible 3 (lower jaw) from Ternifine, Morocco.[6,7]

The claim is also made that the Neandertals had no hooks, nets, or spears for fishing. Jean-Philippe Rigaud (University of Bordeaux), excavating a cave (Grotte XVI) in southern France, has discovered cultural remains of recent modern humans but also of earlier Neandertals dating back to about 75,000 years. He states that the transition between the two groups was very gradual, that they behaved in much the same way, and that they both fished. "In fact, both groups appear to have fished extensively, judging from the abundant remains of trout and pike, among other species. . . . Neandertals may have even smoked their catch, based on evidence of lichen and grass in the Mousterian fireplaces."[8]

The general belief that the Neandertals did not make use of aquatic resources is another evidence of the Neandertals getting a bum rap. Evidence that more modern people made use of such resources is usually found on the seashore. The Neandertals lived during the Ice Age. In parts of France, the seashore is now more than thirty miles inland from where it was during the Ice Age. If the Neandertals had hooks, nets, or spears for fishing, the evidence would be miles out on the ocean floor. To say categorically that they didn't have these things is sheer prejudice.

In the following list of cultural items that have been found in direct association with Neandertal fossils, notice especially (1) the many sites having *bone* tools, tools hafted (with a handle) with a rather exotic bitumen (a type of natural pitch used as an adhesive); (2) bone awls at Petralona, Greece, and Régourdou, France, suggesting the possibility of sewing; and (3) specific space allocation and habitation areas—all things that the Neandertals were not supposed to have.

Stone tools are found in association with most Neandertal sites. I have not listed a site here unless it had other cultural items besides stone tools. Nor have I listed all of the sites that show the controlled use of fire.

We will demonstrate in the section on dating that the evolutionary dating of these sites is meaningless. As a young earth creationist, I do not accept these dates. I first describe the Neandertal fossils found at the site, then the evolutionist's date, and then the cultural items. This is only a partial list of the important Neandertal cultural sites with Neandertal fossils.

1. Arago Cave (Tautavel) skull, skeletal fragments, France, dated at about 400,000 years of age. Excavations show the presence of structured and walled living areas, indicating cognitive and social capacity in Neandertal populations.[9] The site contains Tayacian (contemporary with Acheulean assemblages but without hand axes) and Middle Acheulean tools.[10] Hearths have been found, indicating the ability to control fire.[11]

2. Arcy-sur-Cure caves remains, France, dated at about 34,000 years of age. The site contains jewelry ornaments (*bone*, teeth, and ivory) with Neandertal fossils and iron pyrites with engraving.[12] A Mousterian hearth is surrounded by a circle of blackened and heated blocks. There is evidence of a separation between ground that was littered with debris and clear ground, which suggests an original wall that separated the living area from the damp part of the cave, indicating the socially structured use of space.[13]

3. Biache-Saint Vaast partial skulls, France, dated at about 156,000 years of age. The site contains Mousterian tools, including convergent side scrapers with evidence that they may have been mounted with a handle.[14]

4. Bilzingsleben skull fragments, teeth, Germany, dated about 412,000 years of age. This site has many hearths and has produced the world's largest collection of *bone* artifacts, with workshops for working bone, stone, and wood. The people here made structures similar to those made by Bushmen of southern Africa today. Three circular foundations of bone and stone have been uncovered, 9 to 13 feet across, with a long elephant tusk possibly used as a center post. A 27-foot-wide circle of pavement made of stone and bone may have been an area used for cultural activities with an anvil of quartzite set between the horns of a huge bison. A 15-inch-long piece of an elephant tibia has what appears to be engraving with seven lines going in one direction and twenty-one lines going in another direction. Two other pieces of bone have cut lines that seem to be too regular to be accidental. Archaeologist Dietrich Mania (University of Jena) says, "They are graphic symbols. To us it's evidence of abstract thinking and human language."[15]

5. Boxgrove lower leg bone, England, dated at about 500,000 years of age. The site contains an early Acheulean industry including "flint bifaces [hand axes] and debitage from their manufacture and butchered animal remains."[16] The numerous tool-manufacturing sites are used as proof that the Boxgrove people did not plan ahead and thus were not fully evolved moderns. Archaeologist Wil Roebroeks claims that "modern people would have their tool kit with them."[17] He ignores the fact that modern people do not know how to make stone tools and that it might have been more energy-efficient to make stone tools on the spot than to carry them a distance.

6. Castel di Guido skull fragments, maxilla, femur, Italy, dated at about 450,000 years of age. At the site 5,800 *bone* and Acheulean stone artifacts were discovered. Some bone implements were rather simple. "Other bone implements show a higher degree of secondary flaking and are comparable to the classic forms of stone tools; especially remarkable are several bone bifaces made with bold, large flake removals. The presence and abundance of undeniable, deliberately shaped bone tools make Castel di Guido a truly exceptional site."[18]

7. Ceprano calvarium, Italy, dated at about 800,000 years of age. The second layer above this fossil contains Acheulean choppers and *bone* tools, as well as fossil remains of straight-tusked elephant, rhinoceros, hippopotamus, fallow deer, and Irish elk. Although the literature does not suggest it, there is the possibility that this skull represents a burial from the higher bone tool layer.[19]

8. Fontana Ranuccio teeth, Italy, dated at about 450,000 years of age. The site contains some of the earliest artifacts found in Europe—Acheulean tools, including well-made hand axes, *bone* tools that were flaked, like stone, by percussion, and bifaces (hand axes) made of elephant bone.[20]

9. Gesher Benot Ya'acov femora, Israel, dated at about 600,000 years of age. The site contains a wooden plank with man-made polish. The tools are Middle Pleistocene Acheulean bifaces, flakes, and hand axes very similar to African tools.[21,22]

10. Gibraltar cave and quarry remains; Hortus Cave remains, France; La Quinta rock shelter remains, France; Krapina rock shelter remains, Croatia; and Sima de los Huesos cave, Spain. These sites range in age from 40,000 to 400,000 years. Semicircular or troughlike grooves on premolars and molars at these five sites "strongly suggest that picking one's teeth is indeed one of humanity's oldest habits."[23]

11. Kebara Cave remains, Mount Carmel, Israel, dated at about 60,000 years of age. The site contains Levalloiso-Mousterian tools, with evidence that the Neandertals here may have mounted points on wooden spear shafts with leather thongs.[24]

12. Krapina rock shelter remains, Croatia, dated at about 130,000 years of age. The site contains Mousterian tools, with a few Upper Palaeolithic tools from the upper layers.[25] Jan F. Simek and Fred H. Smith write, "Subtle chronological changes in raw material selection and technology of tool blank production are observed. These changes involve increasingly sophisticated and selective use of lithic materials . . . and are interpreted as reflecting behavioral change among Neandertals rather than between Neandertal and modern human populations."[26]

13. La Chaise Caves remains, France, dated at about 160,000 years of age. The site contains evidence of the presence of structured and walled living areas, indicating cognitive and social capacity.[27]

14. La Ferrassie Rock Shelter remains, France, dated at about 33,000 years of age. The site contains tools of the Charentian Mousterian culture, together with an engraved *bone* found with La Ferrassie 1 and a rectangle of calcareous stones, 3 by 5 meters, carefully laid one beside the other to construct a flat surface for "clearly intentional work."[28,29] A study in *Nature* by Wesley Niewoehner (California State University, San Bernardino) of the La Ferrassie I thumb and

index finger bones reveals that the Neandertal joints and movements were within the modern human range. He and his associates write:

> As there is no significant difference between Neanderthals and modern humans in the locations of their muscle and ligamentous attachments, there remains no anatomical argument that precludes modern-human-like movement of the thumb and index finger in Neanderthals.
>
> The demise of the Neanderthals cannot be attributed to any physical inability to use or manufacture Upper-Palaeolithic-like (Châtelperronian) tools, as the anatomical evidence presented here and the archaeological evidence both indicate that they were capable of manufacturing and handling such implements.[30]

15. La Quina Rock Shelter remains, France, dated at about 64,000 years of age. The site contains *bone* tools such as antler digging picks and highly modified lower ends of wild horse humeri.[31] Richard Klein, a doubter regarding Neandertal art, writes, "The list of proposed art objects from Mousterian/MSA sites depends on the author, but frequently cited examples include a reindeer phalanx and a fox canine, each punctured or perforated by a hole (for hanging?) from La Quina in France; . . . [and] occasional bones with lines that may have been deliberately engraved or incised."[32]

16. Le Lazaret Cave skull fragment, teeth, France, dated at about 156,000 years of age. Richard Klein states that the site contains "clusters of artifacts, bones, and other debris that could mark hut bases or specialized activity areas." Klein adds, "The presence of a structure is suggested by an 11 x 3.5 m concentration of artifacts and fragmented animal bones bounded by a series of large rocks on one side and by the cave wall on the other. The area also contains two hearths. . . . The rocks could have supported poles over which skins were draped to pitch a tent against the wall of the cave."[33]

17. Mauer (Heidelberg) mandible, Germany, dated at about 500,000 years of age. The site contains structured and walled living areas, indicating cognitive and social capacity in pre-Neandertal populations.[34]

18. Neandertal (Feldhofer Cave) skeleton, Germany, dated at about 40,000 years of age. This site was thought to have been destroyed due to limestone mining soon after the first Neandertal fossils were discovered. The site was rediscovered in 1997, however, and more of the original Neandertal skeleton was found, together with fossils of a 44,000-year-old anatomically modern person. "In addition to the fossils, thousands of stone tools and flakes from tool production have been found, representative of both Mousterian and Aurignacian lithic technologies. Burnt and cut bones of non-cave-dwelling animals at the site seem to show evidence of food preparation."[35] A *Nature* article adds that "this is evidence of food preparation and cooking, indicating that the Neanderthals belonged to a settlement."[36]

19. Orgnac 3 Cave teeth, France, dated at about 156,000 years of age. Klein writes that the site contains artifacts and other cultural debris surrounded or accompanied by natural rocks that may represent structure bases in the cave.[37]

20. Pech de l'Aze Cave child's cranium, France, dated at about 45,000 years of age. The site contains a cluster of large limestone blocks clearly assembled by humans but of unknown use.[38]

21. Petralona Cave skull, Greece, dated at about 450,000 years of age. Blackened fire stones and ashes are evidence of the controlled use of fire. It would be impossible for fire in the cave to be of nonhuman origin. Artifacts include stone tools of the early Mousterian culture and *bone* awls and scrapers.[39]

22. Qafzeh, Israel, dated at about 98,000 years of age. The site contains Levalloiso-Mousterian tools, with many of Upper Palaeolithic style, as well as evidence of art or aesthetics and *Glycymeris* shells (a nonedible, scalloplike species) that must have come from at least 50 kilometers away.[40,41]

23. Régourdou Cave skeleton, France, dated at about 70,000 years of age. The site contains *bone* tools, such as an antler digging pick and an awl.[42]

24. Saint-Césaire Rock Shelter skeleton, France, dated at 36,300 years of age. Richard Klein writes, "However, there is a significant problem with the idea that the Neanderthals could not behave like moderns. This is the occasional discovery of artifact assemblages that comprise a blend of Neanderthal/Mousterian and Cro-Magnon/Upper Paleolithic artifact types." After discussing this blend of artifacts at Arcy-sur-Cure, he continues, "Neanderthal authorship of the Châtelperronian is implied even more conspicuously at La Roche à Pierrot Rock-Shelter, Saint-Césaire, west-central France, where a partial Neanderthal skeleton was directly associated with typical Châtelperronian stone tools."[43] Châtelperronian tools are considered to be more "advanced" than typical Mousterian Neandertal tools.

25. Umm el Tlel skull, Syria, dated at about 42,500 years of age.[44] The site contains Mousterian tools hafted with bitumen at very high temperatures. Prior to this, the earliest hafted tools were dated at about 10,000 years of age. A *Nature* report says, "These new data suggest that Palaeolithic people had greater technical ability than previously thought, as they were able to use different materials to produce tools."[45] Simon Holdaway (La Trobe University, Australia) states, "But evidence for hafting in the Middle Palaeolithic may indicate that more complex multi-component forms existed *earlier, so changing our perceptions of the relationships between the two periods*" (emphasis added).[46] That is a remarkable statement. Just a few years ago, we were continually told that the Neandertals had no adhesives.

26. Vértesszöllös skull fragment, teeth, Hungary, dated at about 400,000 years of age. The site contains hearths involving shallow depressions with charred bones,

a continuous living surface, and several thousand pebble tools—Mousterian industry without hand axes.[47,48]

The above cultural items have all been found in direct association with Neandertal fossils. Listed below are cultural items not associated with Neandertal fossils. Since they are, however, in the proper time frame and are associated with Neandertal tools, it is presumed that they are of Neandertal origin.

1. What is beyond question the single most stunning discovery of Neandertal rock art was announced in late 2003. It is described as a human "face-mask" of palm-sized flint that has been reworked and altered. It was found in ice-age deposits at La Roche-Cotard, France, and is considered to be about 32,000 years old. It's identification with the Neandertals is based on its being "side by side with Mousterian tools"[49] in an undisturbed layer eight feet under the surface.

The rock was hand-trimmed to enhance its human appearance by percussion flaking, the same way stone tools were made. It's human appearance was further enhanced "by a shard of animal bone pushed through a hole behind the bridge of the nose creating the appearance of eyes or eyelids." The report adds: "It is clearly not accidental since the bone is fixed firmly in place by two tiny wedges of flint."[50]

Naturally, even the significance of this object as a piece of Neandertal art is being challenged.[51] However, the probability is virtually zero that this stone and bone object is just a natural occurrence.

2. Richard Klein, a doubter regarding Neandertal art, writes, "The list of proposed art objects from Mousterian/MSA sites depends on the author, but frequently cited examples include . . . occasional bones with lines that may have been deliberately engraved or incised, from . . . Molodova I in Ukraine, and a number of other European sites."[52] Other possible examples of art he mentions are numerous pierced animal bones from Prolom II, Crimea, Ukraine; two grooved cave bear teeth from Scalyn, Belgium; a fragment of cave bear molar plate carved and polished to an oval shape and an invertebrate fossil with an engraved line crossing a natural one to form crosses on both surfaces from Tata, Hungary; and a plate of flint cortex with incised semicircles and other lines from Quneitra, Israel. He also lists, at the Acheulean site of Berekhat Ram, Israel, probably more than 300,000 years old, a lump of volcanic rock suggesting a female figure.

3. A flute made from the thighbone of a cave bear, 43,000 to 82,000 years old, found in Divje Babe Cave 1, Slovenia (northern Yugoslavia). It is associated with Mousterian tools and used the same seven-note system as is found in western music.[53]

4. Structured base or living area, La Baume Bonne Cave, France, 156,000 years old. Richard Klein writes, "Concentrations of artifacts and other cultural debris surrounded or accompanied by natural rocks may represent structure bases."[54]

5. Bone tools at Malagrotta, Italy, 450,000 years old. "The presence and abundance of undeniable, deliberately shaped *bone* tools make Castel di Guido a truly exceptional site. Malagrotta, a locality 8 km to the north, has yielded a comparable assemblage, also characterized by bone implements, including a biface, and microlithic stone tools" (emphasis added).[55,56]

The Neandertals may not be the folks next door. But considering their cultural inventory, is there any doubt that they were fully human? The above items alone indicate that the Neandertal culture exceeded that of the Tierra del Fuegans or the Tasmanian Aboriginals in the 1800s. Yet no evolutionist today would even remotely consider denying those people the status of *Homo sapiens sapiens.*

CHAPTER 25

IN LIFE AND DEATH

So Very Human

THE LIFESTYLE OF THE Neandertals can be summed up in just one word—hunting. Studying the Neandertal sites with their collections of the largest game animals gives the overwhelming impression that the Neandertals were occupational hunters. But just because fossils of large animals are found in association with Neandertal fossils at over half of the Neandertal sites does not mean that the Neandertals necessarily hunted and killed those animals.

Let me illustrate. Some years ago, I shot a large bull elk near Steamboat Springs, Colorado. Since an elk is such a large animal, it cannot be hauled out of the woods in one piece like a deer. It must be quartered. During the field dressing and quartering process, I left behind many parts of the elk. I also left behind a 30.06 rifle cartridge from the bullet that killed the elk. Let us suppose that 50,000 years from now, archaeologists are digging in that spot. They discover the elk fossils associated with the 30.06 cartridge. Does that mean that the elk was toting a 30.06 rifle? Obviously not.

Even my secular hero, Louis Leakey, made the mistake of thinking that the first *Australopithecus boisei* (Zinjanthropus) he found was a tool maker because the stone tools and the fossil were found together. Some paleoanthropologists still make that mistake because occasionally stone tools have been found with australopithecines. There is no way of ever proving that those tools were used by those australopithecines. There is always the possibility that humans had been on the scene. However, when those tools (or animals) are found repeatedly in association with human fossils, such as the Neandertals, there is the strong possibility of a relationship. The more often the association occurs, the higher the probability of that relationship.

Since the vast number of associations of Neandertal fossils and large animal fossils occur in caves, there are a number of ways in which that association could occur. The first scenario involves the large grazing animals that never inhabit caves, such as deer, elk, caribou, wild ox, and bison. This is the largest group of animals found in association with the Neandertals, and it is also the group of animals used by hunting peoples in the north for food throughout history. These animals, or parts of them, could be brought into the caves by either humans or carnivores. The larger the animal, however, the more likely it is that its presence in the cave is the result of Neandertal activity—especially when the bones show evidence that the animal was butchered with stone tools. Carnivores tend to feast on large animals at the site of the kill rather than bringing the animals to their dens.

A second scenario involves very large animals that do not possess as much nutritional value yet show up in abundance at Neandertal sites, such as elephant and rhinoceros fossils. These animals almost certainly were transported by the Neandertals to the cave sites and "were used as raw material, as prestige objects or were possibly of symbolic nature."[1] This idea stands in contrast to the views of many paleoanthropologists that the Neandertals, because of limited mental ability, were not into symbolism.

A third scenario includes animals that use caves for hibernation or habitation, such as cave bear, cave hyena, wolf, lynx, fox, wildcat, and other even smaller animals. Not only could these animals be responsible for some of the other animal fossils in the caves, but these animals also show up themselves as fossils at some of the Neandertal sites. It is less likely that all of these animals were the result of Neandertal hunting activities.

In keeping with efforts of many paleoanthropologists to denigrate the Neandertals, some claim that the Neandertals were not a hunting people at all—that they were basically scavengers. This suggestion is so contrary to the evidence that it hardly needs a refutation.

In summary: (1) The largest group of animals found at Neandertal sites are the very same types of animals used by humans for food today; (2) These animals are usually very large grazers, unlikely to be carried to the sites by carnivores; (3) Many show cut marks made by stone tools, indicating that they were butchered; (4) The Neandertals had the thrusting spears, hand axes, and other weapons to effectively hunt these animals; and (5) The Neandertal fossils show the injuries typical of those who handle large animals, such as cowboys. Thus it seems impossible to deny the Neandertals the reputation they so richly deserve—stunning big game hunters.

One writer asks why we see so many bunnies if Neandertals were big game hunters. That remark is silly beyond belief. Just because one is out for big game doesn't mean he can't be opportunistic if he sees a few bunnies—especially if he is hungry. If Mr. Neandertal enjoys a charcoal-broiled wild-ox T-bone steak for dinner, that doesn't mean that he wouldn't enjoy a bunny for breakfast.

There is another possibility. Bunnies are not the easiest game to hunt, especially if one does not have a gun. While Mr. Neandertal was out after wild ox or straight-tusked elephant, Mrs. Neandertal may have been hunting bunnies—with nets! Nets do not preserve well. They have not yet been found with Neandertals, but circumstantial evidence suggests that possibility.

I have hunted deer, antelope, and elk with fine rifles and smaller animals with good shotguns. Yet, as a hunter, I have been only marginally successful. To hunt for one's livelihood using only the simple weapons available to the Neandertals demands a skill, a courage, and an accuracy that few hunters have today. Further, it demands the ability to plan ahead—the very quality that these Neandertal detractors claim Neandertals didn't have. It's obvious that these detractors never went into the woods with a rifle.

Our estimation of the hunting prowess of the Neandertals is enhanced when we realize that many of the animals they hunted were larger varieties of the animals we know today. Presumably, these animals were also more ferocious and better able to defend themselves. French archaeologist Alain Tuffreau states, "For humans to kill such big mammals before bows and arrows were invented, they needed a group and a strategy."[2] There is evidence that the Neandertals kindled fires to drive big game, then rushed in to make the kill with spears. This strategy not only indicates the Neandertal ability to plan ahead but would be impossible without language capability. Neandertals hunted, or defended themselves against, woolly mammoth, elephant, giant long-horned wild ox, giant bison, giant wild horse, giant cave bear, giant deer, giant Irish elk, giant pig, giant hyena, giant warthog, giant cheetah, giant saber-toothed cat, and giant lynx.

Most remarkable is that about half of the Neandertal sites that have fossil animal remains reveal the presence of elephants and woolly mammoths. Some Neandertal groups seem to have subsisted largely on elephant or woolly mammoth meat. Paleontologist Juan Luis Arsuaga (Complutense University, Madrid) writes:

> The elephant is the largest possible game animal on the face of the earth. . . . Beyond the physical capacity of prehistoric humans to hunt elephants, the crux of the polemic is in their mental capacity to develop and execute complex hunting strategies based on seasonally predicable conditions. Planning is powerful evidence for [fully human] consciousness.[3]

In order to demonstrate that the Neandertals were hunters of extraordinary ability, I have listed some of the individual Neandertal sites, especially those revealing the weapons and the strategies used by the Neandertals in their hunting. I have also listed some sites associated with Mousterian artifacts even when Neandertal fossils were not found, since the terms "Mousterian" (regarding tools) and "Neandertal" (referring to human fossils) have been considered almost synonymous. (Lately, however, it is recognized that some Neandertals made other types of tools as well as Mousterian, and that modern humans, at times, made Mousterian-type tools.) To list all of the sites and all of the large game animals found at those sites would be boring, repetitious, and unnecessary.

Although I give the evolutionary dates for these sites, we will show in a later section on dating that these dates are meaningless. As a young earth creationist, I do not accept these dates.

1. Arago Cave (Tautavel), skull, skeletal fragments, France, dated at about 400,000 years of age. Animal fossils found in association are elephant, wild boar, wild ox, red deer (elk), reindeer, ibex, rhinoceros, and wild horse.[4] Fossils indicate that the principal game animal was the wild horse, although bison, cattle, and goats were also eaten.[5]

2. Biache-Saint Vaast, France, dated at about 156,000 years of age. Animal fossils found in association are roe deer, red deer (elk), giant deer, long-horned wild ox, Merck's rhinoceros, steppe rhinoceros, and wild horse. There is also evidence that ten brown bear and cave bear had been skinned and butchered, the meat cut from the bones, and the bones broken open to remove the marrow.[6] Hundreds of flint tools and animal bones indicate that this site was a camp where animals were butchered. "French scientists see evidence in these artifacts of a surprisingly advanced tool kit."[7] Many of the bones were of strong and dangerous young-adult long-horned wild ox. All of this indicates Neandertal ability to plan ahead, as well as language capability.

3. Karain ("The Black Cave"), Anatolia, Turkey, with site levels dated from about 40,000 to about 350,000 years of age. During these times, the Mousterian stone tool assemblages change, but the assemblages of animal fossils change very little. They include fossils of elephant, wild goat, ibex (goat), red sheep, fallow deer, roe deer, red deer (elk), wild boar, long-horned wild ox, wild horse,

hippopotamus, and cave bear.[8] This site indicates that Neandertal tools were not as stereotyped as some suggest.

4. Kebara Cave, Mount Carmel, Israel, dated at about 60,000 years of age. The site contains Levalloiso-Mousterian tools, with evidence that the Neandertals may have mounted points on wooden spear shafts with leather thongs.[9] Stone-tipped spears were used by both Neandertals and modern humans at Kebara and Qafzeh Caves, Israel.[10]

5. La Quina Rock Shelter, France, dated at about 64,000 years of age. Animal fossils found in association are elephant, wild horse, woodland caribou, long-horned wild ox, great cave bear, giant bison, and deer.[11] Arthur Jelinek (University of Arizona) and André Debénath (Université de Perpignan, France) write, "The enormous quantities of animal bone that are concentrated in a depositional sequence at the base of a stepped series of cliffs suggest repeated successful hunting episodes."[12]

6. Feldhofer Cave, Germany, dated at about 40,000 years of age. "In addition to the [human] fossils, thousands of stone tools and flakes from tool production have been found, representative of both Mousterian and Aurignacian lithic technologies. Burnt and cut bones of non-cave-dwelling animals at the site seem to show evidence of food preparation."[13] A *Nature* article adds that "this is evidence of food preparation and cooking, indicating that the Neanderthals belonged to a settlement."[14]

7. Notarchirico-Venosa, Basilicata, femur, Italy, dated at about 500,000 years of age. There is evidence of butchering of big-game animals.[15]

8. Qafzeh, Israel, dated at about 98,000 years of age. Animal fossils found in association are rhinoceros, fallow deer, wild ox, gazelle, and wild horse.[16] As stated above, stone-tipped spears were used by both Neandertals and modern humans here and at Kebara Cave.[17]

9. Saint Brelade Cave, Jersey, Channel Islands, England, dated at about 47,000 years of age. It is believed that these islands were attached to continental Europe during the Ice Age, the time Neandertals hunted woolly mammoth and woolly rhinoceros here, possibly stampeding them off a cliff.[18]

The above hunting items have all been found in direct association with Neandertal fossils. Listed below are hunting items not associated with Neandertal fossils. Since they are in the proper time frame and are associated with Neandertal tools, it is presumed that they are of Neandertal origin.

1. At Ambrona and Torralba, Spain, dated 250,000(?) years ago, were found fossils of forty-seven straight-tusked elephants in association with bifaced and other stone implements that seem to represent planned hunts. Also found are

fossils of horses, steppe rhinoceros, red deer (elk), fallow deer, long-horned wild ox, and lion.[19]

2. At Áridos, Spain, dated 250,000(?) years ago, were found the remains of two elephants that were clearly butchered by humans, and the flint and quartzite stone tools used in butchering were manufactured on-site. There is evidence that the stone tools were resharpened on-site to continue butchering and that the flint was brought from a site two miles away. This seems to be a clear case of human planning by Neandertals.[20]

3. At Schöningen, Germany, dated 400,000 years ago, were found three fir spears, fashioned like modern javelins, cleft at one end to accommodate stone points, that are the world's oldest throwing spears. They are six to seven and one-half feet long and required powerful people to use them. This proves that there were big game hunters at that time and suggests a long tradition of hunting with such tools. It is presumed that Neandertals used them.[21] "If they are what they seem to be, these would be the first known weapons to incorporate two materials, in this case stone and wood. The Neanderthals almost surely used the many stone points found in Mousterian sites for the same purpose."[22] At the same site was found on a bed of black peat a fossilized horse pelvis with a wooden lance sticking out of it.[23]

Neandertal Mortuary Practice

Evidence of mortuary practice includes body positioning, grave construction, placing of artifacts and animal parts in the grave, the arrangement of stones around the grave, and placing of flowers in the grave. Neandertal burials show a strong preference for east-west orientation, an unlikely chance occurrence and a very common practice today. Another indication of intentional burial is that many Neandertal remains are wholly or partially articulated, rather than being torn apart and scattered by scavengers. Further, almost all of the Neandertal and earliest modern human burials have been in caves or rock shelters, with virtually none in open-air sites.

Juan Luis Arsuaga writes, "Finally, there is the question of burial, a symbolic and ritualized behavior if ever there was one, but as we have seen, a topic that stimulates heated debate."[24] He continues:

> Since it cannot be denied that burial of the dead is purposeful and requires planning and consciousness, the only option for those who deny that human species other than our own had the capacity for such activity is to deny that they engaged in it. Actually, according to their arguments, those other species are not human. But the scenario that they propose is that the numerous nearly complete

> Neanderthal bodies that have been found in caves were not buried. They ended up there as a result of actions by other, nonhuman, biological and/or geological agents. Anything will do, from floods that washed the bodies into the caves to lions and hyenas who carried them into their dens.
>
> It is not even necessary to discuss the question of Neanderthal burial, not least because I have spent years arguing that their ancestors at the Sierra de Atapuerca engaged in funerary behavior 300,000 years ago as the find at Sima de los Hueso [*sic*] establishes.[25]

One of the strongest evidences that the Neandertals were fully human, burial, relates to their association with caves. To most people, their association with caves is an indication of just the opposite—that they were "cave men," not fully human. This misunderstanding occurs because most people do not understand what the Neandertals were doing in the caves. Since so many of their remains have been found in caves, it was assumed that they lived in caves because they had not evolved enough to invent more sophisticated dwellings. The public is not aware that Neandertal dwellings have been found.[26] Nor is the public aware that thousands of people across the world live in caves today. When Ralph Solecki (Columbia University) excavated Shanidar Cave, Iraq, he discovered that about eighty Kurds had lived in that cave until 1970, during a time of political unrest.[27]

While it is almost universally believed that the Bible is worthless for scientific research, the book of Genesis sheds light on the activity of early humans regarding the use of caves. The first reference to caves is in Genesis 19:30, which states that Lot and his daughters lived in a cave after fleeing the destruction of Sodom. This is in keeping with the use of caves throughout human history as temporary or permanent shelters. But all other references in Genesis to caves refer to a usage that is virtually unknown today.

Genesis 23:17–20 records a business transaction between Abraham and Ephron the Hittite. Abraham wanted to purchase property in order to bury Sarah.

> So Ephron's field in Machpelah near Mamre—both the field and the cave in it, and all the trees within the borders of the field—was deeded to Abraham as his property in the presence of all the Hittites who had come to the gate of the city. Afterward Abraham buried his wife Sarah in the cave in the field of Machpelah near Mamre (which is at Hebron) in the land of Canaan. So the field and the cave in it were deeded to Abraham by the Hittites *as a burial site* (emphasis added).

Upon his death (Gen. 25:7–11), Abraham was buried in that same cave. In Genesis 49:29–32, Jacob instructs his sons that he too is to be buried in that cave where Abraham and Sarah were buried. We then learn that Jacob buried his wife Leah there and that Isaac and Rebekah were buried there also. Abraham and Sarah, Isaac and Rebekah, and Jacob and Leah were all buried in the cave in the field of Machpelah that Genesis 23:20 tells us Abraham purchased "as a burial site." Only Sarah died in the geographic area of the cave. All of the others had to be transported some distance to be buried there, and Jacob's body had to be brought up from Egypt. It was important then, as it is today, to be buried with family and loved ones.

The Neandertal fossil evidence shows that the Neandertal burial practice is in complete accord with the Genesis record. At least 475 Neandertal fossil individuals have been discovered so far at about 124 sites in Europe, the Near East, and western Asia. This number includes those European archaic *Homo sapiens* fossils that are now called Neandertal or pre-Neandertal. Of these 475 Neandertal individuals, at least 258 of them (54 percent) represent burials—*all of them burials in caves or rock shelters.* Further, caves obviously were used as family burial grounds or cemeteries, as the following sites show:

Krapina Rock Shelter, Croatia—75–82 Neandertals buried. Klein writes, "Arguably, Krapina was a specialized burial site, since it contains relatively limited evidence for actual occupation."[28]

Sima de los Huesos (cave), Spain—33 plus Neandertals interred

Arcy-sur-Cure Caves, France—26 Neandertals buried

La Quina Rock Shelter, France—25 Neandertals buried

Kebara Cave, Mount Carmel, Israel—21 Neandertals buried

Amud Cave, Galilee, Israel—16(?) Neandertals buried

Tabun Cave, Mount Carmel, Israel—12 Neandertals buried

Shanidar Cave, Iraq—9 Neandertals buried

La Ferrassie Rock Shelter, France—"La Ferrassie was a veritable cemetery of eight graves"[29]

Guattari Cave, Monte Circeo, Italy—4 Neandertals buried

Ksar 'Akil Rock Shelter, Lebanon—3 Neandertals buried

Spy Cave remains, Belgium—3 Neandertals buried

Engis Caverns, Belgium—3 Neandertals buried

It is understandable why burial in caves was common in ancient times. Graves in open areas must be marked so that future generations can return to pay homage to their ancestors. However, grave markers or reference points can be changed, destroyed, or moved. Directions to the grave site can become confusing over time. Landscapes can change, and memories of certain features can become clouded. Just as Abraham did not always live in one place, so the Neandertals probably moved seasonally, following herds of game. Since caves are usually permanent, it would have been easy to locate the family burial site if it were in a cave. Descendants could be sure they were at the very spot where their ancestors were buried.

Most anthropologists recognize burial as a very human, and a very religious, act. Richard Klein writes, "Neanderthal graves present the best case for Neanderthal spirituality or religion."[30] But the strongest evidence that Neandertals were fully human and of our species is that at four sites people of Neandertal morphology and people of modern human morphology were buried *together*. In all of life, few desires are stronger than the desire to be buried with one's own people. Jacob lived in Egypt but wanted to be buried in the family cemetery in the cave of Machpelah. Joseph achieved incredible fame and power in Egypt but wanted his bones to be taken back to Israel (Gen. 50:25; Exod. 13:19; Josh. 24:32). Until recently, it was the custom to have a cemetery next to the church so that the church family could be buried together. For centuries, many cities had separate cemeteries for Protestants, Roman Catholics, and Jews so that people could be buried with their own cultural and spiritual kind.

Skhul Cave, Mount Carmel, Israel, is considered to be a burial site of anatomically modern *Homo sapiens* individuals. Yet the Skhul IV and Skhul IX fossil skulls are closer to the Neandertal configuration than they are to modern humans.[31] Qafzeh, Galilee, Israel, is also considered to be a burial site of anatomically modern humans. Qafzeh skull 6, however, is clearly Neandertal in its morphology.[32] Tabun Cave, Mount Carmel, Israel, is one of the classic Neandertal burial sites. But the Tabun C2 mandible is more closely aligned with modern mandibles found elsewhere.[33] The Krapina Rock Shelter, Croatia, is one of the most studied Neandertal burial sites. A minimum of seventy-five individuals are buried there, but the remains are fragmentary, making diagnosis difficult. The addition of several newly identified fragments to the Krapina A skull (now known as Krapina 1) reveals it to be much more modern than was previously thought, indicating that it is intermediate between Neandertals and modern humans.[34,35]

That Neandertals and anatomically modern humans were buried together constitutes strong evidence that they lived together, worked together, intermarried, and were accepted as members of the same family, clan, and community. The false distinction made by evolutionists today was not made by the ancients. To call the Neandertals "cave men" is to give a false picture of who they were and why caves were significant in their lives. The human family is a unified family. "From one man he (God) made every nation of men, that they should inhabit the whole earth" (Acts 17:26).

In comparing the Neandertal burial practice with Genesis, I do not wish to imply that Abraham and his descendants were Neandertals. What the relationship was—if any—between the people of Genesis and the Neandertals we do not know. Young-earth creationists tend to believe that the Neandertals were a postflood people. What is striking is that the burial practice of the Neandertals seems to be identical with that of the postflood people of Genesis. It is not without significance that both Lazarus and Jesus were buried in caves (John 11:38; Matt. 27:60) and that this practice has continued in some cultures up to modern times.

Almost universally, when we bury someone we love we bury them with some item, some treasure, that they love and with flowers. These items are called "grave goods." It is claimed that the Neandertals did not have full human consciousness—a sense of mortality or the possibility of immortality. A major evidence for this is the alleged lack of grave goods in Neandertal burials. But because the Upper Paleolithic people often had rather elaborate grave goods, researchers have simply not noticed that the Neandertals had grave goods—items related to their lifestyle of hunting. The following is a sample of their mortuary practice and grave goods from some Neandertal sites.

1. Amud Cave remains, Upper Galilee, Israel, dated at about 40,000 years of age. An infant was buried in a small niche with a red deer maxilla (upper jaw) lying on its pelvis. Because items in caves are usually disturbed by animals:

> The articulation of Middle Paleolithic hominid skeletons is the major criterion for their designation as intentional burials. In the case of Amud 7, this claim is enhanced by the discovery of a red deer maxilla leaning against the pelvis of the buried hominid. . . . The case of Amud 7 is thus reminiscent of Qafzeh 11, where fallow deer antlers were found associated with a young individual, and of Skhul V, where a suid [wild pig] mandible was associated with the hominid. Both instances have been accepted as burials with possible grave goods by even the more skeptical scholars.[36]

Both these Qafzeh and Skhul individuals are anatomically modern *Homo sapiens*, indicating similar burial practices in the two populations.

2. Dederiyeh Cave infant skeleton, Syria, dated at about 75,000 years of age. An infant lay on its back with arms extended and legs flexed. At its head was a slab of stone, of a rock type rare in the cave deposits, while a triangular piece of sculpted flint lay on the infant's chest (heart), in the most sterile layer of the burial fill.[37,38] The significance of that last phrase is that the sculpted flint was not there by accident.

3. Krapina (rock shelter) remains, Croatia, dated at about 130,000 years of age. There are fossil remains of 75 to 82 individuals, all between the ages of

about 3 and 20.[39] Klein states, "Arguably, Krapina was a specialized burial site, since it contains relatively limited evidence for actual occupation."[40]

4. La Chapelle-aux-Saints (cave) skeleton, France, dated at about 50,000 years of age. The skeleton was described as lying east-west in an intentionally dug grave with his head beneath three or four large fragments of long bone (femur), themselves lying under the articulated bones of a bovid's lower leg.[41]

5. La Ferrassie Rock Shelter remains, France, dated at about 33,000 years of age. "La Ferrassie was a veritable cemetery of eight graves."[42] An engraved bone was found under La Ferrassie I.[43]

6. Le Moustier Rock Shelter remains, France, dated at about 40,300 years of age. The site contained a person that had been buried in the attitude of sleep, with hematite powder sprinkled on the remains.[44] At sites such as Le Moustier, "deliberate burial is the most plausible explanation for the presence of nearly complete or articulated Neanderthal skeletons."[45]

7. Qafzeh, Israel, dated at about 98,000 years of age. Qafzeh II is a child whose left hand rests on a fallow deer's skull and antlers placed across the child's neck. Qafzeh IX is the burial of a young adult with a child laid across its feet. Qafzeh IX is a Neandertal; Qafzeh II is considered to be an anatomically modern *Homo sapiens.*[46] There seems to be no difference in the burial practices of the two populations.

8. Régourdou Cave skeleton, France, dated at about 70,000 years of age. The site is an intentional burial in a cave, with bear bones arranged with the skeleton and the grave covered with a stone slab.[47]

9. Roc de Marsal (cave) skeleton, France, dated at about 55,000 years of age. An elephant tusk is buried with the skeleton.[48]

10. Shanidar Cave remains, Iraq, dated at about 46,000 years of age (upper level) and 75,000 years of age (lower level). There are intentional burials of five individuals in the cave, with flowers placed on at least one body. Both Arsuaga and Hayden refer to these flowers as "grave goods."[49]

11. Sima de los Huesos (Atapuerca) cave, Spain, dated at about 400,000 years of age. Juan Luis Arsuaga writes:

> There is only one known case that suggests any funerary practice before the time of the Neanderthals and modern humans and that is at Sima de los Huesos. Although bodies were amassed one atop the other in a place chose for that purpose, at the Sima they were not *buried,* a process defined as the digging of a grave and the placing of a body within it. . . .[50]
>
> When a member of the group died in the caves at Sierra de Atapuerca or nearby, the body was carried to this hidden niche and deposited there. That is why we came to call this place Sima de los Huesos, which in English means bone pit or depository, in other words, an ossuary [a receptacle to preserve the bones of the dead].[51]

12. Skhul Cave, Skhul IV & IX remains, Mount Carmel, Israel, dated at about 91,000 years of age. The bodies of four individuals (I, IV, V, and VII) appear to have been purposefully arranged with folded and flexed limbs. Skhul V, an adult male, was laid on his back with the mandible of a wild boar placed within his arms, while Skhul IX lay with the skull of a bovid.[52] Skhul IV and IX are Neandertals; the rest are anatomically modern *Homo sapiens*. Again, there seems to be no difference in the burial practices of the two populations.

13. Teshik-Tash Cave child remains, Uzbekistan, CIS, dated at about 50,000 years of age. The site contains a child buried with mountain goat horns surrounding the body.[53]

Perhaps the most poignant illustration of Neandertal mortuary practice is the apparent burial at Antelias Cave, Lebanon, of an eight-month-old fetus with grave goods of fallow deer and ibex,[54,55] appropriate symbols of love and devotion for a hunting people. Since many modern humans abort their fetuses, perhaps a case could be made that in some practices, Neandertals were more "civilized" than many modern humans.

Regarding the significance of the act of burial, allow me to repeat the comment of Richard Klein: "Neanderthal graves present the best case for Neanderthal spirituality or religion."[56] The inclusion of grave goods seems to indicate some thought of an afterlife—life beyond the grave. All of this testifies to the full humanity of the Neandertals.

Although our emphasis has been on death, it is important to know that the Neandertals have a record of tender care for individuals before death. This has been established at Shanidar Cave, Iraq, and at the Bau de l'Aubesier Rock Shelter, France. In *The Proceedings of the National Academy of Science*, Neandertal authority Erik Trinkaus *et al.* write regarding the Neandertals, "These human populations therefore had achieved a level of sociocultural elaboration sufficient to maintain *debilitated* individuals and to provide the motivation to do so" (emphasis added).[57] In life and death, they are so very human.

CHAPTER 26

Stunning Answers from the Land Down Under

I LOVE AUSTRALIA. I love the land and its people. I love their warmth and their accent. (Actually, it's arrogant of me to claim that they have an accent. That implies that I'm normal and they aren't. From their point of view, I'm the one who has the accent. And who's to say which is normal?) But after you read this chapter, you will be aware of another reason why I love Australia.

There is one question regarding the Neandertals that we have not fully addressed—the Neandertal/*erectus* morphology (*erectus* is just a smaller version of Neandertal and the most unique aspect of both is their skull shape). Since we claim that they did not evolve from other primates, the question is, "Are there adequate nonevolutionary explanations for the distinctive morphology of the Neandertals and *Homo erectus*?" We touched on some of the possibilities in chapter 6. There are a number of plausible explanations, and some of them might be related to the biblical record of earth history.

Many people believe that there is a serious conflict between science and the Bible. A classic work on this subject is Andrew White's *A History of the Warfare of Science with Theology*. However, the famous scientist and educator

James B. Conant made this interesting observation on White's book: "The warfare White describes has been for the most part *a series of battles in regard to the interpretation of the past*" (emphasis added).[1]

The overwhelming majority of people working in science and technology deal with the present, not the past. The overwhelming majority of books and journal articles of a scientific nature also deal with the present, not the past. In truth, there is simply no conflict between the Bible and scientific discoveries and observations in the present. The only conflict between science and the Bible involves the evolutionist community's interpretation of the past history of the earth and its life.

While science thrives on observation and experimentation in the present time frame, it has no mechanism to observe the past with the same accuracy as it observes the present. The scientific method (or methods) applies to the past only indirectly. In the absence of historical records, all data regarding the past involves interpretations which may or may not be correct. Not without reason did Oxford scholar R. G. Collingwood say that the study of the past is really a study of the human mind, indicating that there is a high degree of subjectivity in all scientific reconstructions of the past.[2] It is unfortunate that many scientists think that because they are able to make authoritative statements about the present processes of nature, they can also speak dogmatically about the past history of our planet and ignore historical documents such as Genesis.

The failure of the scientific community to recognize the high degree of subjectivity in its interpretations of past events is the major cause of the "warfare" between the Bible and this area of science. While Christians may not always have handled this warfare well, the *cause* of the conflict cannot be laid upon our shoulders. Believing that we have a reliable historical record of two crucial past events, creation and the flood, we Christians accept with gratitude and thanksgiving the contributions of science in our lives today while we challenge the authority of science to make dogmatic statements about the past.

The nature of the scientific method and the nature of biblical revelation are two distinct and separate things. Typical of the confusion that comes from not understanding that fact is this statement by Eugenie C. Scott (National Center for Science Education, an organization founded to combat scientific creationism): "The scientific method is vastly superior to revelation (or other epistemologies) as a means to discover the workings of the natural world."[3]

Scott is absolutely right in what she says, but she doesn't have a clue as to why she is right. She is right, but she is comparing apples with oranges. The purpose of revelation is not to tell us about the "workings of the natural world." Why should God give us revelation of things we can discover for ourselves when we utilize (among other means) the scientific method?

God's purpose in biblical revelation is to give us information on things we could not know by any other means. For this reason God has given us a revelation of two momentous historic events: creation and the flood. Knowledge of these two singular, unrepeatable events is beyond the scope of the scientific method. Scott confuses the issue. She has lumped information of the past through God's revelation with information of the present obtained by the scientific method. She then indicates that she prefers the information provided by the scientific method above the information provided by the Word of God, even though the scientific method is not adequate for the task of dealing with past singularities.

The Ice Age

One of the remarkable features of the biblical revelation, when it is interpreted literally, is its internal consistency in recording past events that help to explain the present world. Nowhere is this consistency seen more clearly than in the ability of the worldwide Genesis flood to provide the only adequate explanation for a great geological mystery: the Ice Age. Few realize that the Genesis flood and the Ice Age are intimately connected in terms of cause and effect. It was the severe disruption of the global climate by the Genesis flood that caused the Ice Age to develop immediately afterward. (The scientific community speaks of four ice ages during the Pleistocene, but the three earlier ones are open to challenge.)

It would be natural to assume that all that is needed to produce an ice age is a series of very cold winters. Not so. In many areas of the world the winters are very cold; yet these areas may have little snow. In contrast to our present climate, the basic requirements for an ice age are (1) much cooler summers and (2) much more snowfall; contributing mechanisms are (3) warmer winters and (4) warmer oceans. This combination is such an unlikely scenario that the Ice Age would not have happened were it not for God's judgment of the flood. The scientific community, limiting itself to present processes, has hard sledding in trying to account for the cause of the Ice Age. Over sixty theories have been proposed to explain the Ice Age, but all have serious defects. From computer simulations and from what we know of atmospheric science, however, the Genesis flood was capable of producing the Ice Age.

Relatively warmer winters were one of the mechanisms that supplied the abundant snowfall necessary for the Ice Age. First the northern oceans had to remain ice free to provide the moisture source for that abundant snowfall. Today, the Arctic Ocean is normally frozen, and colder winters cause major portions of the North Atlantic to freeze as well. Thus cold winters cut off the moisture sources for an ice age. Furthermore, cold air carries very little moisture. Air temperatures in winter had to be warm enough to carry the massive amounts of water that were precipitated as snow. Those who live in northern areas know that very cold weather seldom brings large amounts of snow, whereas some of the heaviest and wettest snowstorms occur in spring.

No matter how heavy the snowfall might have been in winters, if the summers had been warm enough to melt all of the snow, no Ice Age would have occurred. An ice age cannot develop in one year. There must be a heavy accumulation of snow over many years without significant summer melts. Hence very cool summers were a basic necessity for the Ice Age.

I was raised in Fargo, North Dakota. Fargo is in the midst of some of the richest farmland on earth—the Red River Valley, which is an ancient lake bed formed by glacier meltwater. To the east, in Minnesota, there is also much evidence of ancient glaciers. Today, in winter the temperature sometimes drops to forty degrees below zero. In summer, however, one-hundred-degree weather is not uncommon. That is not an ice age–producing climate. The mystery of the Ice Age is this: In the very geographic areas where there is incontrovertible evidence of past continental glaciers, the climate today is such that it is impossible for an ice age to be produced. Further, it stretches the imagination to create a scenario that would have allowed glaciers to be produced there in the past. Certainly, as far as the Ice Age is concerned, the present is not the key to the past. A logical inference is that since the Ice Age was caused by conditions that do not exist today, only historical records such as the book of Genesis can give us clues as to what those past conditions were.

Oxygen isotope and other data give scientific evidence that the deep ocean was warmer during extended periods of the past than it is today. While the scientific community cannot explain this fact, warm oceans are exactly what the Genesis model would predict for the early earth. The waters above the expanse (atmosphere), created on day two of creation week (Gen. 1:6–8), would cause a relatively uniform climate worldwide, including warm oceans and warm polar regions. The oceans would have become even warmer at the time of the flood when hot water, which was a part of the "fountains of the deep," mixed with them. These warm oceans served as the source for the abundant moisture necessary for the heavy snowfall of the Ice Age.

After the flood, the oceans, while gradually cooling, would remain warm for hundreds of years. However, the land surfaces would cool very quickly for the following reasons: (1) The waters above the atmosphere, having contributed to the rains for the flood, would now be gone and their insulating effect would be removed; (2) The intense volcanism of the flood would continue for a time after the flood and would put high amounts of particulate matter and aerosols into the atmosphere, reflecting the sun's rays back into space; (3) Immediately after the flood the earth would be barren and denuded, and barren land is highly reflective of the sun's rays; (4) The heavy cloud cover produced by the warmer oceans would also be highly reflective of the sun's rays; (5) The rapidly building snow cover would itself have a cooling effect on the land. This formula of very cool summers and very heavy snowfall, together with warmer winters, warmer oceans, and cold land surfaces, becomes the secret for the Ice Age. It explains why the Ice Age started immediately after the flood, why there had not been an ice age before that time (evolutionists' claims of pre-Pleistocene ice ages notwithstanding), and why there cannot be an ice age again. It also explains why the scientific community, which rejects the Genesis flood, has never been able to adequately explain the cause of the Ice Age. Old-earth evangelicals who believe in a local Noachic flood are also hard-pressed to explain the cause of the Ice Age and seldom address the issue.

Creationist Michael J. Oard (National Weather Service) has meticulously detailed the cause and nature of the Ice Age in his monograph *An Ice Age Caused by the Genesis Flood*.[4] While those rejecting the Genesis account of early earth history are at a loss to explain the cause of the Ice Age, they are in some agreement with creationist atmospheric scientists such as Michael Oard and Larry Vardiman (Institute for Creation Research) as to what the Ice Age climate was like. And that is what is relevant to our subject.

The Ice Age Climate

During the Ice Age, continental ice sheets covered large sections of North America, northern Europe, and northwest Asia. The ice occurred even on the high mountains of the tropics and in the Southern Hemisphere as well. The climate over much of the earth was cold, damp, and rainy. Thick cloud cover caused by the warm oceans together with heavy post-flood volcanism robbed the earth of much of its sunshine. This condition would have lasted until the slowly cooling oceans gradually approached their present temperature range. It is significant that the book of Job, with its setting after the flood (Job 22:16) and probably before Abraham, has more references in it to snow, ice, and violent weather than any other book of the Bible.

Michael Oard estimates that the time from the end of the flood to when the largest volume of ice and snow covered the land (glacial maximum) was about five hundred years and the time for deglaciation was about two hundred years. He writes, "Thus, the total length of time for a post-flood ice age from beginning to end, is about 700 years."[5] However, Oard cautions that these are difficult figures to come by because "there are too many variables that are poorly known."[6]

The human responses to the lack of sunshine and the harsh climate of the Ice Age would have been: (1) to seek out natural shelters such as caves; (2) to construct shelters out of whatever material was available; and (3) to wear heavy clothing, probably animal skins, to cover much or all of the body.

Since the Neandertals appear to be a postflood/post-Ice Age people, this could help explain why they hunted big game—using the meat for food and the hides for clothing and shelter. It could also explain their use of caves. If much of the ground was snow-covered or frozen, caves would have been an option for shelter and burial. Agriculture would have been virtually impossible.

The Neandertal/*Erectus* Morphology

In the 3-million-year transition period from a Lucy-like creature to modern humans, evolutionists would expect that at the midway point, 1.5 million years ago, humans might still retain a superficial resemblance to their ancestors. But at 100,000 years ago, modern humans had arrived. How does an evolutionist explain the thick skull bones, the heavy browridges, and the prognathism (jutting forward of the upper and lower jaws) of the "classic" Neandertal people 40,000 years ago? A lesser-known but far more serious problem for evolutionists is the Upper Pleistocene *Homo erectus* fossils—the *Homo erectus* Java Solo people 27,000 years ago and the robust *erectus*-like Kow Swamp people of Australia just 9,500 years ago.

It is here that the fatal flaw of evolutionary theory surfaces for all who are willing to face the truth. It involves two problems: (1) According to evolutionary theory, the above-mentioned morphology was not supposed to be present in the Upper Pleistocene; and (2) Since it is present, there is not enough time for this morphology to evolve by evolutionary mechanisms into modern humans in just 40,000 years or less.

Evolutionists have proposed two solutions to this problem. These two solutions are contradictory to each other, but logic has never been the strong suit of evolutionists. These solutions are: (1) regarding the Neandertals, to wipe these individuals out, exterminate them through the invading African Eve people, and/or (2) regarding the late *Homo erectus*, to propose nonevolutionary mechanisms to explain this morphology. The problem, of course, is that if there are nonevolutionary mechanisms to explain this morphology in the Upper Pleistocene, why do we need evolution to explain this morphology in the Lower and Middle Pleistocene?

Nonetheless, the mechanisms I mention below are all nonevolutionary mechanisms that have been offered by evolutionists themselves to explain the Neandertal/*erectus* morphology of fossils in the Upper Pleistocene.

The Ice Age climate with its lack of sunshine because of the heavy cloud cover, the related need for shelter, and the wearing of heavy clothing over much of the body has suggested the disease of rickets to explain the Neandertal/*erectus* morphology. This identification of fossil humans and rickets was made by evolutionist Francis Ivanhoe in a paper in *Nature*. Ivanhoe said that "every Neandertal child skull studied so far shows signs compatible with severe rickets."[7] These include the child remains from Engis (Belgium), La Ferrassie (France), Gibraltar, Pech de l'Aze (France), La Quina (France), Starosel'e (CIS), and Subalyuk (Hungary). Less extreme cases are seen in the child remains from Teshik-Tash (C.I.S.), Shanidar (Iraq), and Egbert or Ksar 'Akil 1 (Lebanon). The rickets skull morphology seen in these children has carried over into the adult Neandertals and other fossil humans. The gross bowing of the long bones of the body, so typical of rickets, is seen in both Neandertal children and adults.

Still another possible explanation of the Neandertal morphology is syphilis. Evolutionist D. J. M. Wright (Guy's Hospital Medical School, London) observed that, "In societies with poor nutrition, rickets and congenital syphilis frequently occur together. The distinction between the two is extremely difficult without modern biochemical, seriological, and radiographic aids."[8]

Based on his examination of the Neandertal collection at the British Museum, Wright found a number of features in the Neandertal's morphology that were compatible with congenital syphilis. These conditions are seen in both adult and child skulls. Wright specifically mentioned the original Neandertal skullcap as well as the Gibraltar II, Starosel'e, and Pech de l'Aze Neandertal remains.

Over the years, the evolutionist literature has suggested a number of conditions—geographical, environmental, pathological, cultural, and dietary—that could produce a Neandertal-like morphology. Richard Klein compares the teeth wear of the Neandertals to that of Eskimos, who tend to use their jaws as clamps: "The forward placement of Neanderthal jaws and the large size of the incisors probably reflect habitual use of the anterior dentition as a tool, perhaps mostly as a clamp or vise." He continues:

> Biomechanically, the forces exerted by persistent, habitual, nonmasticatory use of the front teeth *could account in whole or in part* for such well-known Neanderthal features as the long face, the well-developed supraorbital torus, and even the long, low shape of the cranium. Massive anterior dental loading could further explain the unique Neanderthal occipitodmastoid region which perhaps provided the insertions for muscles that stabilized the mandible and head during dental clamping (emphasis added).[9]

Klein has given a plausible nonevolutionary explanation for most of the unique features of Neandertal morphology: they could be the result of the unique stresses their jaws and teeth were subjected to when used as tools.

Klein also states, "The long, low shape of the Neanderthal cranium with its typically large occipital bun probably reflects relatively slow postnatal brain growth relative to cranial vault growth."[10]

Klein also recognized the effect that Ice Age geographic isolation could have on the development of the Neandertals when he wrote that "some of the European mid-Quaternary fossils clearly anticipate the Neanderthals, while like-aged African and Asian ones do not. Clearly, the implication is that the Neanderthals were an indigenous European development."[11]

On the other side of the world, fossil discoveries in Australia reveal a condition that defies evolution as an explanation. Two populations were living side by side in very recent times. One population had a very modern morphology, and the other had a *Homo erectus*-like morphology. The *erectus*-like fossils include the Mossgiel individual (discovered in 1960 and dated at about 6,000 ya), the forty Kow Swamp individuals (first discovered in 1967, dated at about 10,000 ya), and the Cossack skeletal remains (discovered in 1972 and dated from just a few hundred years ago to about 6,500 ya).

We can sense the evolutionists' bewilderment as they write about these fossils. Jeffrey Laitman (Mt. Sinai School of Medicine) mentions fossil authorities who speak of the "extreme disparities" found between these two groups.[12] Richard G. Klein declares that "the range of variation [between these two groups] is extraordinary and may indicate that Australia was colonized more than once, by very different people."[13] Klein then gives reasons for discounting a direct relationship between the Ngandong Java *Homo erectus* people, which he dates between 27,000 and 53,000 years old, and the Australian *Homo erectus*-like people: (1) The Australian *Homo erectus*-like people are younger in date than very modern Australian people dated earlier; (2) The Australian *Homo erectus*-like people have skulls like the Java *Homo erectus* people, but their body bones are very modern; and (3) He then gives nonevolutionary explanations, especially climate, to explain the difference.[14] Without intending to do so, Klein confirms the thesis of this book—that the differences between the *Homo erectus*-shaped people and modern people are not the result of evolution.

In a 1972 article in *Nature*, probably the most prestigious science journal in the world, Alan G. Thorne (Australian National University) and P. G. Macumber (Geological Survey of Victoria) reported on the Kow Swamp fossils having an *erectus*-like morphology but dating at just 9,500 years ago. Because of the absurdly late date for *erectus*-like fossils, they felt compelled to give a nonevolutionary explanation. In a sure and confident manner they say that "the Kow Swamp series represents an isolated and remnant population."[15] But at the very time they were writing those words, the Cossack skeletal material, with that very same morphology, was being discovered on the west coast of Australia, two thousand miles away. Hence we are not dealing with an "isolated and remnant population." The authors of the Cossack article write that the Cossack discovery "indicates that this morphology was not a regional variant but continental in distribution."[16]

In the same issue of *Nature* in which Thorne and Macumber made their report on the Kow Swamp material, an editorial by an unsigned correspondent (a leading authority in the field) suggested possibilities for the *erectus*-like morphology of the Kow Swamp fossils: (1) these fossils could represent a small inbred community; (2) the thick bones of these fossils could be the result of differential survival—thicker bones would survive intact longer than thinner bones; (3) the thick cranial bones could be the result of a nutritional problem; (4) low-grade anemia; (5) genetic factors; (6) endocranial factors; or (7) a pathological condition.[17] The writer certainly covered all the bases. In doing so, however, he gave away the store. The admission that one or more of these factors could produce a *Homo erectus*-like morphology is also an admission that the concept of human evolution is not needed to explain that morphology—which is what creationists have claimed all along. All of the explanations suggested in that editorial are nonevolutionary.

Many evolutionists have suggested that these diverse Australian populations are the result of two or even three migrations into Australia from elsewhere.[18] But this explanation does not solve the problem; it just pushes the problem back to the Asian land mass. Evolutionists date the first humans in Australia at a bit before 40,000 ya, with some evidence suggesting 60,000 ya. Hence, even if these successive waves of migration were separated by as much as 20,000 years, the differences in morphology cannot be ascribed to evolutionary processes. On the evolution timescale, 20,000 years is nothing. It just means that these two morphologically diverse groups, which may have been separated geographically, had been living as contemporaries on the Asian mainland. Hence the dual migration hypothesis doesn't even address the problem, let alone solve it.

The most common explanation (also a nonevolutionary one) for these Upper Pleistocene *Homo erectus*-like fossils is the idea of cranial deformation. The artificial deformation of bones is well-known in human history. A classic example is the distortion of the feet of Chinese women caused by binding in childhood. In the Americas, a common cause of skull deformation was the strapping of an infant to a cradle board. In South America, Inca infants of noble birth would have their heads bound so as to give the heads a pointed shape. The purpose was to distinguish the nobility from the commoners.

Richard Klein suggests cranial deformation as a likely explanation for the Kow Swamp fossils in his 1989 edition but omits the idea entirely in his 1999 edition, perhaps recognizing that it is not a well-founded argument.[19] *The Catalogue of Fossil Hominids* also suggests cranial deformation as a possibility for the skull of the Kow Swamp 2 individual.[20] As to the particular method of cranial deformation,

Chris Stringer says that the practice of head binding "was certainly responsible for some of the peculiarities in cranial shape amongst the Kow Swamp people."[21] Phillip J. Habgood (University of Sydney) suggests that the method used was one of repeated pressure to the front and back of the infant's cranium. He feels that this method, unlike binding, would allow for the degree of variable deformation that one sees between the Kow Swamp 5 and Kow Swamp 2 individuals.[22] Thorne, while acknowledging the possibility of some form of cranial deformation, likewise sees no evidence of long-term binding in the Kow Swamp fossils.[23]

Freedman and Lofgren, describing the Cossack cranium, say that it is very similar to the Kow Swamp crania but very different from other recent western Australian skulls. They further state that the differences in the fossils do not appear to be due to artificial cranial deformation.[24]

To use cranial deformation as the explanation for the large number of Upper Pleistocene *Homo erectus*-type fossils seems contrived. That explanation is never given for fossils of similar morphology in the Lower and Middle Pleistocene. Why is it valid for one geologic period and not for the others? Nor is that explanation given for the Neandertal and early *Homo sapiens* fossils possessing a somewhat similar morphology. If cranial deformation can produce that morphology, then evolution is not needed.

Textbook illustrations of valid cases of artificial cranial deformation are quite different from the typical *Homo erectus* morphology.[25] I have seen many museum specimens of artificial cranial deformation, and they, too, are very different from a *Homo erectus* morphology. In fact, it is hard to imagine a method of artificial cranial deformation that would result in an *erectus*-like skull. Certainly, no type of cranial deformation could produce the thick cranial walls that are so typical of *erectus*-like individuals.

Without question, the most striking matter of significance to support creationism and challenge evolution is an mtDNA study by Gregory J. Adcock (Australian National University) and associates published in 2001 in the *Proceedings of the National Academy of Science*.[26] They recovered mtDNA from ten ancient Australians. These include:

Lake Mungo 3 skeleton, with gracile morphology similar to that of anatomically modern humans of today, dated at 40,000 to 60,000 years old

3 fossil individuals from Willandra Lakes, with gracile morphology similar to anatomically modern humans of today, dated at a bit less than 10,000 years old

6 Kow Swamp fossil individuals, with robust morphologies outside of the range of contemporary indigenous Australians, dated at 8,000 to 15,000 years old

Regarding the Lake Mungo 3 fossil, the results show that "sequences from the lineage that includes LM3's mtDNA no longer occur in human populations, except as the nuclear Insert on chromosome 11."[27] In other words, Lake Mungo 3's mtDNA is different from all humans living today even though his morphology (shape) is the same as humans of today. In fact, LM3's lineage is even deeper (older) than that of African Eve. Using Out of Africa logic, that would mean that all living humans originated in Australia.[28] To the contrary, the authors state that finding deep lineages in Africa does not prove an African origin for all living humans any more than finding even deeper lineages in Australia proves an Australian origin for all living humans, because the findings are consistent with several different interpretations.[29]

Further, with the exception of Kow Swamp 8 and Lake Mungo 3 (above), "the ancient Aboriginal sequences, including those from individuals with both *robust* and *gracile* morphologies, are within a clade [a group of humans derived from a common ancestor] that includes the sequences of living Aboriginal Australians."[30] In other words, although some of these fossil individuals were *erectus*-like and some were modern in morphology, their mtDNA was virtually the same.

The stunning conclusion is this: "The Ancient mtDNA Sequences Do Not Differentiate *Gracile* and *Robust* Morphologies."[31] They add that the unique mtDNA of LM3 did not survive (except on chromosome 11) but "the lineages of the alleles contributing to this *gracile* phenotype have survived." In contrast, the mtDNA of the Kow Swamp *robust* people has survived but their *robust* morphology has not survived, "implying that the allelic lineages of many of the genes that contribute to this phenotype have been lost." The result is that "lack of association between the survival of nuclear and mtDNA lineages is expected because they have different transmission patterns between generations." After concluding that there must have been genetic exchange between the LM3 people and those with modern lineages, they state, "Similar exchanges between people with other Pleistocene mtDNA lineages, like that of the Feldhofer Neandertal individual, may have occurred."[32] In other words, the alleged differences between the Neandertal mtDNA and modern human mtDNA mean nothing.

An equally stunning observation is as follows:

> The morphologies of the more *robust* ancient individuals are outside the range of living indigenous Australians, but unlike the situation in Europe, there is a consensus that all prehistoric Australian human remains represent part of the ancestry of living Aboriginal Australians.[33]

Robust ancient Australians, dated less than 10,000 years of age, genetically mixed with anatomically modern Australians, also about 10,000 years of age, to produce the anatomically modern Aboriginal Australians of today. There is universal agreement among evolutionists regarding that fact. Then why couldn't robust Neandertals, dated at about 40,000 years of age, genetically mix with anatomically modern Europeans, also dated at about 40,000 years of age, to produce the anatomically modern Europeans of today? Evolutionists say that it didn't happen. If it happened in Australia, why couldn't it happen in Europe? We do not need African Eve.

Do you understand now why I love Australia?

SECTION VI

The Creationist Dating Revolution

The Copernican Revolution put the sun where it belonged. The Creationist Dating Revolution puts the Bible where it belongs.

A million years here, a million years there—pretty soon we're talking about a lot of time.

INTRODUCTION TO SECTION VI

It's About Time

SOME YEARS AGO, I participated in a Creation Week program in the Chicago area. I was picked up at O'Hare airport by Dr. Henry Morris, then president of the Institute for Creation Research in San Diego. He was on his way into downtown Chicago to participate in an informal debate on creation versus evolution on WGN-TV.

The other participant in the talk-show debate was Dr. Peter J. Wyllie, professor of geological sciences at the University of Chicago and editor of the *Journal of Geology*. I mentioned to Dr. Morris that I had met Wyllie at the annual convention of the American Association for the Advancement of Science (AAAS) in Denver just a month before. Since Dr. Morris had not met Wyllie, he asked me to introduce them when Wyllie arrived at the studio.

While Dr. Morris was being made up for the TV cameras, I told Dr. Wyllie how much I appreciated a recent book of his, *The Dynamic Earth*. I then asked him a question about what appeared to be a serious flaw in the basic assumptions of the radiometric dating methods. Wyllie responded by saying that the radiometric dating methods were not his area of expertise. "You may have noticed," he said, "that I did not go into them in my book." I got the clear impression that Wyllie did not feel comfortable discussing the radiometric dating methods and preferred not to answer questions about them. I did not press the matter.

During the debate, I was amazed that Wyllie's entire argument against creation and for evolution was based on the alleged evidence provided by the radiometric dating methods for the age of the earth and its various rock strata. Wyllie implied that the time frame in Genesis was very wrong and could not be trusted. When the program was over, I remarked to Wyllie, "I was a bit surprised that after telling me that the radiometric dating methods were not your area of expertise, you based your entire argument against creation on them." He just smiled.

I asked him if he was aware of the basic assumptions on which the dating methods were based. "No," he replied, "I remember studying about that in graduate school, but I've forgotten what they are."

It really is about *time.* Nothing has been as effective in robbing people of their faith in Jesus Christ and the Bible as the alleged scientific "proofs" that the Earth is very old and, hence, the Bible is wrong. These "proofs" are based on the radiometric dating methods. Even those who know nothing about them, such as Wyllie, have used them with vicious effectiveness. The alleged long ages "proven by the dating methods" have been more of a threat to biblical faith than the theory of evolution. Without long ages, any concept of evolution would be impossible—and absurd.

Time is an amazing entity. It is probably the first entity created by God. We all think we know what time is, but I challenge you to define it. Defining *time* is probably the most difficult problem in philosophy. However, our concern here is not the definition of time but how much of it God created. Did God create billions of years of time or just thousands of years of it?

It's all about time because time is inseparably involved with most miracles, and the creation of the universe is the king-sized miracle—the granddaddy of all miracles. Those who opt for billions of years in the creation of the universe say that a miracle is a miracle whether it takes six days or fifteen billion years. That statement is both very true and very false. It all depends on what the purpose of a miracle is.

If the pupose of God in creation was just to do stuff, then obviously the time involved doesn't matter. However, creation is always used in the Bible as a testimony to God's power and majesty. Hence, he tells us clearly that he created the universe in six days. That those six days are literal days is seen by the fact that in Exodus 20:8–11 the six days of creation are linked to the six days of our workweek.

Our heavenly Father is the master teacher, and he uses the "show and tell" method. He could have created the universe instantaneously. However, he did it in six days to illustrate how we are to order our days of work and rest. In the same way, he created Eve from Adam's body to illustrate that husband and wife are truly one flesh in marriage. In all that God does, he teaches us the great truths of life.

Christ's miracles were not done just to do stuff. Their purpose was to clearly demonstrate who he was—the Messiah and the Son of God. And almost all of his miracles were time-dependent. That is, the miraculous element was the compaction of time from days, weeks, or months into just seconds. When the purpose of a miracle is to witness to the nature and power of God, time is not neutral. Time is on the side of the skeptic and the unbeliever. Let me illustrate.

In Luke 8:22–25, Jesus and his disciples were crossing the Sea of Galilee. Because of the geography of that area, the Sea of Galilee is often subject to sudden and severe storms. Several of the disciples were commerical fishermen who had spent much of their adult lives on this body of water and knew it well. But they had never seen any storm like this one. They cried to Jesus, asleep in the boat, that they were going to drown. Jesus got up and rebuked the winds and the waves, and the storm subsided. However, the response of the disciples tells the real story. Although some of them had spent much time on the Sea of Galilee, the suddenness with which the storm subsided left no doubt in their minds that Jesus was responsible. "Who is this? He commands even the winds and the water, and they obey him" (v. 25).

To the modern mind, however, Luke's account is unacceptable. It must be rewritten to make it believable. "Jesus stood up, rebuked the winds and the waves, *and twelve hours later* the Sea of Galilee was smooth as glass." No miracle. No problem.

Twelve hours, or less, can turn a miracle demonstrating Christ's authority over nature into a strictly natural phenomenon—the storm system moved off, and a fair weather system moved in to take its place. This is what the injection of time into a time-dependent miracle can do.

To a person of genuine, informed faith in God, it may be true that a miracle is a miracle, no matter how long it takes. But miracles are not primarily for the faithful but for the genuine seeker with honest questions. To humanistic skeptics, however, looking for an excuse to continue trusting in themselves, time is all they need to corrupt or destroy God's testimony. This is exactly what has happened to the biblical doctrine of creation. By the injection of fifteen billion years, the greatest of all miracles testifying to the awesome power and majesty of God has been reduced to a naturalistic phenomenon where God is not even needed.

Others say that Genesis 1 simply tells us that God created the universe. It doesn't tell us how. They seem immune to—if not offended by—the statements in Genesis 1 of the method God used to create. God *spoke*—that is, *commanded*—the universe into existence. To the modern mind, this involves two offenses: first, that God's method by which he commanded the universe into existence was *instantaneous* (Psalm 33:9); and second, that his method of creation took the act of creation *outside of scientific investigation or experimentation.* Some prefer a very small god whom they can put in a test tube over a very great God who, by his Word, can command the universe into being.

We are so obsessed with time and process that it is hard to allow God to be the Creator he claims to be. It reminds me of the little boy who said, "If God created the world, he must have a lot of heavy equipment!" We must leave our human concepts behind when we contemplate the God of creation. The English evolutionary paleontologist E. G. Halstead (Reading University) ridiculed Christians, saying that the kingdom of heaven is like being a good dog, obedient to its master. I frankly don't know what else one would do in the presence of a great God who created the universe except to be obedient. I plead "guilty" to wanting to give my God and Savior that kind of obedience. Halstead is dead now. One can only wonder if he has changed his mind. Yes, it is about time—about a God who is not limited by time and about his creatures who are.

However, the day when the greatest miracle in the Bible—the creation of the universe—has been diluted by the injection of time is coming to a close. Thanks to godly scientists committed to the entire Word of God, solid scientific evidence is now coming forth proving that the long ages given by radiometric dating methods are a fiction. The earth is young, and evolution is dead. However, it may take as long as twenty or thirty years for the "coroner" of the scientific establishment to discover that the "body" has assumed room temperature.

The story is told in this section.

CHAPTER 27

A Twenty-First-Century Scientific Revolution

TO AN EVOLUTIONIST, nothing could be more absurd than using the Bible—that religious book full of ancient myths—for direction in scientific research. But what seemed utterly absurd is now proving to be sober fact. A revolution is in the making. This revolution promises to be the undoing of the Darwinian Revolution. The Darwinian Revolution removed the Bible from all serious study of nature. The new Creationist Dating Revolution is putting the Bible back in its proper place as a historical account of the origin of the universe written by the Maker himself.

Ever since the advent of radiometric dating, no "evidence" has been as potent in causing people to think that the Bible is in error and hence not to be trusted. It would be impossible to calculate how many people have turned away from the Bible and from God because science has "proved" that the earth is very old and that Genesis is not trustworthy.

That old earth "proof" is now being revealed for what it is: a fiction. The credit goes to a group of dedicated creationist scientists, known affectionately as the "Gang of Seven," working on the RATE Project (RATE stands for "Radioisotopes and the Age of the Earth") sponsored by the Institute for Creation Research and the Creation Research Society. The project is scheduled for completion in 2005. Solid, rigorous, scientific evidence is being assembled to prove that the earth is indeed very young and that the Bible is accurate in its portrayal of earth history. In terms of the far-reaching implications, the Creationist Dating Revolution could be as revolutionary as the Copernican Revolution and the Darwinian Revolution.

Of the many lines of evidence produced by the RATE project scientists, I will give three that are the most amenable to a nontechnical explanation.

Eyewitness Testimony versus Radiometric Dating

There is no totally independent verification of any of the dating methods. Evolutionists claim otherwise. But since these methods are based on many of the same basic assumptions, it is impossible for any verification to be independent. It's like the case where the town hall tower clock was being set by the clock in the local church steeple, and the church steeple clock was being set by the clock in the town hall tower.

Why would anyone believe something that is unverifiable by independent means? The Bible has been verified by archaeology thousands of times. The Bible continually claims to be eyewitness testimony. The apostle Paul mentions over five hundred witnesses to the resurrection of Jesus Christ, Paul himself being one of them (1 Cor. 15:1–8). The radiometric dating methods have absolutely no independent verification. Isn't that blind faith in a science god created by humans? People are so fearful of "blind faith" in God and the Bible. Isn't faith in the dating methods blind faith in the word of humans? Do we *know* that the earth is four and one-half billion years old? No, we don't.

An amazing situation develops when rocks that were seen to solidify by humans in historic times are then dated by radiometric means. (The age of a rock is defined as the time when that rock *solidifies* from a molten state. The time of *solidification* is the event that is "dated.") This eyewitness testimony is the type of evidence that even people with no scientific background can understand, and the evidence demonstrates that the dating methods are pure science fiction.

The RATE project is collecting many of these illustrations. I will give you just one, but it is representative of the results that are being obtained.

This particular case is the work of Dr. Andrew Snelling, who was associated with Answers in Genesis, Brisbane, Australia. He is one of the "Gang of Seven" RATE scientists. Details of this case are found in the publication *Creation.*[1]

This illustration comes from one of New Zealand's most active volcanoes, Mount Ngauruhoe, located in the center of New Zealand's North Island. This volcanic mountain is an almost perfect cone, rising more than 3,000 feet above the surrounding landscape and 7,500 feet above sea level. Its central crater is 1,300 feet wide.

This volcano is thought to have been active for at least 2,500 years, with more than 70 eruption periods since Europeans first observed its activity in 1839. The last eruption, in 1975, was one of the most violent. Eyewitnesses described loud atmospheric shock waves and reported blocks of volcanic rock 100 feet in length catapulted an estimated two miles into the atmosphere. The smoky eruption plume rose seven to eight miles high.

A project was begun to compare the eyewitness dates of the lava flows with the dates obtained by radiometric dating of those very same rocks. Rock samples were taken from the hardened lava flows of the most recent eruptions, specifically the eruptions on 11 February 1949; 4 June 1954; 30 June 1954; 14 July 1954; and 19 February 1975. These most recent lava flows were clearly visible and easily identified. All of the volcanic rock material samples were from twenty-five to fifty-one years old.

A total of thirteen samples from these eruptions were sent for whole-rock potassium-argon dating to the Geochron Laboratories, Cambridge, Massachusetts, one of the most respected commercial dating laboratories in the world. The samples were sent progressively. One sample from each of the lava flows was sent initially. When these first results were obtained, a second sample from each of the same lava flows was then submitted. In one case, a third sample was later sent in. Additional pieces from two original samples were also submitted to test the consistency of results within samples.

The laboratory was not given any specific information regarding the source of the rock samples, nor were they given any information as to the expected age of the samples. The samples were described only as probably very young with very little argon in them to ensure that the laboratory would take extra care during analysis.

Although the ages of these rocks, based on eyewitness testimony, were between 25 and 51 years old, the 13 rock samples were dated as follows: Four were dated at "less that 270,000 years old," one was dated at "less than 290,000 years old," one was dated at "800,000 years old," three were dated at "1 million years old," one was dated at "1.2 million years old," one was dated at "1.3 million years old," one was dated at "1.5 million years old," and the last one was dated at

"3.5 million years old." All were said to have a margin of error of about 20 percent in either direction.

Mount Nguaruhoe is just one of many cases in which the radiometric dates differ absurdly from known eyewitness dates. The obvious problem is that argon gas was already in these rocks when they solidified. There is no "time zero" for these radiometric clocks. We are told that these dating methods do not work well with young ages (just hundreds of years or less). They only work with such vast ages that there is no way of checking them out. They do not work where we can check them out. The most intelligent reaction to that type of situation is skepticism.

Radiocarbon Becomes a Born-Again Creationist

Radiocarbon, known also as carbon 14 (^{14}C) is different—in fact, it's downright odd. Whereas all other radioactive elements originated, we assume, at creation, carbon 14 is continually "manufactured" in the earth's upper atmosphere. Carbon 14 is an *isotope* of "regular" carbon. (There are slight variations in the atomic weight of some elements because there can be slight variations in the number of neutrons they have in their nucleus. Because the physical and chemical properties of an element are determined largely by the number of electrons it has, a slight variation in the number of neutrons affects—for our purposes—only the weight. These variations of the same element are called *isotopes.*) Although the normal isotope of carbon has an atomic weight of twelve, it can vary all the way from ten to fourteen. Carbon 14 is the particular isotope that concerns us. There is only one atom of it for every trillion (million million) atoms of carbon 12. The total world supply of carbon 14 at any one time would be only about seventy-nine tons.

Carbon 14 is formed as cosmic rays bombard the earth's upper atmosphere with billion-volt energy. The result of this bombardment is the production of neutrons. Oxygen has no affinity for these neutrons, but nitrogen does. Nitrogen has an atomic weight of 14. The reaction of the neutron with nitrogen 14 involves the loss of a proton and an electron to form carbon 14. The weight of the former nitrogen atom remains the same because the loss of the proton is offset by the gain of the neutron. However, the loss of the electron means that the former nitrogen atom no longer behaves like nitrogen. It is now an isotope of carbon, carbon 14, and behaves like carbon. This is a very unstable form of carbon and eventually the atom "decays" back to nitrogen 14. The time it takes for one-half of any given amount to decay back to nitrogen 14 is

5,730 years. This figure is called the "half-life" of the element, and the process of decay is called *radioactivity.*

After being formed in the extreme upper reaches of the atmosphere, the carbon 14 atoms mix with the gases of the lower atmosphere and soon combine with oxygen to become radioactive carbon dioxide. Because plants take in carbon dioxide, this radioactive carbon dioxide finds its way into all plant life. Since all animals either eat plants themselves or feed on animals that do, the radioactive carbon dioxide eventually finds its way into all living matter. During the lifetime of an organism, it builds up what is called an equilibrium condition—as much radioactive carbon dioxide is taken in by eating and breathing as is given off in waste products and in the decay of the radioactive carbon dioxide itself.

When an organism dies, it no longer takes in carbon 14. The equilibrium condition within it then comes to an end. Only the decay of the carbon 14 continues. By knowing the rate of decay, therefore, one can measure the amount of carbon 14 in a sample one wishes to date and, hopefully, compute how long ago the organism was alive. The decay is not a linear decay but an exponential decay and thus is spoken of in terms of the "half-life." It is also obvious that in order to be dated by carbon 14, the sample must contain carbon—such as charcoal, wood, peat, shell, bone, paper, parchment, and cloth.

Carbon 14 dating has its own unique set of assumptions, some of which are quite tenuous. It also needs to be calibrated with known dates obtained by other accurate dating techniques. But these matters are not relevant to our discussion. What is relevant is the very short half-life of carbon 14—just 5,730 years. From the time that carbon 14 dating was first developed in the mid-twentieth century until about 1980, it was recognized that with a short half-life, use of carbon 14 as a dating system was limited to about six half-lives, or 30,000 to 40,000 years. After that, the little remaining carbon 14 in the sample was very difficult to measure, and the margin of error was great. There was also a great deal of "noise" from cosmic rays. Samples that could not be dated by carbon 14 were assumed to be much older, and other dating methods had to be employed.

About 1980, a breakthrough method was developed for measuring the ratio of carbon 14 to carbon 12 with extreme precision in very small samples, using an ion beam accelerator and a mass spectrometer, called the accelerator mass spectrometer (AMS). This new technique improved the sensitivity of the measurements from about 1 percent of the modern value carbon 14/carbon 12 ratio to about 0.001 percent of the modern value ratio. This, theoretically, extended the range of carbon 14 dating from about 40,000 years to about 90,000 years. Obviously, the hope was to date much older material. However, those hoping so were in for the shock of their lives!

In the words of Dr. John Baumgardner (Los Alamos National Laboratory), one of the RATE scientists, "The big surprise, however, was that no fossil material could be found anywhere that had as little as 0.001% of the modern value!"[2] Lest you miss the full impact of that statement, it means that carbon 14 has been found in fossils all the way down to the Cambrian Period, which were considered by evolutionists to be about 600 million years old. That age is now science fiction.

Baumgardner then gives an incredible illustration.

> If one started with an amount of pure ^{14}C equal to the mass of the entire observable universe, after 1.5 million years [a tiny fraction of evolutionist time] there should not be a single atom of ^{14}C remaining! Routinely finding $^{14}C/^{12}C$ ratios on the order of 0.1–0.5% of the modern value—a hundred times or more above the AMS detection threshold—in samples supposedly tens to hundreds of millions of years old is therefore a huge anomaly for the uniformitarian framework [the evolutionary time scale].[3]

Baumgardner also sent a diamond to a radiocarbon laboratory for dating. This had never been done before because evolutionists would consider it absolutely absurd—sheer lunacy. Diamond, formed deep in the Precambrian rocks of the earth, would be almost as old as the earth itself. Further, diamonds have very powerful lattice bonds, so that no biological contamination could find its way into a diamond's interior. Hence, to date diamond by carbon 14 was assumed to be pointless. The report came back: The diamond was about 58,000 years old—an even smaller fraction of evolutionist time.

For those who may question this whole matter, it is important to know that the data used by the RATE scientists is, except for a small portion of it, not their own. It is data that has been confirmed by dating laboratories and published in radiocarbon journals. Evolutionists have been keenly aware of this anomaly since the AMS technique was developed. They are seeking desperately to explain it. Detectable carbon 14 in very old samples has been called "contamination" or "background," which are meaningless terms. Regarding the writings of evolutionists on this matter, Baumgardner says, "Most of these papers acknowledge that most of the ^{14}C in the samples studied appear to be intrinsic to the samples themselves, and they usually offer no explanation for its origin."[4]

The maximum age for any sample with carbon 14 is 250,000 years. That alone is the death knell for evolution. However, the RATE scientists have been doing more than disproving evolution. They have been analyzing the data, removing speculative and evolutionary suppositions, and inserting biblical data in its place. Baumgardner reveals that what they have found "is consistent with the young-earth view that the entire macrofossil record up to the Cenozoic [the highest major segment of the geologic column] is the product of the Genesis Flood and therefore such fossils should share a common ^{14}C age."[5]

The conclusion: "The bottom line of this research is that the case is now extremely compelling that the fossil record was produced just a few thousand years ago by the global Flood cataclysm. The evidence reveals that macroevolution as an explanation for the origin of life on earth can therefore no longer be rationally defended."[6]

Where's the Helium?

A few years ago, a popular fast-food commercial asked, "Where's the beef?" If you are interested in the age of the earth, the most important question you could ask is, "Where's the helium?"

The basement or foundation rocks of the earth—the whitish or grey granites we often see as tombstones—contain flakes of a mineral called biotite, a black mica. If you look at one of these tombstones, you can see these black flecks. In the biotite are microscopic crystals called zircons. Much of the earth's uranium and thorium is in these zircons. Uranium and thorium are, of course, radioactive. A byproduct of uranium and thorium radioactivity is helium (^{4}He). Another form of helium, ^{3}He, is not the result of radioactivity.

"Helium is a lightweight, fast-moving, and 'slippery' atom, not sticking chemically to other atoms," says Dr. D. Russell Humphreys (Sandia National Laboratory, Albuquerque, NM), one of the RATE scientists.[7] Because of these qualities, helium works its way through solids rather rapidly. It squeezes through the spaces between atoms in the crystal lattice of zircons. The leakage rates are so large that old-earth believers had expected that most of the helium produced through radioactive decay had worked its way out of the earth's crust, into the upper atmosphere, and out into space.

The first surprise: Research shows that helium does not escape into space to any great degree. The second surprise: Helium doesn't rise to the top of the atmosphere but spreads throughout the atmosphere from top to bottom. Atmospheric scientist Dr. Larry Vardiman (Institute for Creation Research, and coordinator of the RATE project) has shown that the earth's atmosphere has only about 0.04 percent of the helium it should have if the earth were billions of years old. Vardiman writes:

> If the earth's atmosphere had no helium when it was formed, the current measured column density of helium . . . would have been produced in about 2 million years. This is over 2500 times shorter than the presumed age of the earth. Long-age atmospheric physicists such as [J. C. G.] Walker state that ". . . there appears to be

> a problem with the helium budget of the atmosphere." [J. W.] Chamberlain states that this helium escape problem ". . . will not go away, and it is unsolved."[8]

Vardiman's comment that the helium in the atmosphere "would have been produced in about 2 million years" does not mean that he believes that the earth is 2 million years old. He is pointing out the fallacy in the evolutionist's time scale. Two million years, in the evolutionary scenario, is totally insignificant. That is roughly the time it allegedly took for *Homo erectus* to fully evolve into *Homo sapiens*. Vardiman believes that the helium in the atmosphere is almost entirely primordial—that it was a part of the creation (origin) of the earth and that very little of it is the result of radioactive decay. The origin of ^{3}He is unknown, and most atmospheric scientists believe that it is primordial, so it is not irrational to believe that most ^{4}He is also primordial.

This helium problem is not a new one. Creationists have been writing about it for many years. Nor is the evolutionist's assumption that helium escapes from the atmosphere into outer space new. In 1957, creationist and chemist Melvin Cook (University of Utah) published an article in *Nature* entitled, "Where is the Earth's Radiogenic Helium?" He stated, "Now the experimental evidence from high-altitude studies indicates that hydrogen and helium do not concentrate significantly in the upper atmosphere as has previously been supposed."[9]

Physicist and creationist Robert Gentry (Oak Ridge National Laboratory) has written extensively on the helium and other related problems in the most respected scientific journals. Neither Cook nor Gentry spelled out explicitly the young earth implications of their studies. Both men knew that to do so was the best passport to being rejected for publication. Gentry felt that if he just kept presenting the evidence in the most respected journals, its impact would be recognized. It was! He was fired from Oak Ridge National Laboratory.

Evolutionists have been working on the helium problem without success. They also have been in denial as to the implications of the problem. None of them dare to consider the possibility of a young earth. To even consider that option would mean banishment from the scientific community and deepest humiliation. If true, a young earth would mean the destruction of their worldview and their humanistic religion. The price is far too high. Sometimes it's better to live with the problem.

There are three assumptions that underlie all radioisotope (radiometric) dating: (1) The decay rate of radioactive atoms into daughter products is constant over time; that is, the "clock" has been ticking at a constant rate; (2) The rock or sample being dated is, for all practical purposes, a closed system with little or no contamination from the environment and little or no leaching out of the rock; and (3) The initial amount of isotopes at the time the rock solidified can be known or reasonably estimated. In other words, God is not the only person who is omniscient. Assumptions (2) and (3) are known to be faulty, but the first assumption has seldom been questioned.

In solving the radiometric dating puzzle, Drs. Russell Humphreys and John Baumgardner felt that there were many lines of evidence indicating that large amounts of radioactivity had taken place in the past. Yet Genesis clearly indicates that the earth is young. In solving this seeming contradiction, an obvious starting place was to reexamine the assumption that the decay rate of radioactive atoms into daughter products is constant over time. They also considered the possibility that vast amounts, perhaps billionfold speedups, of nuclear decay might have taken place in the first three days of creation week, at the fall (the curse on nature), and/or during the Genesis flood.

Old-earth people and evolutionists believe that the earth is billions of years old. In that time, an immense amount of helium would have been produced by radioactivity. That helium has not escaped into space and is not in the atmosphere. Where is it? In the zircons! At Fenton Hill, about twenty miles west of Los Alamos, New Mexico, geoscientists from Los Alamos National Laboratory drilled several miles deep into the hot, dry granite rock. Measurements of uranium, thorium, and lead in the zircons from these granites implied 1.5 billion years of nuclear decay at today's rates.

For some reason, there were no reliable studies on the rate at which helium escaped from zircons and no studies at all for the diffusion rate of helium from biotite. This was the RATE scientists' unique contribution to the problem. The RATE project commissioned a world-class noncreationist expert to measure helium diffusion rates in zircon and biotite samples from the Fenton Hill borehole. They also included data from another site, Gold Butte, Nevada.

At Oak Ridge National Laboratory, Gentry and his team calculated how much helium would have been released in those zircons, based on that alleged 1.5 billion years of decay, versus how much helium was still in the zircons today. They discovered that 58 percent of all the helium produced in that alleged 1.5 billion years of decay was still in the zircons. It was clear that the assumption of a constant, slow decay process was wrong. If the earth was as old as evolutionists claim, very little helium would be left.

The RATE group then had to develop theoretical models in order to interpret the data. The evolution model assumed a 1.5-billion-year time span, with continuous radioactive decay resulting in helium production throughout that time. The creation model assumed a time span of about 6,000 years, "with most of the helium produced in one or more bursts of accelerated nuclear decay near the beginning of that time."[10] Humphreys continues, "The data allow us to calculate how long diffusion has been taking place—between 4,000 and 14,000 years! The diffusion rates are nearly 100,000 times higher than the maximum rates the 'Evolution' model could allow. That leaves no hope for the 1.5 billion years."[11]

Since these zircons come from very deep granites, the Precambrian basement rocks of the earth, these rates mean that the earth itself cannot be older than 4,000 to 14,000 years. Further, there seems to be no alternative interpretation of the data—unless it could be shown that the diffusion rates obtained by that world-class authority are somehow flawed.

In the *Proceedings of the Fifth International Conference on Creationism, 2003,* Humphreys and three RATE associates state that the data do not indicate how many acceleration episodes took place or whether they took place during creation week, at the fall, or during the flood. They continue:

> We can say that the "diffusion clock" requires a large amount of nuclear decay to have taken place within thousands of years ago, after the zircons became solid. At whatever time in Biblical history Precambrian rocks came into existence, these data suggest that "1.5 billion years" worth of nuclear decay took place after the rocks solidified not long ago.[12]

Do you remember Archbishop James Ussher? He lived from 1581 to 1665. He's the one responsible for that BC 4,004 date for creation in many of the older Bibles. In evolutionist literature and lectures on evolution, no one is more ridiculed, hated, or humiliated than Ussher. Although his date may be a bit off, he is still more than one million times closer to the date of creation than any evolutionist living today.

The Genesis Flood: The Geophysics of God

Long before John Baumgardner was a member of the RATE team, he was up to other mischief. The 16 June 1997 issue of *U.S. News & World Report* ran a four-page story on Baumgardner.[13] He is in the theoretical division of the Los Alamos National Laboratory, U.S. Department of Energy. The magazine story is about a computer program called Terra that Baumgardner designed. Baumgardner is described as "the world's pre-eminent expert in the design of computer models for geophysical convection." This involves the processes of volcanoes, earthquakes, and the movement of the earth's continental plates.

Although they may not agree with him on Genesis, his peers have high regard for him and his work. Geophysicist Brad Hager (Massachusetts Institute of Technology) is quoted as saying, "There is universal agreement that Terra . . . is one of the most useful and powerful geological tools in existence." Earth scientist Gerald Schubert (University of California, Los Angeles) adds, "As far as the code [Terra] goes, Baumgardner is a world-class scientist."

Are you ready for still another shocker in this chapter? The magazine article continues, "In fact, Baumgardner created Terra expressly to prove that the story of Noah and the flood of Genesis 7:18 . . . happened exactly as the Bible tells it . . . his 'numerical code' [Terra] actually proves the Bible is correct." But the author of the article, to maintain his respectability, had to add this disclaimer: "Or at least in Baumgardner's view it does." If Terra works with everything else, why wouldn't it work with the flood?

One wonders how much richer our lives would be if more people considered the Bible worthy of giving direction in research—as well as in life.

CHAPTER 28

THE DATING GAP

Laying a Really Large Egg

CHAPTER 27 MAY BE too revolutionary for some readers to accept. The reaction could be all the way from "It's too good to be true" to "Anything involving young-earth creation scientists is automatically suspect." I understand. If you are in the second group, this chapter is for you. Could the entire evolutionist establishment be wrong? When a philosophical belief is based on a false foundation, the answer is yes. It has happened before. Former premier Gorbachev, in a speech at Columbia University in 2002, stated that the Soviet communism he served most of his life was pure propaganda.

If recognizing that all of the radiometric dating methods are inherently flawed is too hard a step to take all at once, I offer a smaller step. It is the fact that even if we assume, for the sake of argument only, that evolution is true and that the radiometric dating methods are legitimate, the dates used to establish human evolution are, for all practical purposes, worthless.

Evolution places severe demands on fossils used to support it. A fossil in an evolutionary sequence must have both the proper morphology (shape) to fit that sequence and an appropriate date to justify its position in that sequence. Since the morphology of a fossil cannot be changed, the date is obviously the more subjective element. Yet accurate dating of fossils is so essential that the scientific respectability of evolution is contingent on fossils having appropriate dates.

In past chapters we have seen "The Fake Parade,"which was allegedly supported by proper fossils and dates. Evolutionists now admit that it was false. The newer African Eve scenario also depends heavily on proper dates to deliver it from allegations of racism. Because of severe dating uncertaintics that are seldom mentioned, however, that alleged scenario is similarly false. It is now clear, to those willing to face facts, that to use fossil evidence in support of human evolution is akin to intellectual dishonesty.

Even if we assume that the dating methods are legitimate, it is impossible to give a legitimate fossil sequence to support human evolution. This is because there is a coverage gap involving the dating methods. This gap involves the methods evolutionists believe are the most reliable—radiocarbon and potassium argon (K-Ar). This coverage gap occurs during the very heart of the time period most crucial to human evolution—the period from about 40,000 years ago to about 200,000 years ago on the evolutionist's time scale. It covers the period known as the Middle Paleolithic in Europe and Asia, roughly the same as the African Middle Stone Age (MSA).

This coverage gap lies beyond what is considered the effective range for radiocarbon and prior to what is considered the effective range of potassium-argon. This problem period may be even larger because: (1) Some dating authorities believe that the effective range of K-Ar doesn't begin until about 400,000 years ago; and (2) Many fossils have been found at sites that lack the volcanic rocks necessary for K-Ar dating, and hence they cannot be dated accurately by this or any other method.

In the journal *Science*, Alison Brooks (George Washington University) and her associates state that the origin of anatomically modern humans is "controversial." The problem is the dating of "key hominid fossils from Africa and the Near East." They continue:

> Most of these finds fall in a "gap" between the currently effective ranges of radiocarbon and potassium-argon dating procedures. In the age range from 40 to 200 ka (thousand years ago), many paleoanthropological finds have been *provisionally* dated by correlation with the oxygen isotope record or recognized glacial stages on the basis of stratigraphy, biostratigraphy, or paleoenvironment. Such dating is *uncertain* because of possible microenvironmental variability and problems with low-latitude climatic models of the last 200,000 years (emphasis added).[1]

Brooks *et al.* suggest a new method of dating fossils and archaeological material in this time period. It involves the racemization of amino acids in ostrich eggshell. The amino acid method was developed some time ago for dating bone material at archaeological sites. But because bone is porous, it is subject to groundwater leaching. Hence the method fell into disfavor because it gave questionable dates. However, because ostrich eggshell is very dense, it is thought to be a rather closed system. It is claimed that items found in association with this eggshell can be more accurately dated by the amino acid racemization of ostrich eggshell than by any other method. Ostrich eggshell is common at African sites as well as many Asian sites.

(Amino Acid Racemization—In all living protein, all amino acid molecules have a left-handed "twist." This condition is strong evidence for design because: (1) It is very difficult to duplicate in the laboratory because a normal laboratory mixture would be 50 percent left-handed, 50 percent right-handed; (2) The presence of even one right-handed amino acid in a cell would bring all cellular activity to a halt. It would be like trying to run an engine with 999 round pistons and 1 square piston. At death, the left-handed amino acid molecules begin to convert to a racemic mixture: 50 percent left and 50 percent right. It is believed that under certain conditoins, an organism can be dated by the ratio of left-handed to right-handed molecules it contains.)

Several points are fascinating about this Brooks article. First, creationists have noted an interesting pattern in evolutionist writings regarding dating methods: Shortcomings of the dating methods in current use are not generally acknowledged by evolutionists. Only when evolutionists feel that they have a better method for dating a specific time period do they publicly admit the weaknesses of the methods they have used previously. The result is that the public assumes that the dating methods used at any given time are adequate, whereas dating specialists know this is not true. This case is just the most recent illustration of evolutionists admitting the uncertainties of older dating methods only after "better" ones have been developed.

The second point is that to now claim to have a "better" method for dating human fossils discovered in the future does not correct the inaccurate dates of the human fossils discovered in the past. It is virtually impossible to rectify the human fossil dating errors of the past because: (1) Many of those fossil sites have been destroyed or altered so that reconstruction to allow for redating of fossils after the fact is impossible; (2) To obtain ostrich eggshell that can be shown to have been in unquestioned association with those previously discovered fossils is also virtually impossible; and (3) At many human fossil sites there is no ostrich eggshell to date.

How serious is this dating gap? The period of this gap is the very time during which modern humans are alleged to have evolved from their more primitive human ancestors. At least 445 human fossil individuals fall in this time period and hence are questionably dated. The fossil categories involved are:

Anatomically modern *Homo sapiens*—a minimum of 75 fossil individuals
Neandertals and pre-Neandertals—a minimum of 319 fossil individuals
African and Asian early *Homo sapiens*—a minimum of 32 fossil individuals
African and Asian *Homo erectus*—a minimum of 19 fossil individuals

The third point is that the uncertainty of fossil dates in the Middle Paleolithic (or Middle Stone Age) gap is just the tip of the iceberg. The problem is far more serious than all but a few evolutionists are willing to acknowledge. William Howells (Harvard University) states that the problem involves the entire Middle Pleistocene, from 100,000 to 700,000 years ago. Howells writes, "It cannot be too strongly emphasized how much uncertainty attaches to placement of all but a few of the fossils, absolutely or relatively, especially for the Middle Pleistocene."[2] In fact, Howells's statement "especially for the Middle Pleistocene" sounds as if the problem goes even beyond the Middle Pleistocene. If we include the rest of the fossil individuals that are in the Middle Pleistocene and have dating uncertainties, as Howells suggests, we have a minimum of 626 fossil individuals in the following categories:

Anatomically modern *Homo sapiens*—a minimum of 79 fossil individuals
Neandertals and pre-Neandertals—a minimum of 405 fossil individuals
African and Asian early *Homo sapiens*—a minimum of 54 fossil individuals
African and Asian *Homo erectus*—a minimum of 88 fossil individuals

Even assuming, for the sake of argument, the validity of evolution and of the radiometric dating methods, it is impossible to develop a sequence to support human evolution with 445 to 626 fossils questionably dated. Some might argue that all of these fossils cannot be improperly dated. That may well be true. But how can one demonstrate which fossils are properly dated and which are not?

In very recent years, refinements have been made in both radiocarbon and potassium-argon dating in an effort to cover the gap. However, dating in that period is still far from secure. In 1999, Klein wrote that the potential range for radiocarbon goes out to 40,000 years using conventional methods and 100,000 years using linear accelerators. But he adds:

> The bottom line is that pending fuller, more secure calibration, radiocarbon dates that antedate 22 ky [thousand years] ago are more useful for arranging

> sites relative to one another than for placing them in absolute time. The most general point is that no radiocarbon date implies a true calendar age unless it has been calibrated, and this is currently impossible before 22 ky ago.[3]

A more recent article in *Science* states that radiocarbon dates have only been calibrated back to 24,000 years.[4]

Regarding potassium-argon dating, Klein remarks:

> The very long half life of ^{40}K means that the radiopotassium method has no practical maximum limit (it can be used to estimate the age of the earth) but that in most cases it cannot be used to date rocks younger than a few hundred thousand years old. This limitation occurs because they contain too little ^{40}Ar for accurate measurement and because the statistical error associated with the age estimate therefore may be as large as the estimate itself.[5]

The inability of the radiocarbon and K-Ar methods to cover this time period explains why many alternative dating methods have been devised to attempt to give coverage. None of these are considered to be as accurate as radiocarbon or K-Ar, however, and each of these alternative methods has serious problems of its own.

The concept of human evolution demands precise dating of the relevant fossil material. Evolutionists now admit that the dates for the human fossils in the most significant Middle Paleolithic/Middle Stone Age period and elsewhere are uncertain. This means that there is no such thing as a legitimate evolutionary fossil sequence leading to modern humans. Also, evolutionists cannot make accurate statements regarding the origin of modern humans based on dates of fossils. That they continue to do so reveals that their statements are based on a belief system, not on the practice of a rigorous science.

Not many fossils related to the origin of modern humans have been discovered since 1990, and not many of those recently discovered fossils have been able to be dated by ostrich eggshell. Further, except for the Border Cave, South Africa, site, very few older sites have been redated by ostrich eggshell. This method is not the savior of human evolution for which some had hoped.

The stark reality is that there is no legitimate fossil evidence leading up to modern humans. We are alone. But that's not a problem because we have a God who loves us.

SECTION VII

REALITY!

“There is no use trying,” said Alice; “one can’t believe impossible things.” “I dare say you haven’t had much practice,” said the Queen. “When I was your age, I always did it for half an hour a day. Why, sometimes I’ve believed as many as six impossible things before breakfast.”

—LEWIS CARROLL, *ALICE IN WONDERLAND*

“I don’t want something complicated, Baby; I just want something simple—like the truth.”

—RADIO TALK SHOW SONG

INTRODUCTION TO SECTION VII

"Get Real!"

"GET REAL!" IS AN expression frequently heard. It means "Face the facts," or "Stop living in denial," or "Come back to planet Earth from wherever you are in outer space." Yet those who use the expression seldom define what reality is. And believe it or not, there is a question about it.

The Museum of Modern Art, in the La Jolla section of San Diego, is on one of the most beautiful coastlines in all of Southern California. Yet a sign in that museum states that *true reality* is not to be found outside of the museum. *True reality* is to be found in the art objects within the museum.

Philosophers are divided on whether or not the world outside that museum is real. Some would say that both the art objects in the museum and the world outside the museum—including those things we see, touch, smell, and hear—are all part of objective reality. Others say that true reality exists only in the mind—or elsewhere. What we see, touch, smell, and hear are not physical objects reflecting reality. They are only "sense-data" that are not necessarily related to reality.

In direct conflict with the historic Western worldview in which both God and the world are very real, two popular philosophies consider a part of that historic worldview to be unreal. The first one considers everything to be god and the world to be unreal; the second one considers only the world to be real and God to be unreal—or at best irrelevant.

The first worldview in direct conflict with the historic Western one involves the Eastern pantheistic religions and the recent infestation of those religions known as New Age. These faiths consider everything to be god. There are many variations on this pantheistic theme. However, since there are obvious imperfections in the world, the world—or at least our perception of it—is to some degree unreal. But because the world *seems* very real to us, pantheists liken it to a dream. A dream also seems very real, but when we awaken, we are aware that it was just a dream. In the same way, they say, when we get into god and reach enlightenment, we will recognize that the present world was just a dream.

Genesis 1:1 clearly teaches that God created the universe. By definition, whatever or whoever is eternal is God. Since God is eternal and the universe had a beginning, it is obvious that God and the universe are not the same entity. Further, since God *created* this objective entity called the universe, this act constitutes proof that the universe is real. One cannot create something that is not real. Since we humans would have no mechanism to discern reality from unreality, God created us "in his image." This "image" allows the human mind to recognize the existence and nature of God through his creation of the objective universe. Thus we can be certain that the world we are living in is real (Rom. 1:18–21).

The other worldview in conflict with the historic Western one is cosmic evolution—with biological evolution as an intrinsic part. Here, the universe is all there is. It represents the sum total of reality. The proper philosophic term for this view is *Naturalism.* Since the universe is self-contained and a closed system, God is unreal—or simply irrelevant. At best, God is just an optional part of that system—an add-on. This is why theistic evolution is, biblically, a contradiction in terms.

However, not only does evolution make God unreal, it makes the world itself somewhat unreal by postulating a false view of its origin and the history of life. This is not by accident. Whatever Darwin's scientific motives might have been, it has recently come to light that his greater motivation was a philosophical motive to "ungod" the universe. (See *Charles Darwin and the Problem of Creation,* by Neal C. Gillespie [Chicago: The University of Chicago Press, 1979].) To "ungod" the universe was Darwin's way of attempting to escape reality.

It also recently has been admitted that Darwin's "proof" was actually philosophical reasoning without a great deal of scientific basis. I quote from the most prestigious recent evolutionist, Ernst Mayr (Harvard University): "One must grant Darwin's opponents the validity of two of their objections. First, Darwin produced embarrassingly little concrete evidence to back up some of his most important claims" (*Nature* 248, 22 March 1974, p. 285). The evidence for evolution has *never* been strong. Nor is it strong today.

To live in reality is to recognize the creation of the universe by God, the creation of humans in his image, the entrance of sin into the world through Satan, the death of Christ on the cross as our substitute and as payment for our sins, the triumph of Christ over death by his resurrection from the dead, the availability of eternal life to all who will trust in Christ as their Savior, and the certainty that those who will not do so must pay for their own sins—a horrible and totally unnecessary fate. Reality is recognizing that there is no way we can escape our responsibility to God.

Welcome to reality!

CHAPTER 29

THE PRETEND HUMANS

The Nonhuman Fossil Primates

He's got 'em on the list—he's got 'em on the list;
And they'll none of 'em be missed—they'll none of 'em be missed.

—CHORUS, "THE MIKADO"

IF "AUSTRALOPITHECINE" SOUNDS LIKE "Australian," there is a reason. The Latin word *australo* means "southern." The first fossil australopithecine was so named because it was found in South Africa by Raymond Dart. The word *pithecus* means "ape." Hence the word *Australopithecus* means "southern ape." Many australopithecine fossils have also been found in east Africa, but the name remains. The name "Australia" comes from that same word, *australo*, because Australia is the "southern continent." Other than that, there is no relationship between the australopithecines and Australia. The australopithecines are strictly extinct, nonhuman, African primates. The fossil taxon known as *Homo habilis* is simply a gang of australopithecines in disguise.

Homo habilis: The Little Man Who Isn't There

Necessity is the mother of invention. In the 1960s, human evolution needed a transitional form. Evolutionists needed to bridge the gap between the extinct gracile australopithecines (which everyone acknowledged to be nonhuman) and *Homo erectus* (which everyone acknowledged to be human). The gracile australopithecines were small—although how small they were would not be revealed until 1986, when Tim White found Olduvai Hominid 62. It was an adult, and it was just three feet tall. On the other hand, *Homo erectus* was large—although how large he was would not be recognized until 1984 when Alan Walker (Pennsylvania State University) found the skeleton of a teenage boy who, when fully grown, would have been almost six feet tall (KNM-WT 15,000). As Ian Tattersall put it, "Between *A. africanus* and *H. erectus* there yawned a huge gulf that begged to be filled."[1] And filled it was.

Working at Olduvai Gorge, Tanzania, in the 1960s, Louis and Mary Leakey found the remains of a type of fossil individual that seemed to be a good candidate for human ancestry. Louis gave these fossil individuals some colorful names. There was "Cinderella" (Olduvai Hominid 13), "George" (Olduvai Hominid 16), "Twiggy" (Olduvai Hominid 24, named after a flat-chested English actress because the skull was crushed flat when it was found), and "Johnny's Child" (Olduvai Hominid 7, so named because it was found by Louis's son Jonathan). These fossils consisted of isolated cranial fragments, hand bones, and foot bones. The foot bones seemed to indicate bipedality. The hand bones seemed to indicate manual dexterity. The associated stone tools, formerly attributed to "Zinjanthropus" (see chapter 2), were now ascribed to these newly discovered individuals.

In 1964, Louis Leakey, Phillip Tobias (University of the Witwatersrand, South Africa), and John Napier (University of London) announced in *Nature* a new human ancestor: *Homo habilis*.[2] The name means "handy man" or "man with ability." Since some of the fossils were found in Bed I, they were dated at 1.8 Mya.

From the start, the fossils were the subject of intense controversy. Some felt they were just a mixture of australopithecine and *Homo erectus* fossils. Hence they did not constitute a new taxon. Even those who were sympathetic to the new category recognized that the fossils were a mixture of juvenile and adult material, and juvenile material is very difficult to evaluate because the bones change a great deal as one moves toward adulthood.

Early on, some evolutionists claimed that *Homo habilis* had been launched mainly on the power of Louis Leakey's personality. Those of us who have heard Louis lecture will readily admit that he had the charisma to do it.

The problem of *Homo habilis* was solved in 1986. Tim White, working with Don Johanson and others at Olduvai Gorge, discovered a partial adult skeleton designated Olduvai Hominid 62 and dated at about 1.8 Mya.[3] The cranium and teeth of Olduvai Hominid 62 are very similar to the smaller *Homo habilis* skulls, KNM-ER 1805 and 1813 and Olduvai Hominid 24. Thus it was classified with them. However, it was the *first time* that postcranial material had been found in unquestioned association with a *Homo habilis* skull. The surprise was that the body of this *Homo habilis* adult was not large, as *Homo habilis* was supposed to be. It was actually smaller than the *Australopithecus afarensis* fossil known as Lucy, just a bit more than three feet tall and rather apelike.

Thus there was strong evidence that the category known as *Homo habilis* was not a legitimate taxon but was composed of a mixture of material from two (possibly three) separate taxa—one or more as small as Lucy and another a bit larger. This discovery should also have removed the taxon *Homo habilis* as a legitimate transition between the australopithecines and *Homo erectus*, but it didn't.

With the discovery of Olduvai Hominid 62, Louis Leakey's audacious claim of having found the "missing link," that elusive transitional form between the australopithecines and humans, came crashing down in flames. The reason is clear. With absolutely no proof that the various fossil fragments ascribed to *Homo habilis* actually belonged together, he tried to fill a need by making various bits and pieces into a person. It didn't work. It is another example of the old geological proverb, "If I hadn't believed it, I wouldn't have seen it."

Bernard Wood (George Washington University, Washington, D.C.), perhaps the world's leading authority on evolutionary family trees (known technically as phylogenies), suggests that none of the *Homo habilis* fossils represent human ancestors. He goes on to say, "The diverse group of fossils from 1 million years or so ago, known as *H. habilis*, may be more properly recognized as australopithecines."[4]

Yet more than forty years after the first alleged *Homo habilis* fossils were discovered, paleoanthropologists are still attempting to invent a transitional form out of *Homo habilis*. Richard Leakey describes the problem:

> Of the several dozen specimens that have been said at one time or another to belong to this species, at least half probably don't. But there is no consensus as

to which 50 percent should be excluded. No one anthropologist's 50 percent is quite the same as another's.[5]

The tremendous need for that transitional form between the australopithecines and humans still persists. Milford Wolpoff (University of Michigan), who both needs *Homo habilis* and believes in it, states it well. "The phylogenetic outlook suggests that if there weren't a *Homo habilis* we would have to invent one."[6]

Other Pretend Humans

To give extensive coverage to the australopithecine fossil record is unnecessary. The australopithecines are simply extinct primates. The fact that *sapiens*-like fossils have appeared in the fossil record before the australopithecines and lived as contemporaries with the australopithecines throughout all of their history reveals that the australopithecines had nothing to do with human origins. Australopithecine authority Charles Oxnard (University of Western Australia) concludes, "The genus *Homo* may, in fact, be so ancient as to parallel entirely the genus *Australopithecus*, thus denying the latter a direct place in the human lineage."[7] The composite fossil chart on page 337 shows this to be true.

Other paleoanthropologists have expressed their belief that none of the australopithecines are legitimate human ancestors. Surveying one hundred years of paleoanthropology, Matt Cartmill (Duke University), David Pilbeam (Harvard University), and the late Glynn Isaac (Harvard University) observe, "The australopithecines are rapidly sinking back to the status of peculiarly specialized apes."[8]

The dramatic discovery of KNM-WT 17000, a super-robust australopithecine dated at about 2.5 Mya, has brought chaos to the australopithecine family tree. The super-robust form (*boisei*) was originally thought to have evolved from *robustus* and to have become extinct at about 1 Mya. The fact that super-robust forms were on the scene much earlier than *robustus* came as a shock. Turmoil surrounds both the East African and the South African robust australopithecines. R. A. Foley (Cambridge University) writes of them, "They are almost certainly not human. They are probably not robust, and they are possibly not even australopithecines. Worse still, it may be that they should not be treated as a group at all."[9]

Because of the increasing uncertainty regarding the robust australopithecines, more and more paleoanthropologists are referring to them by their older name, *Paranthropus*, indicating that they belong in a separate extinct genus.

The case for the australopithecines as human ancestors has been based on three claims made for them by evolutionists: (1) They were relatively big brained; (2) They were bipedal; and (3) They appear in the fossil record at the relevant time.

The unique distinction between humans and animals is ignored by most evolutionists. It is that humans are created in the image of God. Besides that spiritual distinction, it is obvious from the charts in this book that the australopithecines do not appear in the fossil record at the relevant time. They are far too late. Furthermore, although brain organization is more important than brain size alone, the significant gap between the cranial capacities of the largest australopithecine and the smallest human, fossil or living, has not been bridged. There is not a smooth transition from nonhuman to human fossils in this regard. And the scientific evidence that the dating methods are a fiction and that the earth is young (section VI) makes such evolution impossible.

The evidence for australopithecine bipedality is controversial. While there is strong evidence that australopithecine locomotion was significantly different from that of humans or other primates, the issue is irrelevant.[10] Bipedality does not prove a human relationship. The birds are bipedal, but no one suggests that they are closely related to humans. Evolutionists make much of the alleged australopithecine bipedality because to make a case for human evolution they must demonstrate the origin of bipedality from a primate stock. Unfortunately, the australopithecine "evidence" comes far too late in the fossil record. As shown by the Laetoli footprints (discussed in chapter 32), when the australopithecines first appear in the fossil record, true humans were already walking.

For many years, paleoanthropologists have commented on the strange lack of chimpanzee and ape fossil ancestors. Sean B. Carroll (University of Wisconsin) observes, "In fact, there are no identified archaic chimpanzee fossils."[11] How strange! Yet evolutionists claim to find more and older alleged human ancestors. They have yet to figure out that what they are finding are extinct species of chimpanzees and apes—not human ancestors.

CHAPTER 30

NEVER DISCUSS THEOLOGY WITH A CHIMP

So God created man in his own image, in the image of God he created him; male and female he created them.

—GENESIS 1:27

We are primates, we are of nature not above nature. Today researchers commonly look to monkeys and apes to try to deduce the attributes that humans held at an earlier stage of their evolutionary history.[1]

—PAT SHIPMAN (PENNSYLVANIA STATE UNIVERSITY)

A RECENT ARTICLE IN *Science* begins with this sentence: "For almost 30 years, researchers have asserted that the DNA of humans and chimps is at least 98.5% identical." Another journal, *NewScientist*, calls this "a figure touted so widely it has almost become a mantra."[2] In the many times you have read or heard that claim, have you ever questioned it?

The science on which that figure was based seems sound. First, it gives a mathematical figure, even including a decimal point, so it has the aroma of accuracy. We moderns are impressed with figures, especially if they contain decimal points. And the phrase "at least 98.5%" suggests that scientists were being cautious on the conservative side. Also, the material being measured is DNA, the very stuff of life. You can't get any more basic than that! Further, since we are told how the scientific community works, we assume that the 98.5 percent figure was not released to the public until it had been checked many times by other scientists.

Yet a thinking person has some legitimate reasons to question that claim. The first reason is logical. Anyone comparing a human with a chimp could sense instinctively that there is far more than a 1.5 percent difference between the two. However, some could say that what a person calls "logical" in this matter might depend on one's philosophical viewpoint.

On a more objective note, the second reason is that the "first draft" of the entire human genome, the project of sequencing or mapping the entire human DNA, was not completed and published until 2001. Two separate teams of investigators published their separate results, one in the journal *Nature* and the other in the journal *Science*. The two results had considerable differences. Further, a project by a five-nation consortium to sequence the full genome of the chimpanzee was not even organized until 2001.

Is it politically incorrect to ask, "If the first draft of the human genome project was not completed until 2001 and the chimpanzee genome project is still in its initial stages, on what basis was the 98.5 percent DNA similarity between humans and chimps published back in the early 1970s?" To my amazement, I am not aware of anyone in the scientific community who has asked that question.

Allow me now to give the *entire* quotation, from the journal *Science*, containing the sentence with which I began this chapter.

> For almost 30 years [from about 1972], researchers have asserted that the DNA of humans and chimps is at least 98.5% identical. Now research reported here last week at the American Society for Human Genetics meeting [in Baltimore, October 2002] suggests that the two primate genomes might not be quite as similar after all. A closer look has uncovered nips and tucks in homologous sections of DNA that weren't noticed in previous studies.[3]

The new figure, 94.5 percent or less, was proposed by the same man who many decades ago had published the 98.5 percent figure, Roy Britten (California Institute of Technology). Britten's original 98.5 percent figure compared only one type of variation, called "single base variation," in which a single "letter" of the genetic code is different in corresponding strands of DNA from humans and chimps.

However, there are two other major types of variation that Britten ignored in his original analysis. These types involve large sections of many "letters" of DNA, called insertions and deletions. When Britten factored these types of variation in, they added 4 percent to the difference between humans and chimps. Further, this difference is based on only about one million DNA bases out of the estimated three billion that make up each of the human and chimp genomes. Britten says of his estimate, "It's just a glance."[4]

I am not a geneticist nor a biochemist, but I am mystified as to why vast sections of DNA that were either insertions or deletions were ignored in the comparison of humans and chimps. It's like comparing elephants and tigers but ignoring the trunk of the elephant because tigers don't have trunks. Obviously, comparing a trunkless elephant to a tiger reduces the differences between them.

Although *NewScientist* states that Britten "ignored" the two other types of variation and *Science* says that these other types "weren't noticed," there are two other possibilities: (1) The added genetic information utilized in the three current evaluations was not available then; and/or (2) Britten made the estimate on preliminary information because he had a political and social agenda, namely evolution.

Scientists are not immune to fads. A few years ago, the fad was to emphasize the similarities between humans and chimpanzees. Robin Dunbar (University of Liverpool), reviewing a book stressing the vast differences between humans and chimps, writes:

> Notwithstanding the enthusiasm in the 1970s and 1980s for the similarities between humans and our primate cousins, both in popular culture and among academics, the fact is that humans are very different from even our ape sister species.[5]

Claims of a close relationship between humans and chimpanzees based on genes are becoming more and more absurd. The mouse genome was sequenced in late 2002. Both humans and mice have about 30,000 genes. An editorial in *NewScientist* states, "What's the difference between Stuart Little and William Shakespeare? Answer (to a very rough approximation): about 300 genes."[6] Alison Abbott, writing in *Nature*, adds "The two genomes, it turns out, a remarkably similar: 99% of mouse genes have a direct human counterpart."[7] The next time your friendly neighborhood evolutionist claims that you are closely related to a chimpanzee, tell him that he is more closely related to a mouse. Could anything be more obvious than that human uniqueness is not to be found in the genes alone?

All science is tentative, and all scientists are fallible—like the rest of us. Because we humans are so thankful for the many blessings of modern science and technology, however, we tend to think of scientists as gods and of science as the conveyer of all truth. Philosophers of science readily admit the limitations of the scientific methods, but working scientists, who seldom study philosophy, seem strangely unable to sense at what point rigorous scientific fact ends and wishful thinking begins.

Conflicted thinking fills the minds of many Christians. On the one hand, even though they know that science is a fallible human activity, some will still accept as authoritative the statements of scientists about our evolutionary origins. At the same time, these same Christians, claiming that Jesus Christ is *Lord*, will equivocate regarding the authority of his clear statement, "Haven't you read . . . that at the *beginning* the Creator made them male and female?" (Matt. 19:4, emphasis added). Since Christ is God, is all-knowing, and proved his authority over all of nature by rising from the dead, to claim that he just adapted his words in a prescientific age or to a prescientific people is an insult to both Christ and the Word of God. He is the one who said, "Heaven and earth will pass away, but my words will never pass away" (Matt. 24:35).

In the real world, it is impossible for two opposite concepts to both be true at the same time. We cannot have evolved from a chimpanzeelike transitional form and at the same time have been created by God in his image. One concept or the other is obviously false.

Who Am I? What Am I?

We humans are unique, and even a thoughtful evolutionist like Juan Luis Arsuaga must acknowledge it:

> We are unique and alone now in the world. There is no other animal species that truly resembles our own. A physical and mental chasm separates us from all other living creatures. There is no other bipedal mammal. No other mammal controls and uses fire, writes books, travels in space, paints portraits, or prays. This is not a question of degrees. It is all or nothing; there is no semi-bipedal animal, none that makes only small fires, writes only short sentences, builds only rudimentary spaceships, draws just a little bit, or prays just occasionally.[8]

What does it mean to be human? Anatomically, we are vertebrates, we are mammals, and we are primates. But because we were created in the image of God, there are light-years of distance between humans and all other members of the animal kingdom. Only humans are a combination of the physical and the spiritual. Only humans have a moral component in their natures and a sense of responsibility toward God. Only humans have the capacity for a relationship with God. Only humans will live forever somewhere—their specific change of address depending on their relationship to God through Jesus Christ.

Evolutionists struggle with three unsolvable problems. (1) What are the unique differences between humans and the other animals? This question can *never* be answered correctly by them because they deny that humans have a spiritual component. (2) What caused a particular line of primates to evolve into humans? That question can *never* be answered correctly by evolutionists because evolution never happened. (3) What does the word *human* mean? Not only do evolutionists not have a generally accepted definition for the Neandertals, for the early *Homo sapiens*, or for *Homo erectus*, they do not even have a definition for humans. Could anything demonstrate more convincingly that evolution does not bring clarity or understanding? It only spawns confusion.

Chris Stringer, speaking of fossil humans, states, "There is an ongoing debate about the concept of modernity [what it means to be fully human], in terms of both morphological and behavioural characteristics."[9]

Ian Tattersall confesses:

> There is no universally agreed definition of what "human" means. The word was invented before people knew anything about the apes, let alone before anybody had any concept that we had close extinct relatives. So "human" is a very elusive term. And we do all tend to use it a little loosely—I certainly tend to use it rather loosely.[10]

In a *Peanuts* cartoon strip, Charlie Brown is talking to his dog, Snoopy. Charlie Brown, deep in thought, says, "You dogs are so lucky. You don't have to worry about things like sin and salvation." Snoopy replies, "Yes, theologically speaking, we dogs are off the hook." Evolution reduces humans, theologically, to the level of dogs by attempting to take us "off the hook" of our responsibility and relationship to God.

It is fascinating that the Word of God gives only one unique distinction between humans and animals—our being created in the image of God. However, that image is not specifically defined. Since the divine image was an integral part of human nature at creation, obviously something of it was retained even after the fall (Gen. 9:6). More of the divine image is restored when a person receives the salvation provided through Christ's death (Col. 3:10). Two of the greatest theologians of the church, John Calvin and Jonathan Edwards, believed that the divine image originally had two components: (1) a moral component having a disposition toward holiness and love and (2) a natural component of reason and will.[11,12] The moral component was largely lost in the fall. The component of reason, although affected by the fall, was retained.

Our being created in the image of God has incredible implications. These implications involve brain size and quality, language ability, and culture. We will discuss the first two now, but since we emphasized human culture in discussing the Neandertals and *Homo erectus*, we will not address it again here.

Brain Size and Quality

The human brain is the most highly organized aggregate of matter in the universe. Scientists originally thought that the brain was organized or designed like a computer. Now we know that the human brain is actually a computer network with 100 million neuron computers. With that incredible degree of design, it doesn't seem that recognizing design in the universe (or recognizing God) should be a problem—but to some it is.

In relation to human body size, the human brain is universally recognized as much larger than is necessary to carry out our physical functions. The reason is this: By creating us in his own image, God intended humans to know him and to have a relationship and intimate fellowship with him (Genesis 2). It is normal for humans to seek to comprehend the God who created them and to contemplate his glory, grandeur, and majesty. Based on God's revelation in the Bible, we are able to think God's thoughts after him. Such a task requires a much larger brain than the size needed for the operation of the human body alone. This fact is certainly adequate to explain our disproportionately large brain in relation to our body size.

Based on Romans 1:18–20, theologian Jonathan Edwards concluded that humans' reasoning ability is qualitatively very much like God's, although quantitatively it is infinitely smaller.[13] Because God made our reasoning ability like his own, he can communicate with us, and we with him. This explains the triumphs of the human mind in every area of human achievement. The fact that many of these accomplishments are used for evil and that the human mind tends to worship itself rather than the God who created it is irrelevant at this point. It is simply impossible not to marvel at the accomplishments of the human mind.

Regarding brain size, Juan Luis Arsuaga observes:

> The average size of the human brain is usually said to be 1,350 cc, but our population is so large and so varied that this figure is more of a convention than anything else. . . . The average brain volume of a human female is less than 1,300 cc while the average for males exceeds 1,400. . . . About ten percent of completely normal modern humans have a brain capacity of less than 1,100 or more than 1,600 cc.[14]

Chinese archaeologist Dr. Li Chi had a cranial capacity of 2000 cc, and Jonathan Swift and Ivan Turgenev each had a capacity of more than

2000 cc. Anatole France and an Australian Aboriginal, Topsy, each had a cranial capacity of under 1000 cc. A. H. Schultz cited a Melanesian with a cranial capacity of 790 cc as the lowest on record for a normal human.[15] The late anthropologist Marvin Harris (University of Florida) claimed that the variability of human cranial capacity starts at 850 cc.[16] However, Stephen Molnar writes, "In fact, there are many persons with 700 to 800 cubic centimeters."[17]

Ralph L. Holloway (Columbia University) observes:

> Microcephaly is a disease characterized by the association of a body of normal size with an abnormally small brain. . . . Humans with microcephaly are quite subnormal in intelligence, but they still show specifically human behavior patterns, including the capacity to learn language symbols and to utilize them.[18]

Normal human brain size ranges from about 800 cc to about 2200 cc. Yet the quality of the brain is the same. The key is that human brains are wired differently and with far more complexity than are animal brains, including primates. In humans, there is a general relation between brain size and body size. For instance, men's brains are generally larger than women's brains because men generally have a larger body size. However, there is no difference in brain quality. (Wives may dispute that statement.) There is a Rubicon that cannot be crossed between the smallest human brain and the largest primate brain.

There can be problems in interpreting whether or not a fossil is human based on brain size. My own rule is that, normally, no human fossil brain should be smaller than the smallest normal human brain today (about 775 cc for fossil adults), but it could be larger than human brains today. Undisputed evidences of culture, such as in burials, can help discriminate between humans and nonhumans. There is a condition in the fossil record known as gigantism. A giant ape could have a brain size approaching the smaller human brain sizes. In this case, items that could help in discriminating between human and ape are the presence or lack of culture, the proportion of face size to brain size, extreme prognathism, a large sagittal crest, and other items common in ape morphology.

The figures in the chart below represent average cranial size. My computations are based on published cranial sizes of fossil adults whose skulls are sufficiently well-preserved to make such measurements; I have made no attempt to discern whether these fossils are males or females. These numbers include the new Neandertal figures, which contain the European group formerly known as archaic *Homo sapiens.*

14 anatomically modern *Homo sapiens* individuals	1531 cc
37 Neandertal and pre-Neandertal individuals	1407 cc
12 African early *Homo sapiens* individuals	1274 cc
2 Asian early *Homo sapiens* individuals	1230 cc
32 Asian *Homo erectus* individuals	997 cc
5 African *Homo erectus* individuals	869 cc

Skull KNM-ER 1470 has a cranial capacity of about 800 cc and may well be human. Without KNM-ER 1470, the average of eight *Homo habilis* (a flawed taxon) skulls is 615 cc, the average for *Australopithecus africanus* is 440 cc, and the average for the Lucy-type *Australopithecus afarensis* is 413 cc.

Language Ability

"Without language, we would be only a sort of upright chimpanzee with funny feet and clever hands. With it, we are the self-possessed masters of the planet," writes Matt Cartmill.[19] Humans can talk. No animal can. Animals can communicate with grunts, howls, growls, songs, whistles, whines, or roars, but this is not language. With language, and its amazing connections to the brain, humans can write and recite a romantic poem about the moon or fly to the moon. We can use our brain to plan and explain elaborate schemes or perform surgery on the brain. The gap between human language and animal communication is vast and absolute. Yet evolutionists think that some animals are essentially like us but with fewer smarts and more fur.

We read stories about chimpanzees who can communicate with humans using American sign language and dolphins who may be able to communicate with humans. But animals only talk in Gary Larson's *Far Side* cartoons, not in the real world.

To the evolutionist, the origin and evolution of language is a black hole. They will never resolve it. The reason is that language did not evolve. Adam and Eve were fully capable of speech the moment they were created (Genesis 2–3). And because of the confusion of languages at Babel, all languages cannot be traced back to a single original language.

The greatest linguistic authority (although he has his detractors), Noam Chomsky (Massachusetts Institute of Technology), believes:

> The deepest structures of language are innate, not learned. We are all born with the same fundamental grammar hard-wired into our brains, and we are preprogramed to pick up the additional rules of the local language.[20]

Humans can speak. Other primates can't. Because evolutionists believe that we evolved from some other primate, they believe that speech *had* to evolve. Belief is a powerful persuader. Hence any tiny shred of "evidence" for the evolution of speech is magnified and made to sound plausible. Evolutionists are fantastic science-fiction writers.

Most who hold to the Out of Africa story question whether the Neandertals had full speech capability. Matt Cartmill lets the cat out of the bag. "Many paleoanthropologists, especially those who like to see Neanderthals as a separate species, accept this story."[21] There is no room in the Out of Africa story for fully human Neandertals.

To deny Neandertals' speech ability is to deny them full human status. Philip Lieberman and others show the very close association between speech and thought. Lack of speech ability implies lack of conceptual ability. This would explain the alleged lack of innovation that is said to characterize the Neandertals and confirm the Great Leap Forward in European culture about 40,000 years ago. We showed earlier that the Neandertals were not culture thin. The Out of Africa view is not so much about evolution as it is about protecting human evolution from the charge of racism. That's why evolutionists can't allow the Neandertals to be our ancestors.

Jared Diamond and Philip Lieberman believe that the *alleged* lack of Neandertal inventiveness implies that they did not have modern speech. Science writer James Shreeve comments on this view:

> Equating language and material inventiveness requires a great leap of another sort. Viewed solely through their stone-tool industries, there are aboriginal societies in Australia and New Guinea that were no more "advanced" than the Neandertals until just a few years ago. Yet these people think and communicate in languages as richly expressive as any others on earth, and use these languages to construct wonderfully inventive myths, stories, and cosmologies. *All this highly complex culture would be invisible to an archaeologist 10,000 years from now* (emphasis added).[22]

Denying the Neandertals modern speech while suggesting that chimpanzees could develop human communication skills is bizarre beyond words.

An amusing incident was reported recently in the *San Diego Union-Tribune.*[23] Researchers at Plymouth University in England gave six Sulawesi crested macaque monkeys a computer for their enclosure at Paignton Zoo. First, reported one of the researchers, Mike Phillips, "the lead male got a stone and started bashing the h—l out of it." Phillips admitted that was something we all want to do with our computers at one time or another. The monkeys were also keen on "defecating and urinating all over the keyboard. They pressed a lot of S's. Obviously, English isn't their first language." However, he stated that the monkeys "are not random generators—they were quite interested in the screen, and they saw that when they typed a letter, something happened. There was a level of intention there."

Apparently, the researchers at Plymouth University still believe in the nineteenth-century evolutionist idea that monkeys or chimpanzees, if given enough time, could produce literature. Let me explain why it will never happen in the real world.

Is "zzyzx" a word? Ninety-five percent of you will say no. The rest of you, who are pathologically suspicious, will say, "Yes, because he wouldn't ask the question if it weren't a word." As is often the case, the minority is right. On Interstate 15, between Los Angeles and Las Vegas, just south of the town of Baker, California, and fifty-five miles south of the California-Nevada state line, is Zzyzx Road. Obviously, someone living on that road, in a remote area of the Mojave Desert, decided to have some fun making a word out of the last three letters of the English alphabet.

At some point, that person decided that "zzyzx" was the most interesting arrangement of x, y, and z and that "zzyzx" would be the name of their road. It may have been just a private dirt road, and he or she may have put up a sign, "Zzyzx Road." At that point, it became a word. However, even that would not make "Zzyzx" an "official" word. Only when San Bernardino County or the State of California tied it into the highway system, perhaps with the famous Route 66 or later with Interstate 15, where it now has an overpass, did "Zzyzx" officially become a word.

Individuals may create unofficial words, but a word becomes "official" when some political or social entity decides that a certain arrangement of letters will refer in its language to a particular concept, place, or thing. The word's universality might be limited. There are many communities named *Norfolk*, but only one in each state. A word might have more than one meaning, but there is universal agreement on what those meanings are. A social entity might determine the meaning of a word just through continual usage, but eventually that word achieves official (dictionary) recognition.

The attachment of meaning to a certain arrangement of letters is called "language convention." It is what makes communication rational, meaningful, and understandable. In *Alice in Wonderland*, Alice objects to her companion's strange use of words. He replies, "Words are not the master of us, we are the master of words." Not so. If we do not follow the language convention, the result is confusion. Words, with their attached meaning, are what make a language.

For this reason, the evolutionary idea that monkeys or chimpanzees can acquire language capability is absurd. The association of the word *banana* with a banana when they see a banana or a picture of one is a far cry from true language acquisition. My cat can do that! Chimpanzees cannot associate a group of letters or the sound of a word with true *abstract* meaning. This is a unique human capacity. It is possible that those monkeys at Plymouth University by chance could type zzyzx. That's not the issue. Would it have any meaning to them? No. Would they ever associate it with something *abstract*? No. You had never heard of the word "zzyzx." You have never been to that particular area of Southern California. Yet in your mind you have made the *abstract* association of a word you never knew with a road you have never seen. That's just one of the many differences between your brain and the brain of a chimpanzee! And chances are it will be a long time before you forget the word "zzyzx."

An Overlooked Principle

Someone might ask, "If a chimpanzeelike animal was not our ancestor, why is it possible for chimpanzees to do all of those things we hear about at the primate center at Emory University in Atlanta?" Do chimpanzees do those things in the wild? No. Is there any indication that, if given a lot more time, they would eventually learn to do those things in the wild? No.

Do you remember your last visit to the circus? Remember the tigers sitting on little stands waiting for a signal from their trainer? Do they do anything like that in the wild? No. What about the tricks done by the elephants? Do they ever do anything like that in the wild? No. What about the tricks your dog does? Would he have done those tricks if you had not spent much time training him? We won't even talk about trying to teach my cat, Abby, to do tricks! I remember my last visit to Sea World. A killer whale pushed a man through the water and high into the air with its nose. It then jumped onto a platform completely out of the water and posed for the audience. (I did notice that they kept feeding it lots of fish to keep its tummy nice and full.) But do killer whales ever do anything remotely like that in the wild? No.

Many animals can be trained by humans to do exotic things that they would never do in the wild. That has absolutely nothing to do with evolution. Instead, it seems to be a principle of nature that a being of higher intelligence can teach a being of lower intelligence to do things that it would never do if left to itself. This is simply a residual effect of the dominion that God gave humans over animals at creation (Gen. 1:26, 28).

However, there is a powerful spiritual analogy of our relationship to the animals. It is that God (a being of higher intelligence) has given humans (beings of lower intelligence) a revelation, the Bible, to teach us truths about God and about life that we would have never known on our own. We would never know the true details of creation, the fall of humans into sin, and the judgment of God through the flood without their being revealed in the Bible. In fact, evolution was designed to counteract the biblical teachings about those very events.

But there is something even more important. We would never know about the true God without the Bible. Let me illustrate. There are literally thousands of religions in the world claiming to teach about God. In spite of their almost infinite variety, there is one common thread going through all of them. It is that humans find deliverance, salvation, improvement, and/or union with god through self-effort, good works, good deeds. That is the universal message of all religions invented by humans. *In that sense, all man-made religions are the same.* In contrast to all human religions, the Bible tells us that our works can *never* save us. Our basic problem is not our lack of good deeds. When we stand before a righteous and holy God, our basic problem is far deeper: It is our abundance of sins and our total lack of true righteousness.

Only one religion is absolutely different and absolutely unique—the gospel of the grace of God, who will freely provide for us both forgiveness and righteousness when we trust in Jesus Christ. This message is of a holy God who loved us enough to die for our sins so that we might live with him forever. This message is found *nowhere* but in the Bible.

Let me illustrate with a personal experience. While attending seminary, I worked part-time at the Dallas post office. Coming home very late one night, I remember seeing that the light at the intersection ahead was green. The next thing I knew, I was in the center of the intersection, and the light was red. Feeling that it was unsafe to stop in the middle of the intersection, I elected to go through. Unfortunately for me, one of the cars waiting for the light to change was marked "Dallas Police Department."

Typical of many students, my wife and I were in a situation of "too much month at the end of the money." I thought that if I went to court and pleaded my case, possibly the judge would reduce the fine. I presented my situation to the judge as I explained it to you. I knew exactly what was coming. The judge looked at me quizzically and asked, "Young man, where were you when the yellow light was on?"

I replied, "Your honor, I don't remember seeing the yellow light."

Laughter broke out in the courtroom. From somewhere, a voice rang out: "Your honor, I move that we excuse this young man on the grounds of his honesty." The judge took my ticket, wrote on it "Not guilty," and handed it to me. I departed—with thanksgiving!

Although I did not protest what the judge did, that judge actually broke the law himself. I had, in essence, admitted that I was guilty. I don't mean to make a great deal about a minor offense, but there is a principle here that is basic to our understanding of what law is. If a person has not broken the law, the law cannot touch him. If a person has broken the law, however, the law is very specific about the penalty. There may be degrees of penalty based on circumstances, but nowhere in any legal system is there a provision for a judge to declare a guilty man innocent without some penalty or some satisfaction of justice.

I often share this story and ask, "What should the judge have done without breaking the law if he had wanted to free me?" Most people seem mystified. The answer is that the judge should have paid my fine himself. Then the law would be satisfied, and I would be free. People reply, "In your dreams! Where will you find a judge like that?"

Where will you find a judge that will actually pay your fine for you? There is only one—the creator of the universe and the judge of all men—Jesus Christ. But there is a difference. He will do it legally! His problem is that he is both holy and loving. His holiness demands justice. His love demands mercy. He satisfied both his holiness and his love by paying the "fine" for me—and for you—on the cross.

The most intelligent thing I ever did in my whole life was to trust in Jesus Christ as my Savior.

Meanwhile, never discuss theology with a chimp!

CHAPTER 31

GENESIS

The Footnotes of Moses

THE PRECISE CIRCUMSTANCES OF the composition of the book of Genesis have been a matter of continual interest for Bible scholars. Since there is strong internal as well as external evidence that Moses wrote Exodus, Leviticus, Numbers, and Deuteronomy, and since the Pentateuch is considered to be a unit, the approach of most conservative scholarship has been that Moses wrote Genesis also.

Nowhere does Scripture say, however, that Moses actually wrote the narratives or the genealogies of Genesis. There is no statement in Genesis referring to Moses as its author, as there clearly is in the other books of the Pentateuch. Not even Christ or the apostles say that Moses actually wrote or spoke the words they quote from Genesis. While accepting the Mosaic authorship of Genesis, conservative scholars have not detailed the means by which Moses received his information.

There are three possible means by which Genesis was composed under the inspiration of the Holy Spirit: (1) Moses received his information by direct revelation; (2) Moses wrote Genesis using material passed on by oral tradition; or

(3) Moses wrote or compiled Genesis using earlier-written documents. Literary items in Genesis make the first possibility seem unlikely. The second possibility also seems remote because of the probability of information being lost or degraded in oral transmission. The third option seems most likely. The majority of evangelical scholars accept some version of the third view but give few details.

Ancient Writing

One of the arguments used by critics of the past century in their attack on the historicity and integrity of Genesis was that the art of writing went back only to the time of David, about 1000 BC. Hence no portion of Genesis could have been in written form before that time. It is now known that these critics were not only wrong but very wrong. By the 1930s, our museums were rich with cuneiform writing on clay tablets dating back to 3500 BC. Excavations of the royal archives at Ebla, in northwest Syria, possibly dating as far back as 2700 BC, reveal that writing at that early date was commonplace. It was not necessary in that era for the average person to know how to read and write, but writing was readily available to everyone through a class of professionals known as scribes. In fact, the ancient Sumerians, Babylonians, and Assyrians seemed unwilling to transact even the smallest items of business without recourse to a written document. This characteristic is dramatically seen at Ebla.

It may surprise some to learn that a clear reference to writing is found in Genesis 5:1: "This is the written account of Adam's line." This suggests that the art of writing was known within the lifetime of Adam, which could make writing virtually as old as the human race. To a creationist, this is not surprising. It is obvious that at the time of their creation, Adam and Eve knew how to speak. Yet language is incredibly complex, and no one understands its origin. The ability to write is more complex than the ability to speak. However, since God created our first parents with the ability to speak, it is reasonable to suggest that he created them with the ability to learn to write as well. A naturalistic, evolutionary origin of language stretches credulity.[1]

Cuneiform writing became the system used by all civilized countries east of the Mediterranean—Assyria, Babylonia, Persia—and by the Hittites, who are mentioned seven times in Genesis, beginning at Genesis 15:20. Cuneiform writing consists of a series of wedge-shaped impressions (*cuneia* means "wedge") made in plastic clay. The Hebrew word for "to write" means "to cut in" or "to dig." Abraham, Isaac, and Jacob all would have written in cuneiform. Cuneiform was not a specific language but a method of writing on clay tablets, and it embraced many languages and dialects.

The clay of the Euphrates Valley is remarkable for its fineness, as fine as well-ground flour. The scribes would mix a bit of chalk or gypsum into the clay to keep the tablets from shrinking or cracking. They were then dried in the sun or in a kiln. These clay tablets are the most imperishable form of writing material known, next to stone. It is possible that the two tablets on which God wrote the Ten Commandments were actually clay tablets (Exod. 32:15–16). The western Asian archaeological record suggests that virtually everything written before Abraham left Ur of the Chaldees, and much after that, was written on clay tablets in cuneiform.

As early as 2350 BC, clay envelopes were used for private clay tablet correspondence and sealed with a private seal. A reference to this seal is found in Job 38:14, which is believed to have been written before the time of Abraham. Judah carried a seal with him and gave it to Tamar (Gen. 38:18, 25). Joseph was given Pharaoh's seal ring (Gen. 41:42), which enabled him to act in an official capacity on behalf of Pharaoh.

Although papyrus was the common writing material in Egypt, cuneiform writing was understood, as the Tell el-Amarna tablets, found in Egypt in 1888, reveal. Among these clay tablets were letters, dated about 1400 BC, from Palestinian officials to the Egyptian government—all written in cuneiform.

Those who do not consider the early chapters of Genesis to be reliable history use oral transmission as the explanation for those chapters of the book. But it is absurd to think that God would entrust his eternal Word to the fragile memory of humans. Scripture teaches the opposite. In Deuteronomy 31:19–21, Moses was given a song to teach to the people. He was specifically commanded to write it down so that it would not be forgotten. God said that forgetting is what the people were disposed to do. Obviously, God has little faith in oral transmission.

The Structure of Genesis

All scholars agree that the most significant and distinguishing phrase in Genesis is "these are the generations of." Commentators of all theological schools divide the book around that phrase, which is found eleven times in Genesis (2:4; 5:1; 6:9; 10:1; 11:10, 27; 25:12, 19; 36:1, 9; 37:2). The translators of the Septuagint (the Greek Old Testament) regarded that phrase as being so significant that they named the book after that term. *Genesis* is the Greek equivalent of the Hebrew *tol^{e}dot*, "generations."

It is common for ancient records to begin with a genealogy or a register documenting close family relationships. Because several of the *tol^{e}dot* phrases in Genesis are followed by genealogies, scholars have almost universally assumed that the *tolᵉdot* phrase serves as an introduction to the section that follows. Hence the major sections of Genesis have been made to begin with the *tolᵉdot.* Since the person named in the *tolᵉdot* does not figure prominently—if at all—in the narrative that follows, the word has taken on the meaning of "descendants" ("these are the descendants of").

Yet the lexicon defines *tolᵉdot* as "history, especially family history" or something associated with origins. This would mean that the term is concerned with ancestors rather than descendants. It also suggests that the phrase looks back to the preceding narrative rather than looking ahead to what follows.

The first use of *tolᵉdot* in Genesis 2:4 ("these are the generations of the heavens and the earth") clearly establishes that this reference at 2:4 is looking back rather than ahead. Nothing following Genesis 2:4 deals with "the heavens and the earth." Many commentators recognize that here *tolᵉdot* looks back, even though they interpret the other occasions where it is used as looking ahead. They fail to see that Genesis 2:4 is the key, and that all of the *tolᵉdot* phrases refer back to the previous material.

James Moffatt, in his translation of the Bible, actually lifted the *tolᵉdot* phrase out of Genesis 2:4 and transferred it to Genesis 1:1 so that it serves as an introduction to the first chapter. Other liberal writers have stated that this phrase was out of place at Genesis 2:4 or that it was put there by a compiler merely to serve as a transition.

The Colophon and Mesopotamian Writings

In 1936, P. J. Wiseman wrote a book titled *New Discoveries in Babylonia about Genesis.* Wiseman seems to have found the key that unlocks the details of the authorship of Genesis. His thesis is that internal clues in Genesis reveal how it was written; that the actual authors of Genesis were Adam, Noah, the sons of Noah, Shem, Terah, Ishmael, Isaac, Esau, Jacob, and Joseph; that the authors, other than Joseph, probably wrote in cuneiform on clay tablets; and that Moses, using these records, was the redactor or editor of Genesis rather than its author.

Wiseman's work was recently edited and reissued by his son, Donald P. Wiseman, a noted evangelical scholar.[2] The younger Wiseman was assistant curator of western Asian antiquities at the British Museum and later professor of Assyriology at the University of London. He is also general editor of the Tyndale Old Testament Commentary series. Donald Wiseman endorses his father's work, as does R. K. Harrison, professor of Old Testament, University of Toronto, who has incorporated it into his monumental *Introduction to the Old Testament.*[3] Although P. J. Wiseman is often cited by evangelical scholars, his remarkable insights into the composition of Genesis are not well-known by the evangelical community.

Wiseman asked this question: How was information recorded and how were documents formulated in ancient Mesopotamia, which was the geographical context of much of the book of Genesis? The heart of Wiseman's contribution to the problem of the formulation of Genesis was his insight in identifying the *tol^e dot* phrases in Genesis with ancient Mesopotamian colophons. A colophon is a scribal device placed at the conclusion of a literary work written on a clay tablet, giving—among other things—the title or description of the narrative, the date or occasion of the writing, and the name of the owner or writer of the tablet.

It is not surprising to the student of ancient Eastern customs that many of their literary habits were precisely the opposite of our own. For instance, the Hebrews commenced their writing on what to us is the last page of the book and wrote from right to left. In ancient Mesopotamia (Iraq), it was the end and not the beginning of the tablet that contained the vital information regarding date, contents, and ownership or authorship. This custom was widespread and persisted for thousands of years.

Perhaps the most striking aspect of the colophon practice was that the name in the colophon was the name of the owner or writer of the tablet. Sometimes the owner would also be the writer. If a person was not able to write, however, he would hire a scribe to do the writing for him. The scribe would include not his own name but the name of his employer—the owner of the tablet. Thus it is impossible to overemphasize the importance of the colophon at Genesis 5:1: "This is the written account of Adam's line." Not only does the Hebrew word *sepher* mean "book" or "a complete writing," but the presence of Adam's name suggests that it was a written account *owned or written by Adam*, not just a written account *about* Adam. Genesis 2:4–5:1 gives evidence of being a firsthand, eyewitness account of the experiences of Adam, possibly written by him on a clay tablet.

Derek Kidner (Tyndale House, Cambridge) understands the impact of the Hebrew word *sepher* at Genesis 5:1.

> The opening, *This is the book* . . . , seems to indicate that the chapter was originally a self-contained unit ("book" means "written account", of whatever length), and the impression is strengthened by its opening with a creation summary, and by the set pattern of its paragraphs.[4]

However, Kidner rejects Wiseman's theory that Genesis contains a series of colophons in which the names given are the names of the original writers or owners of the tablets. "By insisting on a complete succession of named tablets the theory implies that writing is nearly if not quite as old as man."[5]

At the risk of being thought a bit naive, one could ask what is wrong with the art of writing being nearly as old as the human race. Here is where preconceptions enter in. If one believes that God created humans directly as humans, there is nothing at all wrong with the idea. The problem is really with Kidner. He is a theistic evolutionist. His philosophy demands that humans not be that intelligent that early in their history. Therefore, Adam could not have known how to write. It's rather gracious of Kidner to allow Adam to be able to speak. This is not the only occasion when Kidner forsakes solid biblical exegesis because of his preconceived notions about origins.

Colophons also included the date or occasion of their writing. It is easy for us in the twentieth century to miss this fact, because we date our writings by the calendar. Not so the ancients. The creation account (Gen. 1:1–2:4) is dated "in the day that the Lord God created the heavens and the earth" (Gen. 2:4). "In the day" equals "when" and implies that the creation account was written very close to the actual time of creation, not centuries later. In Genesis 37:1–2, Jacob dated his tablet as having been written "when he lived in Canaan." Although by our standards that phrase is not precise, it does reflect a specific period in Jacob's life. Before that time he spent many years working for Laban in Haran. After that period he lived with Joseph in Egypt until his death. Leviticus (although probably not written on clay tablets) is dated as having been written when Moses was "on Mount Sinai" (Lev. 27:34), and Numbers is dated as having been written when Moses was "on the plains of Moab" (Num. 36:13). This type of dating was accurate enough for the people of that era, considering the nature of their society.

The use of colophons persisted almost unchanged for over three thousand years in ancient Mesopotamia and elsewhere. Colophons are found in the Ebla tablets in northwest Syria (2700 BC) and in the Akkadian texts from Ras Shamra (1300 BC). Colophons continued at least until the time of Alexander the Great (333 BC), and they are not unknown today. In one of my English Bibles, at the end of the epistle to the Romans, is this statement: "Written to the Romans from Corinth, and sent by Phoebe, servant of the church of Cenchrea." Readers of *Time* and *Newsweek* will recognize that many of the major news articles have the name of the author and the place of writing at the end of the article, such as, "Written by Susan Smith in Washington." These are all suggestive of colophons.

The Authors of Genesis

The internal evidence suggests that Genesis was written on a series of clay tablets as follows:

Genesis 1:1–2:4 Origin of the heavens and the earth. No author is given. P. J. Wiseman suggests that the author was God himself, who wrote it in the same way he wrote the Ten Commandments, probably on clay tablets. According to its date, as given in the text itself, it was written very soon after the act of creation.

Genesis 2:5–5:2 Tablet written by or belonging to Adam.

Genesis 5:3–6:9a Tablet written by or belonging to Noah.

Genesis 6:9b–10:1 Tablet written by or belonging to the sons of Noah.

Genesis 10:2–11:10a Tablet written by or belonging to Shem.

Genesis 11:10b–11:27a Tablet written by or belonging to Terah.

Genesis 11:27b–25:19a Tablets written by or belonging to Isaac and Ishmael.

Genesis 25:19b–37:2a Tablets written by or belonging to Jacob and Esau. Esau's genealogy may have been added later.

It is significant that the last colophon is at Genesis 37:2a. From Genesis 1 to 11, the Mesopotamian setting and local color are very obvious. From Genesis 12 to 37:2a, that Mesopotamian influence persists. Abraham came from Ur of the Chaldees, Isaac sent back there for his wife, and Jacob got his wife from Haran and worked there for many years. From Genesis 37:2b to the end, however, the setting and local color change dramatically. We are now in Egypt. This section has a strong Egyptian flavor and was probably written by Joseph on papyrus or leather; hence it is without colophons, which are only associated with clay tablets.

Strengthening the arguments presented thus far is the fact that in every case the person named as the owner or writer of the tablet could have written the contents of that tablet from his own personal experience. It is also significant that in every case, the history recorded in the various tablets ceases just prior to the death of the person named as the owner or writer of the tablet.

The Role of Moses

All of the tablets could have come to Moses in the way that family records were normally handed down. Nothing would have been more precious to the patriarchs than their family histories and genealogies. It is possible that there were many sets of these tablets and that each member of a patriarchal family had his own set. Of all the personal items that Noah would have taken on the ark, he would have considered his family histories the most precious and most worthy of preservation.

Because of his education in the household of Pharaoh, Moses had the finest scholarly training of that day. He would have known how to read the languages of the cuneiform tablets, as well as Egyptian. Cuneiform writing was well-known in Egypt because of Egypt's relationship with Mesopotamia. Moses's task would have been first of all to organize the book—under the guidance of the Holy Spirit—into a unified whole. The use of previously written documents in no way does violence to the concept of verbal plenary inspiration. Luke also tells us that he used previously written documents (Luke 1:1–4). It is reasonable to assume that each of the original writers of the tablets was guided by the Holy Spirit as well. By retaining the colophons, Moses clearly indicates the sources of his information. Just as a scholar today documents his sources with footnotes or endnotes, so Moses documented his sources of information with the colophons. These colophon divisions, based on the different sources, constitute the framework of the book of Genesis.[6]

Moses' second task would have been translation. Any tablets written in Mesopotamia would have needed to be translated into Hebrew. If this translation had not been done before Moses' time, Moses would have been qualified to do it. Joseph's records, if written in Egyptian, would also have needed to be translated into Hebrew by Moses.[7]

The third major task for Moses, as the redactor or editor, would have been to bring place names up-to-date for the Israelites of the exodus. Geographic names change, and this updating is seen clearly in Genesis 14:2, 3, 7, 8, 15, and 17. This tablet, written in Abraham's day, had in it many geographic names that had become obsolete in the over four hundred years between Abraham and Moses. It is indicative of Moses' deep regard for the sacred text that he did not remove the old names but just added an explanatory note telling of the new names. Such notations are also seen at Genesis 23:2, 19; and 35:19. Genesis 23:2, 19 also indicate that these notations were made before the Israelites entered Canaan, since Moses had to state where these places were. Had the Israelites already been in the land, these notations would not have been necessary.

Several passages indicate the antiquity of the tablets Moses had in his possession. In Genesis 16:14, regarding the well or spring to which Hagar fled, Moses added this note: "It is still there, between Kadesh and Bered." Genesis 10:19 is one of the most important evidences of the great antiquity of the book of Genesis. This passage, part of Shem's tablet, had to have been written before the destruction of Sodom and Gomorrah. Since these cities were destroyed (Genesis 19) and never rebuilt, and their very location was forgotten, this tablet telling of the settlement of clans near those cities obviously had to be written while those cities were still standing.

Implications of the Evidence

The implications of this evidence for the origin of Genesis are staggering. Rather than Genesis having a late date, as is universally taught in nonevangelical circles, the evidence implies that Genesis 1–11 is a transcript of the oldest series of written records in human history. This is in keeping both with the character of God and with the vital contents of these chapters. It is reasonable to expect that the first humans created by God would have had great intelligence and language capabilities and that God would fully inform them as to their origin.

This research also confirms the idea that the Genesis creation and flood accounts are the original accounts of these events and were not derived from the very different and polytheistic Babylonian accounts.[8] It also supports the fact that monotheism was the original religious belief and not a later evolutionary refinement from an earlier polytheism.

This research further serves to falsify the widespread idea that Genesis 1 and 2 give conflicting accounts of creation. It also suggests that the higher-critical theories on the composition and date of Genesis are factually bankrupt.

Just as God has not left us in doubt about our destiny, so he has not left us in doubt about our origin. We have the footnotes of Moses.

Possible Editorial Insertions by Moses for Clarification

Genesis 10:5 "From these the maritime peoples spread out into their territories by their clans within their nations, each with its own language."

Genesis 10:14 "from whom the Philistines came"

Genesis 14:2, 3, 7, 8, 17 Geographic clarifications.

Genesis 16:14 "it is still there, between Kadesh and Bered."

Genesis 19:37b "he is the father of the Moabites of today."

Genesis 19:38b "he is the father of the Ammonites of today."

Genesis 22:14b "And to this day it is said, 'On the mountain of the Lord it will be provided.' "

Genesis 23:2, 19 Geographic clarifications.

Genesis 26:33 "and to this day the name of the town has been Beersheba."

Genesis 32:32 "Therefore to this day the Israelites do not eat the tendon attached to the socket of the hip, because the socket of Jacob's hip was touched near the tendon."

Genesis 35:6, 19, 27 Geographic clarifications.
Genesis 25:20 “and to this day that pillar marks Rachel’s tomb.”
Genesis 36:10–29 Esau’s genealogy probably added later.
Genesis 47:26 “—still in force today—”
Genesis 48:7b “that is, Bethlehem.”

CHAPTER 32

Reality in the Human Fossil Record

HUMAN EVOLUTION HAS BEEN falsified. That is, it has been proven by objective, scientific evidence to be false. In fact, human evolution has been falsified in at least three ways. First, human evolution—and all evolution—has been falsified by the recent evidence that the dating methods are a fiction and that the earth is young (chapter 27). With a young earth, all evolution is dead! Since this development is quite new, its full impact has yet to be seen.

Second, human evolution has been falsified in that virtually every chart of human evolution since 1990 has question marks or dotted lines at the most crucial point—the transition from the australopithecines to true humans. That is why they needed *Homo habilis*. Unfortunately, he's the little man who isn't really there. Although evolutionists are still claiming that human evolution is a fact, Dr. Meave Leakey says, "It may never be possible to say exactly what evolved into what."[1] Those confident human-evolution "trees" of the 1980s have gone the way of the dinosaur. Instead, we have human evolution "bushes" with question marks at the crucial points.

To illustrate, let me refer you to a publication readily available to you: *National Geographic*, the February 1997 issue. This issue has australopithecine and human family "bushes" by two of the most respected paleoanthropologists. The "bush" by Phillip Tobias has nine question marks in it. The "bush" by Bernard Wood has fifteen question marks in it. In fact, from *Homo sapiens*, *Homo erectus*, and *Homo rudolfensis* (KNM-ER 1470), Wood has question marks all the way down with no certain ancestors whatsoever. The certainty with which evolutionists state that human evolution is a fact is a reflection of their philosophical belief, not of the objective evidence. The evidence for human evolution is not solid enough to convict a known thief of petty larceny.

Third, human evolution is falsified by the charts in this book, going back to the first edition published in 1992. However, I am not so arrogant as to think that my book was responsible for falsifying human evolution. All I did was reveal the dirty little secret that paleoanthropologists have known all along. They admit that they are not certain of our distant ancestors. I pointed out that neither is there any evolution from *Homo erectus*, to early *Homo sapiens*, to the Neandertals, and/or to modern *Homo sapiens*. The thesis of this book is that these contrived evolutionary groups were actually living as contemporary humans created by God.

The composite master chart is the most important chart. It is a composite of all of the other charts in this book, all based on the evolutionist dates for the fossils. On this composite chart the full impact of the contemporaneousness of the various fossil groups is revealed—the time relationship of the *Homo* (human) fossils to each other and to the australopithecines. Evolutionists divided the various human fossils into different groups (*Homo sapiens*, Neandertals, early *Homo sapiens*, and *Homo erectus*) based on their belief that humans had an evolutionary history. These are false distinctions. They merely reflect the genetic and geographical diversity of the human family over time. As can be seen, even when we accept the evolutionists' categories for the fossils and their dates, the fossils still do not demonstrate human evolution.

Regarding the chart of anatomically modern *Homo sapiens* and *sapiens*-like fossils, the fossils on this chart dated older than 100,000 years are not called modern *Homo sapiens* by evolutionists—for obvious reasons. While I cannot prove that they are modern *Homo sapiens*, I can say that when we compare these older fossils with the comparable bones of modern humans, they are virtually identical. This means that there are fossils that are indistinguishable from modern humans that extend all the way back to 4.5 Mya (million years ago) on the evolutionary time scale. Because the lower portion of the anatomically modern *Homo sapiens* fossils chart is the more significant, and the more controversial, we will deal specifically with some of the fossils in that lower section.

KNM-ER 1470

The skull and femur known as KNM-ER 1470 and 1481 are dated at about 1.9 Mya on the evolution time scale. KNM-ER 1481 is a completely modern leg bone. The skull, found at the same level as the leg bone but not in association with it, has been the source of controversy ever since it was found in 1972. Richard Leakey first called it human, then it became *Homo habilis* (but it is much larger than the others in that group), and now many call it *Homo rudolfensis.* However, comparisons suggest that skull 1470 is more modern than any of the *Homo erectus* fossils—even the Kow Swamp material, which is only about 10,000 years old.

Not only could skull 1470 possibly qualify for human status based on cranial shape, size (almost 800 cc), and wall thickness, but there is evidence on the inside of the skull of a Broca's area, the part of the brain that controls the muscles for producing articulate speech in humans.

> The two foremost American experts on human brain evolution—Dean Falk of the State University of New York at Albany and Ralph Holloway of Columbia University—usually disagree, but even they agree that Broca's area is present in a skull from East Turkana known as 1470. Philip [sic] Tobias, . . . renowned brain expert from South Africa, concurs. . . . So, if having the brains to speak is the issue, apparently *Homo* has had it from the beginning.[2]

Soon after casts were available, I purchased one of skull 1470 from the National Museums of Kenya, Nairobi. As I studied it, I sensed that there might be a problem with the reconstruction of the face. The original fossil had been found in hundreds of pieces and was assembled over a six-week period by Alan Walker, Bernard Wood, and Richard Leakey's wife, Meave. The skull was far too large for an australopithecine. It cried out, "*Homo*!" However, the face had a bit of an australopithecine slant to it. Pictures taken before plaster was used to fill in the missing pieces reveal that the face of the fossil is rather free floating. It is attached to the skull only at the top, with nothing to stabilize the slant of the face. Further, the maxilla (upper jaw) is not attached to the rest of the face.

Others have also questioned the reconstruction of skull 1470. On several occasions, Richard Leakey protested that the skull was assembled in the only way possible. But it seems that Leakey was not being straightforward. Science writer Roger Lewin, associated with Leakey on several projects, tells a different story regarding skull 1470.

> One point of uncertainty was the angle at which the face attached to the cranium. Alan Walker remembers an occasion when he, Michael Day, and Richard Leakey were studying the two sections of the skull. "You could hold the maxilla [upper jaw] forward, and give it a long face, or you could tuck it in, making the face

short," he recalls. "*How you held it really depended on your preconceptions.* It was interesting watching what people did with it." Leakey remembers the incident too: "Yes, if you held it one way, it looked like one thing [human]; if you held it another, it looked like something else [australopithecine]."[3]

There is no question that bias intervened in the reconstruction of skull 1470. Tucked under, the skull would look much like a modern human. Instead, the face was given an australopithecine slant to make it look more like a transitional form. Having said all that, I still have put a question mark by it. What we can say at the present time is that there is no *compelling* reason why it could not be human.

The Olduvai Circular Stone Structure

At the bottom of Bed I, the lowest bed at Olduvai Gorge, Tanzania, a circular stone structure was found that could only have been made by true humans. This object, fourteen feet in diameter, is considered to be the world's oldest man-made structure. Technically, it is an artifact rather than a fossil. Mary Leakey discovered it on the oldest of the occupation sites (or living floors) at Olduvai during the 1961–62 digging season.

Leakey quickly realized that there was a pattern in the distribution of these stones, an intentional piling of stones on top of each other. The stones themselves are lava rocks that are not indigenous to what was a lakeshore when the structure was built. The several hundred rocks were brought from a source some miles away. Other stones on the occupation site are scarce and haphazardly scattered.

What was this structure? No one knows. Was it a habitation hut, a hunting blind, a weapons pile, or a shelter of some kind? The people of the Okombambi tribe in southwest Africa construct such shelters today. They make a low ring of stones with higher piles at intervals to support upright poles or braces. Over these poles are placed skins or grasses to keep out the wind. Turkana tribesmen living in the desert of northern Kenya make similar shelters.

The living floor where this structure was found was dated by evolutionists at 1.8 Mya (K-Ar). Revisions published in late 1991 indicate that it may be as old as 2 Mya (Laser-fusion 40Ar/39Ar).[4] Mary Leakey also found the usual stone and bone waste, as well as Oldowan tools. These tools are normally considered very primitive by evolutionists. However, Leakey reports that a similar type of stone chopper is used today by the remote Turkana tribesmen to break open the nuts of the doum palm.[5]

The conceptual ability required to make such structures, the physical ability to carry hundreds of large stones several miles, and the fact that similar structures are made by humans today constitute strong evidence that true humans were on the scene at Olduvai at about 2 Mya on the evolution time scale.

The Laetoli Footprints

Beginning in 1978, associates of Mary Leakey discovered a series of what appear to be human footprint trails at site G, Laetoli, thirty miles south of Olduvai Gorge, in northern Tanzania. The stratum above the footprints has been dated at 3.6 Mya, while the stratum below them has been dated at 3.8 Mya (K-Ar). These footprint trails rank as one of the great fossil discoveries.

Mary Leakey described the footprints as "remarkably similar to those of modern man."[6] Three parallel trails are seen, made by three individuals, with one individual partially walking in the footprints of another. The trails contain a total of sixty-nine prints extending a length of about thirty yards. The prints were made in fresh volcanic ash spewed out by Mount Sadiman to the east. A unique combination of circumstances—ash fall, rain, time to harden, and another ash fall that hardened and protected them—caused these amazing prints to be preserved.

These footprint trails have produced a large body of literature. Virtually everyone agrees that the footprints are strikingly like those made by modern humans. In spite of that fact, evolutionists have ascribed these footprints to the Lucy-type hominid known as *Australopithecus afarensis*. This taxon includes mandibles found elsewhere at Laetoli by Mary Leakey as well as fossils found by Donald Johanson in the Afar region of Ethiopia. The assumption, based on the somewhat similar ages of the fossils in the two different localities and the belief that *afarensis* was bipedal, is that the *afarensis* fossils represent the type of individuals who made the Laetoli footprint trails. Not only is this totally unprovable, but Lucy was only three feet tall, while in size these footprints are more like those made by modern humans.

The footprint trails at Laetoli appear to have been made by individuals who were barefoot, probably habitually unshod. The specialist who has conducted the most extensive recent study of these footprints is Russell H. Tuttle (University of Chicago). When Tuttle began his study, he discovered that very few studies have been done on habitually unshod peoples. Studies done on the footprints of shod people would not necessarily be applicable to the Laetoli prints.

As part of Tuttle's investigations, he observed the Machiguenga Indians in the rugged mountains of Peru. The Machiguengas are a habitually barefoot people. More than seventy individuals from ages seven to sixty-seven, both male and female, constituted his study. He concludes, "In sum, the 3.5 million-year-old footprint trails at Laetoli site G resemble those of habitually unshod modern humans. None of their features suggest that the Laetoli hominids were less capable bipeds than we are."[7]

Elsewhere, Tuttle writes, "In discernible features, the Laetoli G prints are indistinguishable from those of habitually barefoot *Homo sapiens*."[8] He is especially struck by the similarity of the Laetoli prints to those of the Machiguenga Indians: "Casts of Laetoli G-l and of Machiguenga footprints in moist, sandy soil further illustrate the remarkable humanness of Laetoli hominid feet in all detectable morphological features."[9]

Not only did Tuttle reject the notion that the Laetoli footprints were made by *Australopithecus afarensis*, but he found that the former work on those footprints by J. T. Stern Jr. and Randall L. Susman (State University of New York, Stony Brook) was flawed.

> In any case, we should shelve the loose assumption that the Laetoli footprints were made by Lucy's kind, *Australopithecus afarensis*. The Laetoli footprints hint that at least one other hominid roamed Africa at about the same time.[10]

> My studies on the Laetoli footprints provide no support for the apish model of Stern and Susman, who, in fact, waffled from their initial position on the basis of undocumented rumors about faults in the casts that they had studied.[11]

If the Laetoli footprints are so much like those of modern humans, why would Tuttle talk about the existence of "one other hominid" in East Africa, one whose identity is totally unknown? Why not ascribe those footprints to humans? Tuttle is honest enough to give us the reason: "If the G footprints were not known to be so old, we would readily conclude that they were made by a member of our genus, *Homo*."[12]

The real problem—the only problem—is that to ascribe those fossil footprints to *Homo* does not fit the evolutionary time scale. According to the theory of evolution, those footprints are too old to have been made by true humans. This is a classic case of interpreting facts according to a preconceived philosophical bias.

Interpreting the Laetoli footprints is not a question of scholarship; it is a question of logic and the basic rules of evidence. We know what the human foot looks like. There is no evidence that any other creature, past or present, had a foot exactly like the human foot. We also know what human footprints look like. But we will never know for sure what australopithecine footprints look like, because there is no way of associating "beyond reasonable doubt" those extinct creatures with any fossil footprints we might discover. On the one hand, we have very positive identification of the human foot with the Laetoli footprints. On the other hand is the total absence of the information needed to make any identification of those prints with australopithecines. Juries deal with that kind of problem continually. The human mind deals with that kind of logic every day. Were it not for the darkness evolution casts on the human mind, there would be no question as to which category those Laetoli footprints should be assigned.

The fossil at the bottom of the chart of anatomically modern *Homo sapiens*-like fossils is an old friend, Kanapoi KP 271. We discussed that elbow bone in chapter 5. William Howells had the very same problem with that fossil that Russell Tuttle has with the Laetoli footprints. According to evolution, that elbow bone is just too old to be human. Allow me to quote Howells again:

> The humeral fragment from Kanapoi, with a date of about 4.4 million, could not be distinguished from *Homo sapiens* morphologically or by multivariate analysis by Patterson and myself in 1967 (or by much more searching analysis by others since then). We suggested that it might represent *Australopithecus* because at that time allocation to *Homo* seemed preposterous, *although it would be the correct one without the time element* (emphasis added).[13]

Evolutionists refuse to call extremely old fossils by their proper names. The reason is to protect the theory of evolution. Hence it is obvious that we are dealing not with science but with something more like quicksilver. There may be many terms we could use in referring to the evolutionist's methodology, but my mother, who would not let me get by with very much, certainly had a name for it.

The facts of the human fossil record are as follows. First, fossils that are indistinguishable from modern humans can be traced all the way back to 4.5 Mya, according to the evolution time scale. That suggests that true humans were on the scene before the australopithecines appear in the fossil record.

Second, *Homo erectus* demonstrates a morphological consistency throughout its two-million-year history. The fossil record does not show *erectus* evolving from something else or evolving into something else.

Third, anatomically modern *Homo sapiens*, Neandertal, early *Homo sapiens*, and *Homo erectus* all lived as contemporaries at one time or another. None of them evolved from a more robust to a more gracile condition. In fact, in some cases (Neandertal and early *Homo sapiens*) the more robust fossils are the more recent fossils in their respective categories.

Fourth, all of the fossils ascribed to the *Homo habilis* category are contemporary with *Homo erectus*. Thus *Homo habilis* not only did not evolve into *Homo erectus*, it could not have evolved into *Homo erectus*.

Fifth, there are no fossils of *Australopithecus* or of any other primate stock in the proper time period to serve as evolutionary ancestors to humans. As far as we can tell from the fossil record, when humans first appear in the fossil record they are already human. It is this abrupt appearance of our ancestors in morphologically human form that makes the human fossil record compatible with the concept of special creation. This fact is evident even when the fossils are arranged according to the evolutionist's dates for the fossils, although we believe that dating to be grossly in error. In other words, even when we accept the evolutionist's dates for the fossils, the results do not support human evolution. The results, in fact, are so contradictory to human evolution that they effectively falsify the theory. This is the true condition of the human fossil record.

The human fossil record, like the fossil record in general, has failed to furnish evidence for evolution. Evolutionists, understandably, are reluctant to admit it. One way they now handle the problem is to claim that the fossils really are not important. Indicative of this new trend is an incredible statement by Mark Ridley (Oxford University): "No real evolutionist, whether gradualist or punctuationist, uses the fossil record as evidence in favor of the theory of evolution as opposed to special creation. This does not mean that the theory of evolution is unproven."[14] The heading of Ridley's article reads, "The evidence for evolution simply does not depend upon the fossil record."[15]

Ridley then presents us with a classic case of revisionist history. He tells us not only that the fossils are unimportant as evidence for evolution but that from Darwin's time on they never were important:

> The gradual change of fossil species has *never* been part of the evidence for evolution. In the chapters on the fossil record in the *Origin of Species* Darwin showed that the record was useless for testing between evolution and special creation because it has great gaps in it. The same argument still applies (emphasis added).[16]

Now allow me to quote Darwin:

> But just in proportion as this process of extermination has acted on an enormous scale, so must the number of intermediate varieties, which have formerly existed, be truly enormous. Why then is not every geological formation and every stratum full of such intermediate links? Geology assuredly does not reveal any such finely graduated organic chain; and this, perhaps, is the most obvious and serious objection which can be urged against the theory. The explanation lies, as I believe, in the extreme imperfection of the geological record.[17]

Ridley claims that the fossil evidence has never been a part of the evidence for evolution and that Darwin recognized that the fossil evidence did not discriminate between creation and evolution because of the gaps in the fossil record. In contrast, Darwin said that the number of intermediate forms fossilized in the rocks must be enormous, that they have not been found, that their absence is the most serious objection one could have against his theory, and that they have not been found only because of the imperfection of the fossil record. Darwin felt that the fossil evidence was so important, and the lack of transitions was such a serious threat to his theory, that he devoted an entire chapter in the *Origin* to "The Imperfection of the Geological Record." Ridley has forgotten his Darwin.

For 150 years, evolutionists have paraded the fossils they have found as evidence for evolution. They promised more and better fossils in the future, hoping that luck and the tooth fairy would validate their hopes. In the early 1970s, when it became obvious that we had a more than adequate sampling of the fossil record, the grim reality dawned that those transitional fossils were not to be found. The punctuated equilibria model of evolution was then invented to explain why they were not found. However, it is imperative to emphasize that the punctuated equilibria model does not remove the *need* for transitional fossils. It just explains why those transitions have not been found. Certainly, the punctuated equilibria theory is unique. It must be the only theory ever put forth in the history of science that claims to be scientific but then explains why evidence for it cannot be found.

The popular myth is that the hominid fossil evidence virtually proves human evolution. The reality is that this evidence has been a disappointment to evolutionists and is being de-emphasized. In actuality, the human fossil evidence falsifies the concept of human evolution. The Bible, the Word of the living God, clearly declares that humans were specially created. The human fossil evidence is completely in accord with what the Scriptures teach.

Charts of the Human Fossil Record

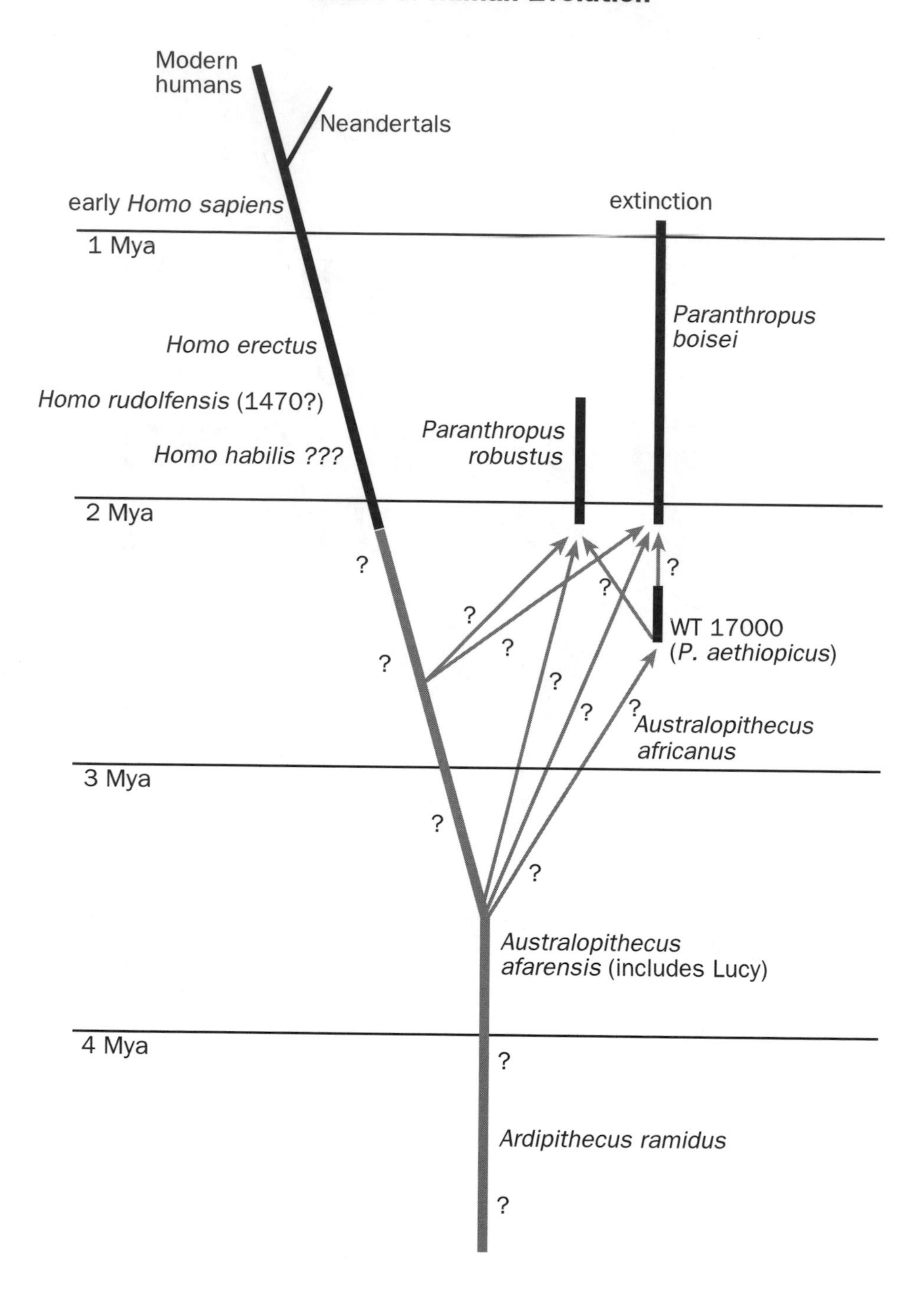
Chart of Human Evolution
Modern humans
Neandertals
early Homo sapiens
extinction
1 Mya
Paranthropus boisei
Homo erectus
Homo rudolfensis (1470?)
Paranthropus robustus
Homo habilis ???
2 Mya
?
?
?
?
WT 17000 (P. aethiopicus)
?
?
?
?
?
Australopithecus africanus
3 Mya
?
?
Australopithecus afarensis (includes Lucy)
4 Mya
?
Ardipithecus ramidus
?

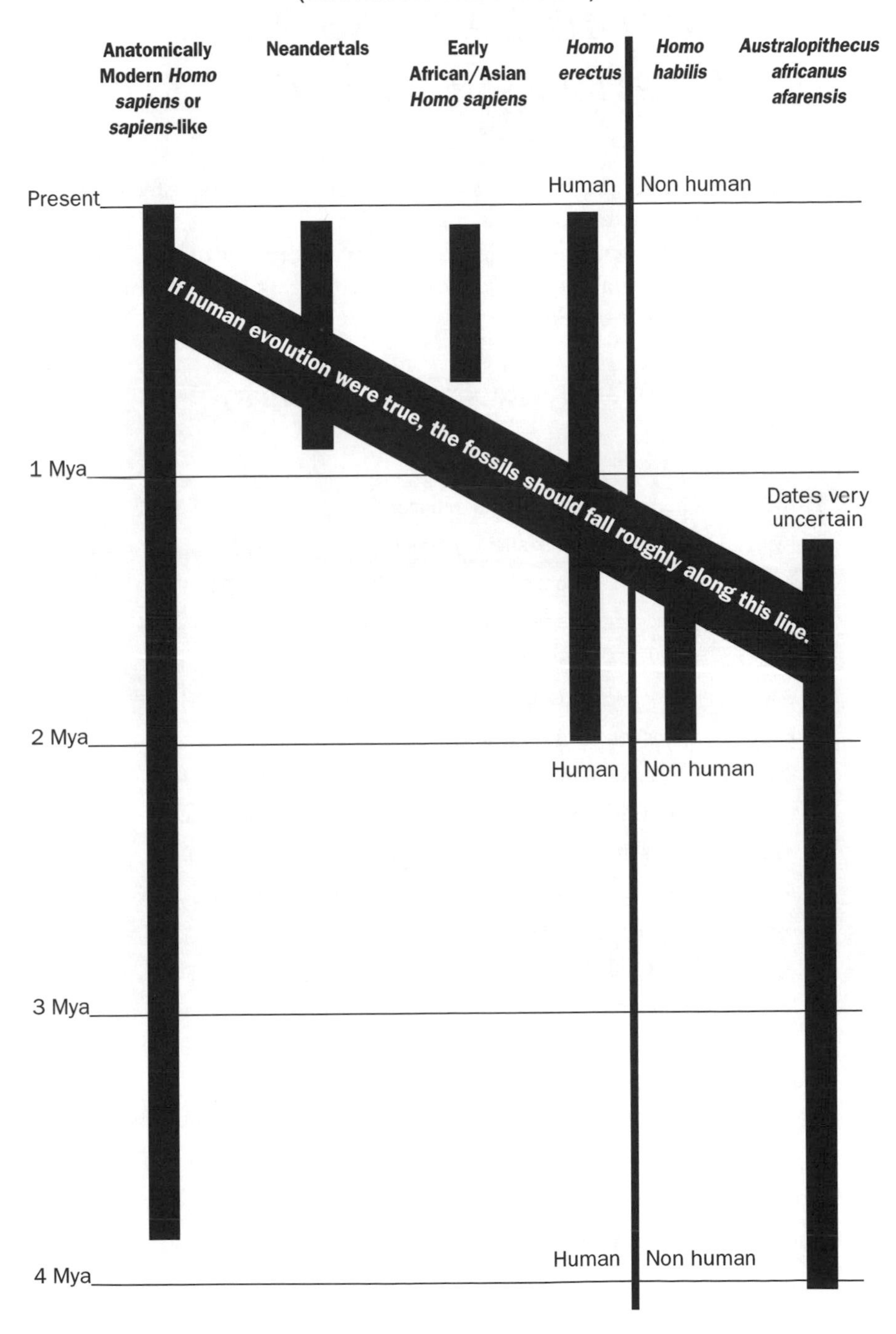
Composite Human Fossil Chart
(Evolution Time Scale)
Anatomically Modern Homo sapiens or sapiens-like
Neandertals
Early African/Asian Homo sapiens
Homo erectus
Homo habilis
Australopithecus africanus afarensis
Human
Non human
Present
1 Mya
2 Mya
3 Mya
4 Mya
If human evolution were true, the fossils should fall roughly along this line.
Dates very uncertain
Human
Non human
Human
Non human

Anatomically Modern *Homo Sapiens* and *Sapiens*-like Fossils

Fossils dated more recently than 28,000 years of age are not listed. When a fossil has a range of dates, the median age is given.

Key: T—Tools Associated; A—Artifacts Associated; B—Burial; C—Cannibalism; ?—Date Uncertain; F—Evidence of Use of Fire; ( ) Minimum Number of Fossil Individuals at the Site

Upper Pleistocene—Evolutionary Time Scale

28 Kya	Chauvet Cave footprints, France (1) A[1]
30 Kya	Cro-Magnon (rock shelter) remains, France (5+) T A B[2]
30 Kya	Mladec (Lautsch) cave remains, Czech Republic (98+) T[3]
30 Kya ?	Maszycka Cave remains, Poland (16+) T C[4]
33 Kya	Abri-Pataud Rock Shelter remains, France (13) T F B[5]
33.8 Kya	Velika Pecina (Great Cave) skull fragments, Croatia (1) T F[6]
35 Kya	Springbok Flats (Tuinplaas) partial skeleton, South Africa (1) T B[7]
35 Kya	Darra-i-Kur remains, Afghanistan (1+)[8]
36 Kya	Fish Hoek (Peer's Cave) skeleton, Cape Province, Africa (1) T[9]
36 Kya	Yamashita-cho 1 (cave) femur and tibia, Okinawa (1) T F[10]
37.4 Kya	Ziyang skullcap and maxilla, Suchuan Province, China (1) T[11]
40 Kya	Lake Mungo skeletons, Australia (2) T A B F[12]
40 Kya ?	Niah Cave skull, partial skeleton, Borneo, Indonesia (1)[13]
40 Kya ?	Brno remains, Czech Republic (3) T B[14]
40 Kya ?	Combe-Capelle Rock Shelter remains, France (1) T B[15]
40 Kya	Mumbwa Cave radius fragments, Zambia (1) T[16]
43 Kya ?	Salawusu (Ordos) remains, Inner Mongolia, China (7?) T[17]
43.2 Kya	Veternica Cave skullcap, Croatia (1) T B? F[18]
44 Kya	Feldhofer Cave skeleton, Germany (1) T F[19]
49 Kya	Equus Cave mandible, teeth, South Africa (1?)[20]

50 Kya	Diré Dawa Cave mandible fragment, Ethiopia (1) T[21]
51 Kya ?	Bacho Kiro Cave mandibles, Bulgaria (2) T A B F[22]
60 Kya	Lake Mungo 3 skeleton, Australia (1) T A B F[23]
67 Kya	Liujiang (cave) skull, skeletal bones, Guangxi Province, China (1)[24]
70 Kya	Dar-es-Soltan Cave 2 skulls, teeth, Morocco (3) T[25]
70 Kya ?	Die Kelders Cave 1 phalanges, teeth, South Africa (1?) T A F[26]
70 Kya ?	Border Cave skull, skeletal remains, Swaziland, South Africa (3) B?[27,28]
80 Kya	Taramsa Hill child skeleton, Egypt (1) T B[29]
84 Kya	Sea Harvest phalanx, tooth, South Africa (1)[30]
91 Kya	Skhul Cave remains, Mount Carmel, Israel (8+) T B[31]
98 Kya	Qafzeh Cave individuals, Israel (19+) T B[32]

Middle Pleistocene

100 Kya ?	Wadjak (caves) skulls, mandibles, Java, Indonesia (2)[33]
105 Kya	Klasies River Mouth Caves skull, skeletal remains, South Africa (5) T F[34,35]
110 Kya	Chouhu partial cranium, Anhui Province, China (1)[36]
117 Kya	Langebaan Lagoon footprints, South Africa (1) T[37]
120 Kya	Tabun Cave, Tabun C2 mandible, Israel (1) T B[38]
130 Kya	Mumba Rock Shelter dental remains, Eyasi, Tanzania (1) T[39]
130 Kya	Omo-Kibish I and III skulls, partial skeleton, Omo Valley, Ethiopia (2) T[40]
135 Kya	Zouhrah Cave mandible, tooth, Morocco (1)[41]
165 Kya	Singa cranium, Sudan (1) T[42]
200 Kya	Nahoon Point footprints, South Africa (1)[43]
400 Kya	Rhodesian Man (Broken Hill/Kabwe) leg bones, Zambia (2+) T A[44,45]
700 Kya	Java Man (*Pithecanthropus I*) leg bone, Java, Indonesia (1)[46]
700 Kya	Sondé tooth, Java, Indonesia (1) (see page 112)[47]

Lower Pleistocene/Pliocene

1.6 Mya	Koobi Fora Footprints, Kenya (1)[48]
1.7 Mya	Gomboré IB-7594, arm bone, Melka Kenturé, Ethiopia (1)[49]
1.76 Mya	Olduvai Hominid 48, clavicle, Tanzania (1) T[50]
1.8 Mya	KNM-ER 813, foot and leg bone fragments, Kenya (1)[51]
1.8 Mya	Olduvai Stone Structure, Tanzania T (see pages 329–30)[52]
1.85 Mya?	KNM-ER 1590, partial cranium, dentition, Kenya (1)[53]
1.89 Mya	KNM-ER 1472, femur, Kenya (1)[54]
1.89 Mya?	KNM-ER 1481, femur, lower leg bone fragments, Kenya (1)[55]
1.9 Mya?	KNM-ER 1470, cranium, Kenya (1)[56]
3.5+ Mya	KNM-KP 271 upper arm bone fragment, Kanapoi, Kenya (1)[57]
3.5+ Mya	KNM-KP 29285 tibia ends, Kanapoi, Kenya (1)[58]
3.75 Mya	Laetoli footprint trails, Tanzania (3)[59]

Neandertal and Pre-Neandertal Fossils

Key: T—Stone Tools Associated; A—Other Artifacts Associated; B—Burial; C—Cannibalism; F—Evidence of Use of Fire; ( ) Minimum Number of Neandertals at the Site

When a fossil has a range of dates, the median age is given; dates of discoveries before 1950 are uncertain.

Upper Pleistocene—Evolutionary Time Scale

17.6 Kya	Banyoles (quarry) mandible, Girona, Spain (1)[1]
24.5 Kya	Lagar Velho 1 child skeleton, Portugal (1) T A B[2]
28 Kya	Vindija Cave remains, Croatia (12) T C?[3]
29 Kya	Mezmaiskaya Cave fragments infant skeleton, Georgia, CIS (1) B[4]
30 Kya	Zafarraya Cave mandible and femur, Spain (2) T[5]
31 Kya	Figueira Brava Cave fragments, Portugal (1?) T[6]
33 Kya	La Ferrassie Rock Shelter remains, France (8) T A B[7]
33 Kya	Hahnöfersand frontal bone, Germany (1)[8]
34 Kya	Arcy-sur-Cure caves remains, France (26) T F A B[9]
35 Kya	Shukbah Cave remains, Israel (2) T F[10]
35+ Kya	Starosel'e remains, Ukraine, CIS (2) T[11]
36.3 Kya	Saint-Césaire Rock Shelter skeleton, France (1) T F B[12]
38 Kya	Ksar 'Akil Rock Shelter remains, Lebanon (3) T B[13]
39 Kya	Combe-Grenal Rock Shelter remains, France (2) T F C[14]
40 Kya ?	Karain ("The Black Cave") cranium, Anatolia, Turkey (1) T[15]
40 Kya	Feldhofer Cave (original Neandertal) skeleton, Germany (1) T F B[16,17]
40 Kya	Amud Cave remains, Israel (16?) T B[18,19]
40 Kya	Monte Circeo remains, Guattari Cave, Italy (4) B T[20,21]
40 Kya	Hortus Cave remains, France (12+) T C[22]
40 Kya	Fond-de-Foret Cave femur, tooth, Belgium (1) T[23]
40 Kya	Saint-Brais Cave tooth, Switzerland (1)[24]

40 Kya	Marillac Cave remains, France (2) T[25]
40 Kya	Antelias Cave fetus skeleton, Lebanon (1) T B[26]
40.3 Kya	Le Moustier Rock Shelter remains, France (2) T B[27, 28]
42.5 Kya	Umm el Tlel skull, Syria (1) T A[29]
42.5 Kya	Denisova Cave teeth, Altayskiy Kray, Russia (Siberia) CIS (1) T[30]
42.5 Kya	Okladnikov Cave teeth, Altayskiy Kray, Russia (Siberia) CIS (1) T[31]
44 Kya	Kulna Cave maxilla, Czech Republic (1) T[32]
45 Kya	Archi 1 child's mandible, Italy (1)[33]
45 Kya	Contrada Ianni child's skull fragment, Italy (1) T[34]
45 Kya	Pech de l'Azé Cave child's cranium, France (1) T F?[35]
45 Kya	Angles sur l'Anglin Rock Shelter tooth, France (1) T[36]
45 Kya	Beegden 1 femur, Netherlands (1)[37]
45 Kya	Carigüela Cave remains, Piñar, Spain (3) T[38]
45 Kya	Columbeira Cave tooth, Portugal (1) T[39]
45 Kya	Le Petit-Puymoyen Cave remains, France (5+) T[40]
45 Kya	Le Placard Cave, Placard 22 child's tooth, France (1) T[41]
45 Kya	Salemas (quarry) remains, Portugal (2) T[42]
45 Kya	Stetten 3 humerus, cave deposits, Germany (1) T[43]
45 Kya	Vergisson (cave) teeth, France (3+) T[44]
46 Kya	Shanidar Cave, upper level, remains I and III, Iraq (2) T F B[45]
47 Kya	Saint Brelade cave teeth, Jersey, Channel Islands, England (1+) T F?[46]
50 Kya	Gibraltar (cave & quarry) remains (2) T[47,48]
50 Kya	La Chapelle-aux-Saints (cave) skeleton, France (1) T A B[49]
50 Kya	Teshik-Tash Cave child remains, Uzbekistan, CIS (1) T B[50]
55 Kya	Sipka (cave) child's mandible, Czech Republic (1) T F B[51]
55 Kya	Chateauneuf-sur-Charente (cave) remains, France (3) T F[52]
55 Kya	Roc de Marsal (cave) skeleton, France (1) T B[53]
60 Kya	Kebara Cave remains, Mount Carmel, Israel (21) T F B[54]
64 Kya	La Quina Rock Shelter remains, France (25) T A B[55,56]
65 Kya	Karain ("The Black Cave") teeth, Anatolia, Turkey (1) T[57]
68 Kya	Caminero partial skeleton, Castaigne Cave, France (1) T[58]
70 Kya	Engis Caverns remains, Belgium (3) B C?[59,60]
70 Kya	Kiik-Koba ("Wild Cave") remains, Ukraine, CIS (2) T B[61]

70 Kya	Molare (cave) Shelter, child's mandible, Italy (1) T F[62]
70 Kya	Spy (cave) remains, Belgium (3) T B[63]
70 Kya	Dzhruchula (cave) tooth, Georgia, CIS (1) T[64]
70 Kya	Genay cranial fragments, France (1) T[65]
70 Kya	La Naulette Cave remains, Belgium (1)[66]
70 Kya	Leuca (cave) child's tooth, Italy (1) T F[67]
70 Kya	Ochoz (cave) remains, Czech Republic (2) T[68]
70 Kya	Ohaba-Ponor (cave) toe bone, Romania (1) T[69]
70 Kya	Wildscheuer (cave) remains, Germany (2) T[70]
70 Kya	Régourdou Cave skeleton, France (1) T A B[71,72]
75 Kya	Dederiyeh Cave infant skeleton, Syria (1) T B[73]
75 Kya	Monsempron remains, France (4+) T F[74]
75 Kya	Shanidar Cave, lower level remains, Iraq (7) T F B[75,76]
80 Kya	Subalyuk Cave remains, Hungary (2) T[77]
80 Kya	Soulabe-las-Maretas (cave) teeth, France (1+) T[78]
80 Kya	Uluzzo (cave) child's tooth, Italy (1) T[79]
90 Kya	Montgaudier Cave remains, France (4+) T[80]
90 Kya	Quinzano (cave) skull fragment, Italy (1) T[81]
90 Kya	Santa Croce Cave femur, Bisceglie, Italy (1) T[82]
90 Kya	Lezetxiki Cave humerus, Spain (1) T[83]
90 Kya	Malarnaud Cave remains, France (1+)[84]
91 Kya	Skhul Cave, Skhul 4 and 9 remains, Mount Carmel, Israel (2) T B[85,86]
98 Kya	Qafzeh Cave, Qafzeh 6 and 9, Israel (2) T A B[87,88]

Neandertal Fossils—Middle Pleistocene

100 Kya	Sala skull fragment, Slovak Republic (1)[89]
100+ Kya	Cava Pompi (caves) cranial and leg fragments, Fofi, Latium, Italy (1)[90]
105 Kya	Gánovce skull fragments and endocast, Slovak Republic (1) T[91]
110 Kya	Moula-Guercy cave remains, France (6) T F C[92]
120 Kya	Tabun Cave remains, Mount Carmel, Israel (12) T B[93]
120 Kya	Taubach child's tooth, Germany (2?) T[94]
125 Kya	Saccopastore crania, Italy (2) T[95]
125 Kya	Cova Negra (cave) skull fragment, Spain (1) T[96]
130 Kya	Krapina Cave remains, Croatia (75–82) T F B C[97,98]
156 Kya	Biache-Saint Vaast partial skulls, France (2) T A F[99]
156 Kya	Le Lazaret Cave skull fragment, teeth, France (3) T A F[100]
156 Kya	Orgnac 3 Cave teeth, France (1+) T A[101]
160 Kya	La Chaise Caves remains, France (18) T[102]
160 Kya	Fontéchevade Cave partial skulls, France (2) T[103]
179 Kya	Bau de l'Aubesier (rock shelter) mandibles, teeth, France (15) T[104]
190 Kya	Montmaurin Cave remains, France (4) T[105,106]
215 Kya	Pontnewydd Cave skull, dental fragments, Wales, UK (4+) T[107]
225 Kya	Karain ("The Black Cave") mandible, hand, arm, and leg fragments, Turkey (1) T[108]
230 Kya	Ehringsdorf/Weimar remains, Germany (9) T C[109,110]
250 Kya	Casal de' Pazzi skull fragment, Italy (1) T[111,112]
300 Kya	Reilingen partial skull, Germany (1)[113,114]
300 Kya	Steinheim skull, Germany (1)[115,116]
300 Kya	Zuttiyeh Cave (Galilee Man) partial skull, Israel (1) T[117]
325 Kya	Karain ("The Black Cave") vertebra, leg fragments, Anatolia, Turkey (1) T[118]
350 Kya ?	Azykh Cave (Azykhskaya Peshchera) mandible, Azerbaijan, CIS (1) T[119,120]
400 Kya	Sima de los Huesos (Atapuerca) cave remains, Spain (33) B[121]
400 Kya	Vértesszöllös skull fragment, teeth, Hungary (2) T F[122]
400 Kya	Swanscombe skull, England (1) T[123,124]
400 Kya	Arago Cave (Tautavel) skull, skeletal fragments, France (13+) T F[125,126]
400 Kya	Altamura Cave skull, skeleton, Italy (1)[127]

412 Kya	Bilzingsleben skull fragments, teeth, Germany (2–3) T A F[128,129]
450 Kya	Castel di Guido skull fragments, maxilla, femur, Italy (1) T A[130,131,132]
450 Kya	Fontana Ranuccio teeth, Italy (1) T A[133]
450 Kya	Petralona Cave skull, Greece (1) T F A B[134,135]
500 Kya	Boxgrove lower leg bone, England (1) T[136,137]
500 Kya	Mauer (Heidelberg) mandible, Germany (1)[138,139]
500 Kya	Notarchirico-Venosa, Basilicata, femur, Italy (1) T[140,141]
500 Kya	Visogliano Rock Shelter mandible fragments, two teeth, Trieste, Italy (1) T[142]
600 Kya	Gesher Benot Ya'acov femora, Israel (1+) T A[143]
800 Kya	Ceprano calvarium, Italy (1) B?[144,145]
800 Kya	Gran Dolina TD6 cave site, Atapuerca, remains, Spain (6) T C[146]

Neandertals with Unknown Dates

Barakai Cave child mandible, Russia, CSI (1) T[147]

Brüx skullcap, Slovak Republic (1)[148]

Me'arat Shovakh (cave) tooth, Galilee, Israel (1) T[149]

Masloukh Cave tooth, Lebanon (1) T[150,151]

Rivaux fragments, southern France (1?)[152]

Zaskolnaya 2 immature mandible, Crimea (1)

Early African/Asian *Homo Sapiens* Fossils

Key: T—Tools Associated; A—Artifacts Associated; B—Burial; C—Cannibalism; ?—Date Uncertain; F—Evidence of Use of Fire; x—Assignment to this category is tentative; ( ) Minimum Number of Early African/Asian *Homo sapiens* Fossil Individuals at the Site

When a fossil has a range of dates, the median age is given.

Holocene/Upper Pleistocene—Evolutionary Time Scale

5 Kya	Cape Flats skull and skeleton fragments, Cape Province, South Africa (3) T[1]
25 Kya	Upper Cave 101 skull, Zhoukoudian, China (1)[2]
65 Kya ?	Haua Fteah (Great Cave) partial mandibles, Libya (2) T[3]
70 Kya	Border Cave humerus, Swaziland, South Africa (1)[4,5]
84 Kya	Mugharet el 'Aliya cave (Tangier) maxilla, tooth, Morocco (3) T[6]

Middle Pleistocene

103 Kya	Jiande remains, China (1?)[7]
105 Kya	Klasies River Mouth Caves skeletal fragments, South Africa (2+) T F[8]
107 Kya	Xujiayao (Hsu-c'hia-yao) skull fragments, Shanxi Province, China (2) T[9]
120 Kya	Laetoli Hominid 18, skull, Tanzania (1) T[10,11]
130 Kya	x Omo-Kibish II skull, Omo Valley, Ethiopia (1)[12]
130 Kya	Mapa (Maba) skull, Guangdong Province, China (1) T[13]
150 Kya	Eliye Springs skull, West Turkana, Kenya (1)[14]
150 Kya	Xindong/Xingdong (Zhoukoudian Cave) tooth, China (1)[15]
160 Kya	Herto skulls, fragments, Middle Awash, Ethiopia (10) T[16,17]
180 Kya	Chaoxian (Yinshan) cave occipital, maxilla, Anhui Province, China (1?)[18]

185 Kya	Dingcun (Ting-T'sun) child skull fragments, teeth, Shanxi Province, China (1) T[19]
193 Kya	Tongzi (Tonzhi) teeth, Guizhou Province, China (1?)[20]
195 Kya	Changyang maxilla, tooth, Hubei Province, China (1) T[21]
200 Kya	Jebel-Irhoud (cave/mine) skulls, mandible, Morocco (4) T[22]
200 Kya	Rabat (Kébibat) quarry skull fragment, maxilla, mandible, Morocco (1)[23]
205 Kya	Dali (Tali) cranium, Shaanxi Province, China (1) T[24]
250 Kya	Wadi Dagadlé maxilla, Djibouti (1)[25]
250 Kya	Jinniushan (Yinkou) partial skeleton, Liaoning Province, China (1) T F?[26]
259 Kya	Florisbad partial cranium, Orange Free State, South Africa (1) T[27]
270 Kya	KNM-ER 3884 cranium, Kenya (1)[28]
300 Kya	Hoedjiespunt remains, Saldanha Bay, South Africa (1?) T[29]
300 Kya	KNM-ER 999 femur, Kenya (1)[30]
325 Kya	Garba 3 cranial fragments, Melka-Kunturé, Ethiopia (1) T[31]
350 Kya	Salé cranium, maxilla fragment, Morocco (1)[32]
400 Kya	Ndutu skull, Tanzania (1) T[33,34]
400 Kya	Rhodesian Man (Broken Hill/Kabwe) skull, Zambia (2+) T A[35,36]
400 Kya	Miaohoushan cave teeth, femur, Liaoning Province, China (1) T F[37]
450 Kya ?	Cave of Hearths mandible, arm bone fragments, Transvaal, South Africa (1) T F[38]
450 Kya	Eyasi cranial fragments, Tanzania (3) T[39]
450 Kya ?	Narmada skullcap, clavicle, Narmada Valley, India (1?) T[40]
500 Kya	Saldanha (Hopefield/Elandsfontein) cranium, South Africa (1) T[41]
600 Kya	x Bodo skulls, Middle Awash, Ethiopia (2) T[42]
600 Kya ?	Ain Maarouf (El Hajeb) femoral shaft, Morocco (1)[43]

Homo Erectus Fossils (Including *Homo ergaster*)

Key: T—Tools Associated; A—Artifacts Associated; B—Burial; C—Cannibalism; ?—Date Uncertain; F—Evidence of Use of Fire; #—Called by some *Homo ergaster*; x—Assignment Tentative; ( ) Minimum Number of Individuals at the Site

When a fossil has a range of dates, the median age is given.

Holocene/Upper Pleistocene—Evolutionary Time Scale

6 Kya	Mossgiel cranium, New South Wales, Australia (1) T F B[1]
6.5 Kya	Cossack skull, mandible, limb fragments, Western Australia (1)[2]
7 Kya ?	Lake Nitchie people, New South Wales, Australia (2+) T B[3]
9.5 Kya	Kow Swamp people, Victoria, Australia (40+) T A F B[4]
9.5 Kya	Cohuna cranium, Victoria, Australia (1)[5]
9.5 Kya ?	Coobool Creek/Crossing people, New South Wales, Australia (29+)[6]
12 Kya	Talgai cranium, Queensland, Australia (1)[7]
27 Kya	Sambungmacan, 3 calvaria, 1 tibia, Java, Indonesia (3–4) T[8,9]
30 Kya ?	Willandra Lakes WHL 50 cranium, New South Wales, Australia (1)[10]
35 Kya	Témara (Smugglers' Cave) cranial fragment, mandible, Morocco (2) T[11]
40 Kya	Java Solo (Ngandong) people, Java, Indonesia (14) T A[12,13]

Middle Pleistocene

130 Kya ?	Hazorea 2, 4, and 5, skull and mandible fragments, Israel (2) T[14]
130 Kya	Laetoli Hominid 29, mandible, Tanzania (1) T?[15]
175 Kya	Hexian (cave) skull fragments, Anhui Province, China (2)[16]
300 Kya	Zhoukoudian (Peking Man) remains, Locality 1, China (50+) T F[17,18]

400 Kya	Lainyamok maxilla, teeth, femur, Kenya (1) T[19]
400 Kya ?	Xichuan teeth, Henan Province, China (1+)[20]
400 Kya ?	Nanzhao (Xiaohuashan) tooth, Henan Province, China (1)[21]
400 Kya ?	Yiyuan (Qizianshan) partial cranium, skeletal fragments, China (1)[22]
400 Kya ?	Bailongdong (White Dragon Cave) teeth, Hubei Province, China (1) T[23]
400 Kya ?	Jianshi (Longgudong cave) teeth, Hubei Province, China (1)[24]
400 Kya ?	Luc Yen teeth, Tiger Cave, Vietnam (1)[25]
450 Kya ?	SM3 skullcap, Poloyo Village, Solo River, Central Java (1)[26]
450 Kya	Kapthurin (KNM-BK 66, 67, and 8518) mandibles, limb fragments, Kenya (2) T[27]
475 Kya	Tham Khuyen Cave teeth, Vietnam (1+)[28]
480 Kya	Lang Trang Caves teeth, Vietnam (3)[29]
500 Kya	Kedungbrubus mandible, Java, Indonesia (1)[30]
600 Kya	Tangshan Cave (Nanjing) crania, tooth, Jiangsu Province, China (3)[31]
600 Kya	*Pithecanthropus* IX skull, Sangiran, Java (1)[32]
600 Kya	Thomas Quarries 1 and 3 skulls, mandible, Morocco (2) T[33]
600 Kya	Sidi Abderrahman cave (Casablanca) mandible fragments, Morocco (1) T[34]
600 Kya ?	Ain Maarouf (El Hajeb) femoral shaft, Morocco (1)[35]
650 Kya	Lantian (Chenchiawo) mandible, Shaanxi Province, China (1) T[36]
650 Kya	Yunxian (Longgudong Cave) teeth, Hubei Province, China (1)[37]

Lower Pleistocene/Pliocene

700 Kya	Java Man (*Pithecanthropus* I) skullcap, Java, Indonesia (1) T[38,39]
700 Kya	Sangiran 17 (*Pithecanthropus* VIII) cranium, Java, Indonesia (1)[40,41]
700 Kya	Yayo (Koro Toro) skull fragment, Chad (1)[42]
700 Kya	Olduvai Hominid 22, right mandible, Tanzania (1)[43]
750 Kya	Ternifine (Tighenif/Palikao) partial skull, mandibles, Algeria (3) T[44]
790 Kya	Olduvai Hominid 2, skull fragments, Tanzania (1) T[45]

900 Kya	Olduvai Hominid 23, mandible fragment, Tanzania (1) T[46]
900 Kya	Olduvai Hominid 34, leg bones, Tanzania (1) T[47]
950 Kya ?	Luonan cave (Donghe) tooth, Shaanxi Province, China (1)[48]
1 Mya	Olduvai Hominid 28, femur, hip bone, Tanzania (1) T[49]
1 Mya`	Olduvai Hominid 12, cranial fragments, Tanzania (1) T[50]
1 Mya	Olduvai Hominid 29, teeth, phalanx, Tanzania (1) T[51]
1 Mya	Buia cranium, pelvic fragments, teeth, Eritrea (1)[52]
1 Mya	Olduvai Hominid 51, left mandible with teeth, Tanzania (1)[53]
1 Mya	Gomboré II cranial fragment, Melka-Kunturé, Ethiopia (1) T[54]
1 Mya	Bouri remains, Middle Awash, Ethiopia (4) T[55]
1 Mya	Yunxian (EV 9001, 9002) crania, Hubei Province, China (2) T[56,57]
1.15 Mya	Lantian (Gongwangling) skull, tooth, Shaanxi Province, China (1) T F[58]
1.2 Mya	Olduvai Hominid 9 (Chellean Man) cranium, Tanzania (1) T? A?[59,60]
1.23 Mya	x Olduvai Hominid 36, lower arm bone, Tanzania (1)[61]
1.3 Mya	'Ubeidiya cranial fragments, teeth, Israel (1) T[62,63]
1.35 Mya	Omo L-996–17, cranial fragments, Ethiopia (1)[64]
1.4 Mya	Konso left mandible, molar, Ethiopia (2) T[65]
1.5 Mya	Java (Sangiran/Trinil) people, Java, Indonesia (43+)[66]
1.52 Mya	KNM-ER 992, mandible with teeth, Kenya (1) T[67]
1.55 Mya	KNM-ER 803, skeletal fragments, Kenya (1)[68]
1.58 Mya	Olduvai Hominid 15, teeth, Tanzania (1)[69]
1.58 Mya	KNM-ER 3883, skull, Kenya (1) T[70]
1.6 Mya	KNM-ER 737, femur, Kenya (1)[71]
1.6 Mya	KNM-ER 820, juvenile mandible with teeth, Kenya (1)[72]
1.6 Mya	KNM-ER 1466, cranial fragment, Kenya (1) T[73]
1.6 Mya	KNM-WT 15000, skeleton, Kenya (1)[74]
1.7 Mya	KNM-ER 807A, maxilla fragment with teeth, Kenya (1)[75]
1.7 Mya	KNM-ER 1808, skeleton, Kenya (1)[76]
1.7 Mya	KNM-ER 1821, cranial fragment, Kenya (1)[77]
1.7 Mya	KNM-ER 730, skull fragments, Kenya (1) T[78]
1.7 Mya	Dmanisi skull (D2280) and mandibles (D211, D2600) CIS (3) T[79,80,81]
1.78 Mya	KNM-ER 3733, cranium, Kenya (1)[82]

1.78 Mya	KNM-ER 1809, femur, Kenya (1)[83]
1.79 Mya	KNM-ER 1507, juvenile left mandible, Kenya (1)[84]
1.8 Mya	Yuanmou teeth, Yunnan Province, China (1) T F[85,86]
1.8 Mya	Swartkrans SK-15, 18a, and 18b, skeletal fragments, South Africa (1) T F[87]
1.8 Mya	Swartkrans SK-27, cranium, South Africa (1)[88]
1.8 Mya	Swartkrans SK-84, SKX-5020, thumb, finger fragments, South Africa (2) T F[89]
1.85 Mya ?	Swartkrans SK-45, 847, skull and maxilla fragments, South Africa (5?) T F[90]
1.9 Mya	# Longgupo (Dragon Hill) Cave mandible, tooth, Sichuan Province, China (1) T[91]
1.9 Mya	Modjokerto Infant skull, Djetis Beds, Java, Indonesia (1)[92]
1.9 Mya	Olduvai Hominid 60, molar showing toothpick use, Tanzania (1)[93]
1.9 Mya	KNM-ER 2598, cranial fragment, Kenya (1)[94]
1.95 Mya	KNM-ER 3228, hip bone, Kenya (1)[95]

Homo Erectus/Homo Habilis–Contemporaneousness

Evolution Time Scale

Each entry (except Dmanisi) is one individual

	***Homo habilis* Fossils**	***Homo erectus* (See *H. e.* chart for details)**
1.5 Mya	Olduvai Hominid 13 cranial fragments, Tanzania[1]	KNM-ER 992
		KNM-ER 803
		O. H. 15
		KNM-ER 3883
1.6 Mya		KNM-ER 737
	Olduvai Hominid 16 skull fragments, Tanzania[2,3]	KNM-ER 820
		KNM-WT 15000
		KNM-ER 1466
1.7 Mya		KNM-ER 1808
		KNM-ER 730
		KNM-ER 807
		KNM-ER 1821
	Dmanisi skulls (D2282. D2700), and fragments, Georgia, CIS (3),[4,5,6]	
*	Sterkfontein Stw 53 skull, South Africa[7]	
	Olduvai Hominid 6 skull fragments, Tanzania[8]	
	Olduvai Hominid 7 skull, hand, Tanzania[9]	
	Olduvai Hominid 8 foot bones, Tanzania[10]	KNM-ER 3733
	Olduvai Hominid 35 tibia, fibula, Tanzania[11]	KNM-ER 1507
1.8 Mya	Olduvai Hominid 49 radius, Tanzania[12]	SK-15, 18a, 18b
	Olduvai Hominid 62 skeletal fragments, Tanzania[13]	SK-84
		SKX-5020
	Omo L 894–1 cranial fragments, Ethiopia[14]	SK-45,847

	KNM-ER 1805 cranium, mandible, Kenya[15]	
	KNM-ER 1813 cranium, teeth, Kenya[16]	
	Olduvai Hominid 4 teeth, Tanzania[17]	KNM-ER 1809
1.9 Mya	Olduvai Hominid 24 cranium, Tanzania[18]	
	Olduvai Hominid 52 cranial fragment, Tanzania[19]	Java (Djetis)
	KNM-ER 1802 mandible, tooth, Kenya[20]	Longguppo
	Swartkrans SK-68 tooth, South Africa[21]	KNM-ER 2598
		KNM-ER 3228
2.0 Mya	KNM-WT 15001 cranial fragment, Kenya[22]	

* Date uncertain

Notes

Chapter 1: "Show Me Your Fossils; I'll Show You Mine"

1. David Van Reybrouck, "Imaging and imagining the Neanderthal: the role of technical drawings in archaeology," *Antiquity* 72 (March 1998): 62.

2. David W. Frayer, "Naming Our Ancestors," review of *Naming Our Ancestors*, edited by W. Eric Mielke and Sue Taytlor Parker, *American Journal of Physical Anthropology* 98, no. 2 (October 1995): 235.

3. M. Braun, J. J. Hublin, and P. Boucher, "New reconstruction of the Middle Pleistocene skull of Steinheim," *American Journal of Physical Anthropology* Supplement 26 (1998): 113.

4. Carl C. Swisher III, Garniss H. Curtis, and Roger Lewin, *Java Man* (New York: Scribner, 2000), 12.

5. James Shreeve, *The Neandertal Enigma* (New York: William Morrow and Company, Inc., 1995), 102.

6. Eric Delson, ed., *Ancestors: The Hard Evidence* (New York: Alan R. Liss, Inc., 1985), 1.

7. Ibid., 1–2.

8. Bernard Wood, "A Gathering of Our Ancestors," *Nature* 309 (17 May 1984): 208.

9. "Old Bones Week," *Discover* (June 1984): 69.

10. Ian Tattersall and Niles Eldredge, "Fact, Theory, and Fantasy in Human Paleontology," *American Scientist* 65 (March–April 1977): 207.

11. Roger Lewin, *Bones of Contention* (New York: Simon and Schuster, 1987), 24.

12. Ellen Ruppel Shell, "Flesh and Bone," *Discover* (December 1991): 41.

13. Ibid.

14. Louis S. B. Leakey, *Adam's Ancestors*, 4th ed. (New York: Harper & Row, 1960), v–vi.

15. Becky A. Sigmon and Jerome E. Cybulski, eds., *Homo erectus: Papers in Honor of Davidson Black* (Toronto: University of Toronto Press, 1981), 5.

16. Roger Lewin, "Ancestors Worshipped," *Science* 224 (4 May 1984): 478.

17. Delson, *Ancestors*, 4.

18. Ibid., 4–5.

19. Tim D. White, Gen Suwa, and Berhane Asfaw, "*Australopithecus ramidus*, a new species of early hominid from Aramis, Ethiopia," *Nature* 371 (22 September 1994): 306–12.

20. Tim D. White, Gen Suwa, and Berhane Asfaw, "*Australopithecus ramidus*, a new species of early hominid from Aramis, Ethiopia," *Nature* 375 (4 May 1995): 88.

21. Ann Gibbons, "Glasnost for Hominids: Seeking Access to Fossils," *Science* 297 (30 August 2002): 1464.

22. Ibid., 1465.

23. Ibid., 1464.

Chapter 2: A Fairy Tale for Grown-ups

1. Sherwood L. Washburn, *Abstracts of the 71st Annual Meeting, American Anthropological Association, Washington, DC*, 1972, 121.

2. Bernard Wood, "A date with Java Man," review of *Java Man*, by Garniss Curtis, Carl Swisher, and Roger Lewin, *NewScientist* (17 February 2001): 54.

3. Carl Zimmer, "Great Mysteries of Human Evolution," *Discover* (September 2003): 33–43.

4. Carl Sagan, *Velikovsky's Challenge to Science*, cassette tape 186–74, produced by the American Association for the Advancement of Science, Washington, DC

5. G. A. Kerkut, *The Implications of Evolution* (New York: The Macmillan Company, 1960), 7.

6. Andrew Hill, "The gift of Taungs," review of *Hominid Evolution*, edited by D. V. Tobias, *Nature* 323 (18 September 1986): 209.

7. Jeffrey H. Schwartz, "Another Perspective on Hominid Diversity," *Science* 301 (8 August 2003): 763.

8. Andrew Hill, "The Myths of Human Evolution," review of *The Myths of Human Evolution*, by Niles Eldredge and Ian Tattersall, *American Scientist* 72 (March–April, 1984): 189.

9. Ian Tattersall, "Outdoor Man," review of *Ecc Homo*, by Noel T. Boaz, *Nature* 388 (14 August 1997): 638.

10. Gibbons, "Glasnost for Hominids," 1467.

11. Jonathan Marks, "Gorillas in the midst," review of *Primate Origins and Evolution*, by R. D. Martin, *Nature* 344 (15 March 1990): 205.

12. Norman Macbeth, *Darwin Retried: An Appeal to Reason* (Boston: Gambit Incorporated, 1971).

13. Phillip E. Johnson, *Darwin on Trial* (Washington, DC: Regnery Gateway, 1991), 81.

14. Ibid., 83.

15. Van Reybrouck, "Imaging and imagining the Neanderthal," 63.

16. Louis S. B. Leakey, "Finding the World's Earliest Man," *National Geographic* (September 1960): 421.

17. F. Clark Howell, *Early Man*, rev. ed. (New York: Time-Life Books, 1968), 41.

18. Tom Waters, "Almost Human," *Discover* (May 1990): 44.

19. Misia Landau, *Narratives of Human Evolution* (New Haven: Yale University Press, 1991).

20. Misia Landau, "Human Evolution as Narrative," *American Scientist* 72 (May–June 1984): 264.

21. Jared Diamond, "The Great Leap Forward," *Discover* (May 1989): 50–60.

22. Richard Leakey, *The Sixth Extinction* (New York: Random House, 1996).

Chapter 3: Dead Reckoning

1. Ian Tattersall and Jeffrey Schwartz, *Extinct Humans* (New York: Westview Press, 2000).

2. Douglas J. Preston, "Four Million Years of Humanity," *Natural History* (April 1984): 12.

3. Boyce Rensberger, "Bones of Our Ancestors," *Science* 84 (April 1984): 29. This publication has since been combined with the Time-Life publication *Discover.*

4. Richard E. F. Leakey, "The Search For Early Man," cassette tape interview, produced by the American Association for the Advancement of Science, Washington, D. C., 1973.

5. John Reader, "Whatever Happened to Zinjanthropus?" *NewScientist* (26 March 1981): 802.

6. Constance Holden, "The Politics of Paleoanthropology," *Science* 213 (14 August 1981): 737.

Chapter 4: Monkey Business in the Family Tree

1. Sagan, *Velikovsky's Challenge to Science.*

2. Vincent Sarich, Creation-Evolution debate (North Dakota State University, Fargo, ND, 28 April 1979).

3. Since the revision of German orthography in 1901, the h in Neanderthal is omitted in the vernacular name; thus, Neandertal. According to the International Code for Zoological Nomenclature, however, the name of the species *Homo neanderthalensis* (King, 1864) must continue to be written with the letter h following the letter t, as well as in the newer name *Homo sapiens neanderthalensis* (Campbell, 1964). Many writers still continue to use the older spelling for the vernacular.

4. Kenneth A. R. Kennedy, *Neanderthal Man* (Minneapolis: Burgess Publishing Company, 1975), 33. The quotation is from Boule, but Kennedy gives no reference.

5. Jerold Lowenstein, Theya Molleson, and Sherwood Washburn, "Piltdown Jaw Confirmed as Orang," *Nature* 299 (23 September 1982): 294.

6. Ronald Miller, *The Piltdown Men* (New York: St. Martin's Press, 1972).

7. Charles Blinderman, *The Piltdown Inquest* (Buffalo: Prometheus Books, 1986).

8. Stephen Jay Gould, "The Piltdown Conspiracy," *Natural History* (August 1980): 8–28.

9. John Winslow and Alfred Meyer, "The Perpetrator at Piltdown," *Science* 83 (September 1983): 32–43.

10. Wilbur M. Smith, "In the Study," *Moody Monthly* (March 1954): 26.

11. The Piltdown hoax never ceases to surprise. After eighty-three years, the one man who up to now seemed to be above suspicion has been added to the list of suspects: Sir Arthur Keith himself. See Frank Spencer, *Piltdown: A Scientific Forgery* (New York: Oxford University Press, 1990). There is still no smoking gun. Spencer's evidence, like all the rest, is entirely circumstantial. Spencer speculates that Keith and Dawson worked together. Dawson's motive was to gain entrance to the prestigious Royal Society. Keith's motive was to distract attention from the smaller-brained Java Man (*Pithecanthropus I*) that had been discovered about twenty years earlier and to strengthen his own position that the brain of man evolved first, before bipedalism. No doubt the list of suspects will continue to grow.

Chapter 5: Looks Aren't Everything

1. Vincent Sarich, Creation-Evolution debate.

2. William W. Howells, "*Homo erectus* in human descent: ideas and problems," in Sigmon and Cybulski, *Homo erectus*, 70–71.

3. F. B. Livingstone, "Gene flow in the Pleistocene," abstract, *American Journal of Physical Anthropology* Supplement 12 (1991): 117.

4. T. C. Partridge, "Geomorphological Dating of Cave Openings at Makapansgat, Sterkfontein, Swartkrans, and Taung," *Nature* 246 (9 November 1973): 75–79.

5. Karl W. Butzer, "Paleoecology of South African Australopithecines: Taung Revisited," *Current Anthropology* 15, no. 4 (December 1974): 382.

6. Ibid., 411.

7. Ibid., 404.

8. Phillip V. Tobias, "Implications of the New Age Estimates of the Early South African Hominids," *Nature* 246 (9 November 1973): 82.

9. Ibid.

10. A. Walker, R. E. Leakey, J. M. Harris, and F. H. Brown, "2.5–Myr Australopithecus boisei from west of Lake Turkana, Kenya," *Nature* 322 (7 August 1986): 517–22.

11. Ian Tattersall, Eric Delson, and John Van Couvering, eds., *Encyclopedia of Human Evolution and Prehistory* (New York: Garland Publishing, 1988), 571.

12. Ibid.

13. Richard G. Klein, *The Human Career: Human Biological and Cultural Origins* (Chicago: University of Chicago Press, 1989), 113.

14. Ibid.

15. Bryan Patterson, Anna K. Behrensmeyer, and William D. Sill, "Geology and Fauna of a New Pliocene Locality in North-western Kenya," *Nature* 226 (6 June 1970): 918–21.

16. Bryan Patterson and W. W. Howells, "Hominid Humeral Fragment from Early Pleistocene of Northwestern Kenya," *Science* 156 (7 April 1967): 65. Originally the stratum was thought to be Pleistocene, but it was later determined to be Pliocene Age. See note 15 above.

17. Ibid., 66.

18. Henry M. McHenry, "Fossils and the Mosaic Nature of Human Evolution," *Science* 190 (31 October 1975): 428.

19. David Pilbeam, *The Evolution of Man* (New York: Funk and Wagnalls, 1970), 151. The describer is W. W. Howells, not F. Clark Howell.

20. McHenry, "Fossils and the Mosaic Nature," 428.

21. Brigette Senut, "Humeral Outlines in Some Hominoid Primates and in Plio-Pleistocene Hominids," *American Journal of Physical Anthropology* 56 (1981): 275.

22. One of the oldest in the sense that it is one of the oldest fossils capable of a legitimate diagnosis.

23. Howells, "*Homo erectus* in human descent," 79–80.

Chapter 6: With a Name like Neandertal, He's Got to Be Good

1. Technically, Neandertal fossils on Gibraltar were discovered earlier, but their importance was not understood, and they have remained in relative obscurity.

2. Donald Johanson and James Shreeve, *Lucy's Child* (New York: William Morrow & Company, 1989), 49.

3. Erik Trinkaus, "Hard Times Among the Neanderthals," *Natural History* 87, no. 10 (December 1978): 58. See also R. L. Holloway, "The Neandertal Brain: What Was Primitive," abstract, *American Journal of Physical Anthropology* Supplement 12 (1991): 94.

4. Trinkaus, "Hard Times," 58.

5. Valerius Geist, "Neanderthal the Hunter," *Natural History* 90, no. 1 (January 1981): 30.

6. Ibid.

7. "Stone tips on ancient hunting," *Science News* (1 July 1989): 13.

8. Geist, "Neanderthal the Hunter," 34.

9. Jared Diamond, "The Great Leap Forward," *Discover* (May 1989): 50–60.

10. Ibid., 55.

11. "How Neanderthals Chilled Out," *Science News* (24 March 1990): 189.

12. Geist, "Neanderthal the Hunter," 36.

13. Christopher B. Stringer, "Fate of the Neanderthal," *Natural History* (December 1984): 12.

14. N. Mercier, H. Valladas, J. L. Joron, J. L. Reyss, F. Leveque, and B. Vandermeersch, "Thermoluminescence dating of the late Neanderthal remains from Saint-Césaire," *Nature* 351 (27 June 1991): 737–39.

15. Michael H. Day, *Guide to Fossil Man*, 4th ed. (Chicago: University of Chicago Press, 1986), 128–29.

16. Ibid., 134.

17. Ranier Berger and W. F. Libby, *Radiocarbon Journal*, vol. 8 (1966): 480.

18. Klein, *Human Career*, 1st ed., 281–82.

19. Ibid., 279.

20. Geist, "Neanderthal the Hunter," 34.

21. Klein, *Human Career*, 1st ed., 283.

22. J. Lawrence Angel, "History and Development of Paleopathology," *American Journal of Physical Anthropology* 56, no. 4 (December 1981): 512.

23. Francis Ivanhoe, "Was Virchow Right about Neandertal?" *Nature* 227 (8 August 1970): 577–79.

24. D. J. M. Wright, "Syphilis and Neanderthal Man," *Nature* 229 (5 February 1971): 409.

25. Michele A. Miller, "New Discoveries at the First Neanderthal Site," *Athena Review* 2, no. 4 (2001): 5.

26. "Germans unearth hoard of Neanderthal remains," *Nature* 407 (7 September 2000): 9.

Chapter 7: Java Man: The Rest of the Story

1. Pat Shipman's recent work, *The Man Who Found the Missing Link* (New York: Simon & Schuster, 2001) is the only biography of Dubois. However, it does not meet the need. It reads more like a historical novel than a biography; Shipman often uses her imagination to describe what Dubois must have thought and done. The work is far too subjective. She does not even give a modern assessment of the Trinil fossils.

2. Bert Theunissen, *Eugene Dubois and the Ape-Man from Java* (Dordrecht: Kluwer Academic Publishers, 1989). Theunissen is on the staff of the Institute for the History of Science, University of Utrecht. The book, originally published in Dutch in 1985, was the author's Ph.D. dissertation. The English translation is an expanded version.

3. G. H. R. von Koenigswald, *Meeting Prehistoric Man*, trans. Micheal Bullock (New York: Harper and Brothers, 1956), 38–39.

4. Theunissen, *Eugene Dubois*, 49.

5. Ibid., 38.

6. Alan Houghton Brodrick, *Early Man* (London: Hutchinson's Scientific and Technical Publications, 1948), 85.

7. Theunissen, *Eugene Dubois*, 44, 68.

8. Brodrick, *Early Man*, 85.

9. Theunissen, *Eugene Dubois*, 121.

10. Ibid., 122.

Chapter 8: Java Man: Keeping the Faith

1. For a definitive exposition of this popular eighteenth- and nineteenth-century concept, see Arthur O. Lovejoy, *The Great Chain of Being* (Cambridge: Harvard University Press, 1964).

2. See Ernst Mayr, "Evolution and God," *Nature* 248 (22 March 1974): 285–86.

3. See Neal C. Gillespie, *Charles Darwin and the Problem of Creation* (Chicago: University of Chicago Press, 1979). Gillespie shows that Darwin's purpose was to "ungod" the universe.

4. Ibid., 137.

5. Theunissen, *Eugene Dubois*, 79–127.

6. Robert F. Heizer, ed., *Man's Discovery of His Past* (Englewood Cliffs, NJ: Prentice Hall, 1962), 138.

7. Ibid., 135–36.

8. Theunissen, *Eugene Dubois*, 68, 77.

9. Ibid., 158.

10. Michael H. Day and T. I. Molleson, "The Trinil Femora," in Michael H. Day, ed., *Human Evolution*, vol. XI, Symposia of the Society for the Study of Human Biology (London: Taylor and Francis, Ltd., 1973): 127–54. See also the comment by William W. Howells, *American Journal of Physical Anthropology* 81, no. 1 (January 1990): 133–34.

11. Von Koenigswald, *Meeting Prehistoric Man*, 34.

12. Klein, *Human Career*, 1st ed., 185.

13. Von Koenigswald, *Meeting Prehistoric Man*, 32.

14. Theunissen, *Eugene Dubois*, 154–55.

Chapter 9: Wadjak Man: Not All Fossils Are Created Equal

1. Theunissen, *Eugene Dubois*, 41, 43.

2. Von Koenigswald, *Meeting Prehistoric Man*, 30.

3. Sir Arthur Keith, *The Antiquity of Man*, rev. ed. (London: Williams and Norgate, 1925), 2:439.

4. Carleton S. Coon, *The Origin of Races* (New York: Alfred A. Knopf, 1962), 399.

5. Theunissen, *Eugene Dubois*, 176, and Keith, *The Antiquity of Man*, 2:440.

6. Theunissen, *Eugene Dubois*, 44, 148.

7. Keith, *The Antiquity of Man*, 2:440–41.

8. Theunissen, *Eugene Dubois*, 72.

9. C. Loring Brace, "Creationists and the Pithecanthropines," *Creation/Evolution* 19 (Winter 1986–1987): 16. *Creation/Evolution* is a publication of the American Humanist Association. Brace's article is the text of his presentation at the University of Michigan debate.

10. Michael H. Day, *Guide to Fossil Man*, 1st ed. (Cleveland: World Publishing Company, 1965), 247.

11. Theunissen, *Eugene Dubois*, 147.

12. Day, *Guide to Fossil Man*, 1st ed., 247.

13. Oakley, Campbell, and Molleson, eds., *Catalogue of Fossil Hominids* Part III (London: Trustees of the British Museum, 1975), 115.

14. Dirk Albert Hooijer, "The Geological Age of Pithecanthropus, Meganthropus, and Gigantopithecus," *American Journal of Physical Anthropology* 9, no. 3 (September 1951): 275.

15. Ibid.

16. Ibid., 278.

17. Ibid.

18. Kenneth Oakley, *Frameworks for Dating Fossil Man* (Chicago: Aldine Publishing Company, 1964), 7.

19. Sir Karl Popper, *The Logic of Scientific Discovery* (New York: Basic Books, 1959).

20. Werner Heisenberg, *Physics and Beyond*, trans. Arnold J. Pomerans (New York: Harper & Row, Publishers, 1971), 63.

Chapter 10: The Selenka-Trinil Expedition: A Second Opinion

1. M. Lenore Selenka and Max Blanckenhorn, *Die Pithecanthropus-Schichen auf Java* (Leipzig: W. Engelmann, 1911), 342.

2. A. G. Tilney, "Pithecanthropus: The Facts" (Stoke, England: Evolution Protest Movement, n.d.).

3. Sir Arthur Keith, "The Problem of *Pithecanthropus*," *Nature* 87 (13 July 1911): 49–50.

4. Alan Houghton Brodrick, *Man and His Ancestry*, rev. ed. (Greenwich, CT: Fawcett Publications, 1964), 127.

5. Coon, *Origin of Races*, 376.

6. George Grant MacCurdy, *Human Origins* (New York: D. Appleton and Company, 1924), 1:316.

7. Theunissen, *Eugene Dubois*, 127.

8. Von Koenigswald, *Meeting Prehistoric Man*, 36.

9. Theunissen, *Eugene Dubois*, 164.

Chapter 11: *Homo Erectus*: A Man for All Seasons

1. Bruce Bower, "Human Ancestors Make Evolutionary Change," *Science News* 127 (4 May 1985): 276.

2. Howells, "*Homo erectus* in human descent," 72

3. C. C. Swisher III, G. H. Curtis, T. Jacobs, A. G. Getty, A. Suprijo, Widiasmoro, "Age of the Earliest Known Hominids in Java, Indonesia," *Science* 263 (25 February 1994): 1118–21.

4. C. C. Swisher III, W. J. Rink, S. C. Antón, H. P. Schwarcz, G. H. Curtis, A. Suprijo, Widiasmoro, "Latest *Homo erectus* of Java: Potential Contemporaneity with *Homo sapiens* in Southeast Asia," *Science* 274 (13 December 1996): 1870–74.

5. Tattersall and Schwartz, *Extinct Humans*, 160–63.

6. Richard G. Klein, *The Human Career: Human Biological and Cultural Origins*, 2d ed. (Chicago: University of Chicago Press, 1999), 270–75.

7. R. L. Susman, J. T. Stern Jr., and M. D. Rose, "Morphology of KNM-ER 3228 and O. H. 28 innominates from East Africa," abstract, *American Journal of Physical Anthropology* 60, no. 3 (February 1983): 259.

8. Tattersall, Delson, and Van Couvering, *Encyclopedia*, 67.

9. Delson, *Ancestors*, 298.

10. Klein, *Human Career*, 2nd ed., 181.

11. R. C. Walter, P. C. Manega, R. L. Hay, R. E. Drake, and G. H. Curtis, "Laser-fusion 40Ar/39Ar dating of Bed 1, Olduvai Gorge, Tanzania," *Nature* 354 (14 November 1991): 145.

12. A. G. Thorne and P. G. Macumber, "Discoveries of Late Pleistocene Man at Kow Swamp, Australia," *Nature* 238 (11 August 1972): 316–19.

13. Ibid., 316.

14. A. G. Thorne, "Mungo and Kow Swamp: Morphological Variation in Pleistocene Australians," *Mankind* 8, no. 2 (December 1971): 87.

15. Thorne and Macumber, "Discoveries of Late Pleistocene Man," 316, 319; and Alan G. Thorne and Milford H. Wolpoff, "Regional Continuity in Australasian Pleistocene Hominid Evolution," *American Journal of Physical Anthropology* 55, no. 3 (July 1981): 337–49.

16. Oakley, Campbell, and Molleson, *Catalogue of Fossil Hominids* Part III, 200.

17. "Talgai Skull," *Science News* 93 (20 April 1968): 381.

18. Tattersall, Delson, and Van Couvering, *Encyclopedia*, 67.

19. L. Freedman and M. Lofgren, "Human Skeletal Remains from Cossack, Western Australia," *Journal of Human Evolution* 8 (1979): 285.

Chapter 12: *Homo Erectus*: All in the Family

1. Michael Day, "Homo turmoil," *Nature* 348 (20/27 December 1990): 688.

2. Ibid.

3. Milford H. Wolpoff, Wu Xin Zhi, and Alan G. Thorne, "Modern Homo sapiens Origins: A General Theory of Hominid Evolution Involving the Fossil Evidence From East Asia," *The Origins of Modern Humans*, ed. Fred H. Smith and Frank Spencer (New York: Alan R. Liss, Inc., 1984), 465–66.

4. Franz Weidenreich, "The skull of *Sinanthropus pekinensis*," *Palaeontol Sinica* (n. s. D, No. 10, 1943): 246, quoted in Wolpoff *et al.*, 466.

5. William S. Laughlin, "Eskimos and Aleuts: Their Origins and Evolution," *Science* 142 (8 November 1963): 644.

6. Gabriel Ward Lasker, *Physical Anthropology* (New York: Holt, Rinehart and Winston, Inc., 1973), 284.

7. Edmund White and Dale Brown, *The First Men* (New York: Time-Life Books, 1973), 14.

8. "*Homo erectus* never existed?" *Geotimes* (October 1992): 11.

9. Susman, Stern, and Rose, "Morphology of KNM-ER 3228," 259.

10. Donald C. Johanson and Maitland A. Edey, *Lucy: The Beginnings of Humankind* (New York: Simon and Schuster, 1981), 144.

11. Stephen Molnar, *Races, Types, and Ethnic Groups* (Englewood Cliffs, NJ: Prentice-Hall, 1975), 57.

12. Erik Trinkaus, public lecture on the Neandertals (Colorado State University, Fort Collins, 3 December 1984).

13. Brace, "Creationists and the Pithecanthropines," 23.

14. Harry L. Shapiro, *Peking Man* (New York: Simon and Schuster, 1974), 125.

15. Sigmon and Cybulski, *Homo erectus*, 227.

16. Ronald W. Angel, "Fire Use," *Science* 284 (30 April 1999): 741.

17. Kathy A. Svitil, "Leonardo of the Pleistocene," *Discover* (October 2003): 18.

18. Hou Yamei, Richard Potts, Yuan Baoyin, Guo Zhengtang, Alan Deino, Wang Wei, Jennifer Clark, Xie Guangmao, and Huang Weiwen, "Mid-Pleistocene Acheulean-like Stone Technology of the Bose Basin, South China," *Science* 287 (3 March 2000): 1622–26.

19. Peter Hadfield, "Gimme shelter," *NewScientist* (4 March 2000): 4.

20. M. J. Morwood, P. B. O'Sullivan, F. Aziz, and A. Raza, "Fission-track ages of stone tools and fossils on the east Indonesian island of Flores," *Nature* 392 (12 March 1998): 173–76.

21. Robert Kunzig, "Erectus Afloat," *Discover* (January 1999): 80.

22. Chris B. Stringer and Rainer Grün, "Time for the last Neanderthals," *Nature* 351 (27 June 1991): 701.

23. M. D. Leakey, "Primitive Artifacts from Kanapoi Valley," *Nature* 5062 (5 November 1966): 581.

24. Lawrence H. Robbins, "Archeology in the Turkana District, Kenya," *Science* 176 (28 April 1972): 360.

25. Klein, *Human Career*, 2nd ed., 335–37.

26. Eileen M. O'Brien, "What Was the Acheulean Hand Ax?" *Natural History* (July 1984): 20–23.

27. Ibid., 23.

Chapter 13: The Voyage

1. Charles Darwin, *Voyage of the Beagle*, Everyman's Library (London: J. M. Dent & Sons, 1959), 196–97. Originally published in 1826, 1836, and 1839.

2. Ashley Montagu, *Man: His First Two Million Years* (New York: Dell Publishing Co., Inc., 1969), 143–44.

3. Darwin, *Voyage of the Beagle*, 219.

4. Ibid., 202–3.

5. Ibid., 194–95.

6. Ibid., 204.

7. Ibid., 205–6.

8. Janet Browne, *Voyaging: Charles Darwin, A Biography*, vol. 1 (New York: Alfred A. Knopf, 1995), 247–48.

9. Ibid., 249.

10. Ibid., 353.

11. Ibid., 244.

12. Francis Darwin, ed., *Life and Letters of Charles Darwin*, seventh thousand Revised. (London: John Murray, 1888), 3:126.

13. Ibid., 3:127–28.

Chapter 14: Tasmanian Devils

1. Jared Diamond, "Ten Thousand Years of Solitude," *Discover* (March 1993): 56.

2. Jerry Bergman, "Nineteenth Century Darwinism and the Tasmanian Genocide," *Creation Research Society Quarterly* 32 (March 1996): 190.

3. James Bonwick, *Daily Life and Origin of the Tasmanians* (London: Samson Low, Son and Marston, 1870), 27. Cited in Bergman, "Nineteenth Century Darwinism," 190.

4. William Knighton, *Struggles for Life* (London: Williams and Norgate, 1886), 272. Cited by Bergman, "Nineteenth Century Darwinism," 191.

5. Charles Darwin, *The Origin of Species* and *The Descent of Man*, The Modern Library (New York: Random House, Inc., n.d.), 542.

6. Ibid., 543.

7. Diamond, "Ten Thousand Years," 57.

8. Ibid.

9. Bonwick, *Daily Life and Origin*, 70. Cited in Bergman, "Nineteenth Century Darwinism," 194.

10. Desmond King-Hele, *Erasmus Darwin: Grandfather of Charles Darwin* (New York: Charles Schribner's Sons, 1963), 75. Cited in Bergman, "Nineteenth Century Darwinism," 192.

11. Bergman, "Nineteenth Century Darwinism," 193.

12. Jared Diamond, "In Black and White," *Natural History* 97 (October 1988): 9.

13. Diamond, "Ten Thousand Years," 57.

14. Ibid., 51.

Chapter 15: The Elephant in the Living Room

1. A. E. Wilder-Smith, *Evolution, Theistic Evolution, or Creation?* transcription of a 1981 lecture (CLP Tapes, P.O. Box 15666, San Diego, CA 92115). Used by permission.

2. Stephen Jay Gould, *Evolution and Human Equality*, public lecture given at The College of Wooster, Wooster, Ohio, in 1987. Videotape distributed by *Natural History* magazine, 2002.

3. Pat Shipman, *The Evolution of Racism* (Cambridge: Harvard University Press, 2002), 117–18.

4. Gould, *Evolution and Human Equality.*

5. Jonathan Marks, "Race and Human Evolution," *Journal of Human Evolution* 33, no. 4 (October 1997): 525.

6. Olivia P. Judson, "The Book of Revelation," review of *The Collected Papers of W. D. Hamilton*, *Nature* 416 (7 March 2002): 17–18.

Chapter 16: Rhodesian Man: The Fastest-Aging Man in the World

1. Arthur Smith Woodward, "A New Cave Man from Rhodesia, South Africa," *Nature* 108, no. 2716 (17 November 1921): 371–72.

2. Day, *Guide to Fossil Man*, 1st ed., 154.

3. Ibid.

4. Coon, *The Origin of Races*, 622.

5. Richard G. Klein, "Geological Antiquity of Rhodesian Man" *Nature* 244 (3 August 1973): 311–12.

6. Coon, *The Origin of Races*, 622.

7. Woodward, "A New Cave Man," 371.

8. Christopher B. Stringer, "An Archaic Character in the Broken Hill Innominate E. 719," *American Journal of Physical Anthropology* 71 (September 1986): 115.

9. Ibid.

10. Klein, "Geological Antiquity," 311–12.

11. Ian Tattersall, *The Last Neanderthal*, rev. ed. (Boulder, CO: Westview Press, 1999), 67–68.

Chapter 17: African Eve: A Woman of No Importance

1. H. Valladas, J. L. Reyss, J. L. Joron, G. Valladas, O. Bar-Yosef, and B. Vandermeersch, "Thermoluminescence dating of Mousterian 'Proto-Cro-Magnon' remains from Israel and the origin of modern man," *Nature* 331 (18 February 1988): 614–16.

2. Y. Rak and B. Arensburg, "Kebara 2 Neanderthal Pelvis: First Look at a Complete Inlet," *American Journal of Physical Anthropology* 73 (1987): 227–31. See also Yoel Rak, "On the Differences between Two Pelvises of Mousterian Context from the Qafzeh and Kebara Caves, Israel," *American Journal of Physical Anthropology* 81:3 (March 1990): 323–32.

3. Stephen Jay Gould, "A Novel Notion of Neanderthal," *Natural History* (June 1988): 20.

4. Chris Stringer, "The Dates of Eden," *Nature* 331 (18 February 1988): 565.

5. Stringer and Grün, "Time for the last Neanderthals," 702.

6. Rebecca L. Cann, Mark Stoneking, and Allan C. Wilson, "Mitochondrial DNA and human evolution," *Nature* 325 (1 January 1987): 31–36.

7. Philip Awadalla, Adam Eyre-Walker, and John Maynard Smith, "Linkage Disequilibrium and Recombination in Hominid Mitochondrial DNA," *Science* 286 (24 December 1999): 2524–25.

8. Henry Gee, "Statistical cloud over African Eden," *Nature* 355 (13 February 1992): 583.

9. Marcia Barinaga, "'African Eve' Backers Beat a Retreat," *Science* 255 (7 February 1992): 687.

10. S. Blair Hedges, Sudhir Kumar, Koichiro Tamura, and Mark Stoneking, "Human Origins and Analysis of Mitochondrial DNA Sequences," *Science* 255 (7 February 1992): 737–39.

11. Allan C. Wilson and Rebecca L. Cann, "The Recent African Genesis of Humans," *Scientific American* (April 1992): 68.

12. Ibid.

13. Marcia Barinaga, "Choosing a Human Family Tree," *Science* 255 (7 February 1992): 687.

14. Wilson and Cann, "Recent African Genesis," 68.

15. Ibid., 72.

16. Marks, "Race and Human Evolution," 525.

17. Matt Cartmill, "The Third Man," *Discover* (September 1997): 62.

18. Alan G. Thorne and Milford Wolpoff, "The Multiregional Evolution of Humans," *Scientific American* Special Edition 13:2 (2003): 48.

19. Kate Wong, "Sourcing Sapiens," *Scientific American* (August 2003): 24.

20. Richard G. Klein with Blake Edgar, *The Dawn of Human Culture* (New York: John Wiley & Sons, Inc., 2002), 250.

21. James M. Bowler, Harvey Johnston, Jon M. Olley, John R. Prescott, Richard G. Roberts, Wilfred Shawcross, and Nigel A. Spooner, "New ages for human occupation and climatic change at Lake Mungo, Australia," *Nature* 421 (20 February 2003): 837–40.

22. Emma Young, "Mungo Man has his say on Australia's first humans," *NewScientist* (22 February 2003): 15.

23. J. C. Ahern and F. H. Smith, "The Transitional Nature of the late Neandertal Mandibles from Vindija Cave, Croatia," *American Journal of Physical Anthropology* Supplement 16 (1993): 47.

24. Tattersall, Delson, and Van Couvering, *Encyclopedia*, 241.

25. Christopher Stringer and Clive Gamble, *In Search of the Neanderthals* (New York: Thames and Hudson, Inc., 1993), 179–80.

26. Tattersall, Delson, and Van Couvering, *Encyclopedia*, 56.

27. Kenneth P. Oakley, Bernard G. Campbell, and Theya I. Molleson, eds., *Catalogue of Fossil Hominids* Part II (London: Trustees of the British Museum—Natural History, 1971), 209.

28. Milford Wolpoff and Rachel Caspari, *Race and Human Evolution* (New York: Simon & Schuster, 1997), 177, 182.

29. Nancy Minugh-Purvis, Jakov Radovcic, and Fred H. Smith, "Krapina 1: A Juvenile Neandertal from the Early Late Pleistocene of Croatia," *American Journal of Physical Anthropology* 111 (2000): 393–424.

30. Marcellin Boule and Henri V. Vallois, *Fossil Men* (New York: The Dryden Press, 1957), 281.

31. F. H. Smith, A. B. Falsetti, and M. A. Liston, "Morphometric analysis of the Mladec postcranial remains," *American Journal of Physical Anthropology* 78, no. 2 (February 1989): 305.

32. M. H. Wolpoff and J. Jelinek, "New discoveries and reconstructions of Upper Pleistocene hominids from the Mladec cave, Moravia, CCSR," *American Journal of Physical Anthropology* 72, no. 2 (February 1987): 270–71.

33. N. S. Minugh, "The Mladec 3 child: Aspects of cranial ontogeny in early anatomically modern Europeans," *American Journal of Physical Anthropology* 60, no. 2 (February 1983): 228.

34. Fred H. Smith, "A Fossil Hominid Frontal from Velika Pecina (Croatia) and a Consideration of Upper Pleistocene Hominids from Yugoslavia," *American Journal of Physical Anthropology* 44 (January 1976): 130–31.

35. Oakley, Campbell, and Molleson, *Catalogue of Fossil Hominids* Part II, 342.

36. Tattersall, Delson, and Van Couvering, *Encyclopedia*, 56, 87.

37. Klein, *Human Career*, 1st ed., 236–37.

38. G. A. Clark, "Neandertal Genetics," *Science* 277 (22 August 1997): 1024.

39. Kathy A. Svitil, "Leonardo of the Pleistocene," *Discover* (October 2003): 18.

40. Ian Tattersall, "Will We Keep Evolving?" *Time* (10 April 2000): 97.

Chapter 18: Splitters versus Lumpers

1. Milford H. Wolpoff, Wu Xin Zhi, and Alan G. Thorne, "Modern Homo sapiens Origins: A General Theory of Hominid Evolution Involving the Fossil Evidence From East Asia," in *The Origins of Modern Humans*, eds. Fred H. Smith and Frank Spencer (New York: Alan R. Liss, Inc., 1984), 465–66.

2. Ian Tattersall, "Once we were not alone," *Scientific American* Special Edition 13, no. 2 (2003): 20–27.

3. Bernard Wood, "Lessons from lemurs," review of *The Monkey in the Mirror*, by Ian Tatttersal, *NewScientist* (9 March 2002): 52.

4. Susan C. Antón and E. Indriati, "Earliest Pleistocene Homo in Asia: craniodental comparisons of Dmanisi and Sangiran," *American Journal of Physical Anthropology* Supplement 34 (2002): 38.

Chapter 19: "We Are All African": A Study in Political Correctness

1. Tim D. White, Berhane Asfaw, David DeGusta, Henry Gilbert, Gary D. Richards, Gen Suwa, and F. Clark Howell, "Pleistocene *Homo sapiens* from Middle Awash, Ethiopia," *Nature* 423 (12 June 2003): 742–47.

2. J. Desmond Clark, Yonas Beyene, Giday WoldeGabriel, William K. Hart, Paul R. Renne, Henry Gilbert, Alban Defleur, Gen Suwa, Shigehiro Katoh, Kenneth R. Ludwig, Jean-Renaud Boisserie, Berhane Asfaw, and Tim D. White, "Stratigraphic, chronological and behavioural contexts of Pleistocene *Homo sapiens* from Middle Awash, Ethiopia," *Nature* 423 (12 June 2003): 747–52.

3. Chris Stringer, "Out of Ethiopia," *Nature* 423 (12 June 2003): 692–95.

4. Ann Gibbons, "Oldest Members of *Homo sapiens* Discovered in Africa," *Science* 300 (13 June 2003): 1641.

5. "Our ancestors the hippo eaters," *NewScientist* (14 June 2003): 3.

6. James Randerson, "The dawn of Homo sapiens," *NewScientist* (14 June 2003): 4–5.

7. Roger Lewin, "Where, When, and How," *NewScientist* (14 June 2003): 5.

8. Michael D. Lemonick and Andrea Dorfman, "The 160,000–Year-Old Man," *Time* (23 June 2003): 56–58.

9. John Noble Wilford, "Fossilized skulls may offer glimpse of direct ancestors," *San Diego Union-Tribune*, 12 June 2003.

10. Rick Gore, "The First Europeans," *National Geographic* 192:1 (July 1997): 101.

11. Stringer, "Out of Ethiopia," 692–93.

12. Tim D. White *et al.*, "Pleistocene *Homo sapiens*," 742, 745.

13. Stringer, "Out of Ethiopia," 693–695.

14. Randerson, "Dawn of Homo Sapiens," 5.

15. Gibbons, "Oldest Members of *Homo Sapiens*," 1641.

16. Lewin, "Where, When, and How," 5.

17. Lemonick and Dorfman, "The 160,000–Year-Old Man," 57–58.

18. Wilford, "Fossilized skulls."

19. Tim D. White *et al.*, "Pleistocene *Homo sapiens*," 742, 745.

20. Clark *et al.*, "Stratigraphic, chronological and behavioural contexts," 750.

21. Stringer, "Out of Ethiopia," 692.

22. Gibbon, "Oldest Members of *Homo Sapiens*," 1641.

23. Randerson, "Dawn of Homo Sapiens," 4.

24. Lewin, "Where, When, and How," 5.

25. Lemonick and Dorfman, "The 160,000–Year-Old Man," 57.

26. Wilford, "Fossilized Skulls."

27. Mildred Cho and Maren Grainger-Monsen, "Dilemmas of a Divisive Concept," *Science* 300 (18 April 2003): 434.

28. Wilford, "Fossilized skulls."

Chapter 20: A Cave in Spain Makes It All Very Plain

1. Ian Tattersall, "Recognizing hominid species in the late Pleistocene," abstract, *American Journal of Physical Anthropology* 81, no. 2 (February 1990): 306.

2. William Howells, "*Homo erectus*—Who, When, and Where: A Survey," *Yearbook of Physical Anthropology* 23 (New York: Alan R. Liss, Inc., 1980), 15.

3. Klein, *Human Career*, 1st ed., 225.

4. Juan Luis Arsuaga, *The Neanderthal's Necklace*, trans. Andy Klatt (New York: Four Walls Eight Windows, 2002), 271–72.

5. Ibid., 224.

6. James Shreeve, "Infants, Cannibals, and the Pit of Bones," *Discover* (January 1994): 40.

7. Chris Stringer, "Secrets of the Pit of the Bones," *Nature* 362 (8 April 1993): 501–2.

8. Klein, *Human Career*, 2nd ed., 268–69.

9. John Noble Wilford, "Our ancestors lived in Europe 800,000 years ago," *San Diego Union-Tribune*, January 22, 2003.

10. F. Clark Howell, *Early Man* (New York: Time-Life Books, 1968): 104.

11. Sigmon and Cybulski, *Homo erectus*, 8.

12. Jeff Hecht, "Long, narrow skulls reveal the colonisation of America," *NewScientist* (6 September 2003): 17.

13. *San Diego Union-Tribune*, September 14, 2003.

Chapter 21: "Otherizing" the Neandertals

1. Diamond, "Great Leap Forward," 50.

2. Ibid., 59.

3. Ibid.

4. Ibid.

5. Ibid., 60.

6. Ibid.

7. B. Arensburg, A. M. Tillier, B. Vandermeersch, H. Duday, L. A. Schepartz, and Y. Rak, "A Middle Palaeolithic human hyoid bone," *Nature* 338 (27 April 1989): 759–60. Bracketed material added for clarity. See also B. Arensburg, L. A. Schepartz, A. M. Tillier, B. Vandermeersch, and Y. Rak, "A Reappraisal of the Anatomical Basis for Speech in Middle Palaeolithic Hominids," *American Journal of Physical Anthropology* 83, no. 2 (October 1990): 137–46.

8. Diamond, "Great Leap Forward," 55.

9. Ibid., 58.

10. Ibid., 55.

11. Paul Mellars, *The Neanderthal Legacy* (Princeton: Princeton University Press, 1996), 366.

12. Richard G. Klein, "Middle Paleolithic People," review of *The Neanderthal Legacy*, by Paul Mellars, *Science* 272 (10 May 1996): 822–23.

Chapter 22: Neandertals, Tasmanians, and the Wild and Woolly West

1. Harris H. Wilder, *The Pedigree of the Human Race* (New York: Henry Holt, 1926), 341. Cited in Bergman, "Nineteenth Century Darwinism," 194.

2. Ann Shepherd, *The Last Tasmanian. Instructors Guide for the CRM/McGraw-Hill Film* (Del Mar, CA: 1990), 4. Cited in Bergman, "Nineteenth Century Darwinism," 195.

3. Diamond, "Great Leap Forward," 50–60.

4. Diamond, "Ten Thousand Years," 48–57.

5. Gerald R. Smith, *Fishes of the Pliocene Glenns Ferry Formation, Southwest Idaho: Papers on Paleontology No. 14* (Ann Arbor: Museum of Paleontology, University of Michigan, 1975), 66–67.

6. Marvin L. Lubenow, "Reversals in the Fossil Record: The Latest Problem in Stratigraphy and Evolutionary Phylogeny," *Creation Research Society Quarterly* 13 (March 1977): 185–90.

7. Klein, *Human Career*, 1st ed., 161.

8. Diamond, "Ten Thousand Years," 54.

9. James Peoples and Garrick Bailey, *Humanity: An Introduction to Cultural Anthropology* (Minneapolis: West Publishing Company, 1994), 115.

Chapter 23: Technical Section: mtDNA Neandertal Park—A Catch-22

1. Patricia Kahn and Ann Gibbons, "DNA From an Extinct Human," *Science* 277 (11 July 1997): 176.

2. Matthias Krings, Anne Stone, Ralf W. Schmitz, Heike Krainitzki, Mark Stoneking, and Svante Pääbo, "Neandertal DNA Sequences and the Origin of Modern Humans," *Cell* 90 (11 July 1997): 19–30.

3. Igor V. Ovchinnikov, Anders Götherström, Galina P. Romanova, Vitally M. Kharitonov, Kerstin Lidén, and William Goodwin, "Molecular analysis of Neanderthal DNA from the northern Caucasus," *Nature* 404 (30 March 2000): 490–93.

4. Gregory J. Adcock, Elizabeth S. Dennis, Simon Easteal, Gavin A. Huttley, Lars S. Jermiin, W. James Peacock, and Alan Thorne, "Mitochondrial DNA sequences in ancient Australians: Implications for modern human origins," *Proceedings of the National Academy of Science* 98, no. 2 (16 January 2001): 537–42.

5. Tomas Lindahl, "Instability and decay of the primary structure of DNA," *Nature* 362 (22 April 1993): 713; Svante Pääbo, "Ancient DNA," *Scientific American* (November 1993): 92.

6. Ann Gibbons, "Ancient History," *Discover* (January 1998): 47.

7. Kary B. Mullis, "The Unusual Origin of the Polymerase Chain Reaction," *Scientific American* (April 1990): 56.

8. Daniel Koshland Jr. and Ruth Levy Guyer, "Perspective," *Science* 277 (22 December 1989): 1543, cited by Paul Rabinow, *Making PCR* (Chicago: The University of Chicago Press, 1996), 5–6.

9. Tomas Lindahl, "Recovery of antediluvian DNA," *Nature* 365 (21 October 1993): 700.

10. Kahn and Gibbons, "DNA From an Extinct Human," 176–77.

11. "Games and theories," *NewScientist* (14 June 2003): 50.

12. Philip Awadalla, Adam Eyre-Walker, and John Maynard Smith, "Linkage Disequilibrium and Recombination in Hominid Mitochondrial DNA," *Science* 286 (24 December 1999): 2524–25.

13. Evelyn Strauss, "mtDNA Shows Signs of Paternal Influence," *Science* 286 (24 December 1999): 2436.

14. Ibid.

15. Ibid.

16. Eleftherios Zouros, Kenneth R. Freeman, Amy Oberhauser Ball, and Grant H. Pogson, "Direct evidence for extensive paternal mitochondrial DNA inheritance in the marine mussel *Mytilus*," *Nature* 359 (1 October 1992): 412, 414.

17. Philip E. Ross, "Crossed Lines," *Scientific American* (October 1991): 32.

18. Strauss, "mtDNA Shows Signs of Paternal Influence," 2436.

19. Ann Gibbons, "Calibrating the Mitochondrial Clock," *Science* 279 (2 January 1998): 28.

20. Ibid., 29.

21. Krings *et al.*, "Neandertal DNA Sequences," 25.

22. G. A. Clark, "Neandertal Genetics," *Science* 277 (22 August 1997): 1024.

23. Jonathan Marks, "Chromosomal evolution in primates," in *The Cambridge Encyclopedia of Human Evolution*, eds. Steve Jones, Robert Martin, and David Pilbeam (Cambridge: Cambridge University Press, 1992), 302.

24. Krings *et al.*, "Neandertal DNA Sequences," 27.

25. Kahn and Gibbons, "DNA From an Extinct Human," 177.

26. Kate Wong, "Ancestral Quandary," *Scientific American* (January 1998): 32.

27. James Randerson, "It's official: Neanderthals and humans didn't date," *NewScientist* (17 May 2003): 14.

28. Adcock *et al.*, "Mitochondrial DNA sequences," 537–42.

29. Krings *et al.*, "Neandertal DNA Sequences," 19–30.

30. Adcock *et al.*, "Mitochondrial DNA sequences," 537–42.

31. Personal communication. Used by permission.

32. Personal communication. Used by permission.

33. Krings *et al.*, "Neandertal DNA Sequences," 20, 22, 26.

34. Hendrik N. Poinar, Matthias Höss, Jeffrey L. Bada, and Svante Pääbo, "Amino Acid Racemization and the Preservation of Ancient DNA," *Science* 272 (10 May 1996): 864.

35. Personal communication. Used by permission.

36. Krings *et al.*, "Neandertal DNA Sequences," 20–21.

37. Ibid., 21–22.

38. Ibid., 22.

39. Gregory J. Adcock, Elizabeth S. Dennis, Simon Easteal, Gavin A. Huttley, Lars S. Jermiin, W. James Peacock, and Alan Thorne, "Human Origins and Ancient Human DNA," *Science* 292 (1 June 2001): 1656.

40. Krings *et al.*, "Neandertal DNA Sequences," 22.

41. Randerson, "It's official," 14.

42. Alison Abbott, "Anthropologists cast doubt on human DNA evidence," *Nature* 423 (29 May 2003): 468.

43. Guido Barbujani and Giorgio Bertorelle, "Were Cro-Magnons too like us for DNA to tell?" correspondence, *Nature* 424 (10 July 2003): 127.

44. Karl J. Niklas, "Turning over an old leaf," *Nature* 344 (12 April 1990): 588.

45. Christopher B. Stringer, "The Legacy of *Homo sapiens*," *Scientific American* (May 1993): 138.

46. Kenneth A. R. Kennedy, review of *Continuity or Replacement: Controversies in Homo sapiens Evolution*, eds. Gunter Bräuer and Fred H. Smith, *American Journal of Physical Anthropology* 89, no. 2 (October 1992): 271–72.

47. Personal communication. Used by permission.

48. Xinzhi Wu and Frank Poirier, *Human Evolution in China* (New York: Oxford University Press, 1995), 113.

49. Robert Foley, "Talking Genes," a review of *Genes, Peoples and Languages* by Luigi Luca Cavalli-Sforza, *Nature* 377 (12 October 1995): 493–94.

Chapter 24: The Neandertal Next Door (Maybe)

1. Paul Bahn, "Better late than never," review of *Timewalkers: The Prehistory of Global Colonization*, by Clive Gamble, *Nature* 369 (16 June 1994): 531.

2. Arsuaga, *The Neanderthal's Necklace*, 32.

3. Jean-Jacques Hublin, Fred Spoor, Marc Braun, Frans Zonneveld, and Silvana Condemi, "A late Neanderthal associated with Upper Palaeolithic artifacts," *Nature* 381 (16 May 1996): 224–26.

4. "Early Music," *Science* 276 (11 April 1997): 205.

5. Tim Folger and Shanti Menon, ". . . Or Much Like Us?" *Discover* 18, no. 1 (January 1997): 33.

6. Rick Gore, "The First Europeans," *National Geographic* 192, no. 1 (July 1997): 101.

7. A. Ascenzi, I. Biddittu, P. F. Cassoli, A. G. Segre, and E. Segri-Naldini, "A calvarium of late *Homo erectus* from Ceprano, Italy," *Journal of Human Evolution* 31, no. 5 (November 1996): 422.

8. Kate Wong, "Paleolithic Pit Stop," *Scientific American* (December 2000): 18–19.

9. Brian Hayden, "The cultural capacities of Neandertals: a review and re-evaluation," *Journal of Human Evolution* 24, no. 2 (February 1993): 136.

10. Day, *Guide to Fossil Man*, 4th ed., 49.

11. "Fire One," *Scientific American* 248, no. 4 (April 1983): 75.

12. Hublin *et al.*, "A late Neanderthal," 224–26. Paul G. Bahn, "Neandertals emancipated," *Nature* 394 (20 August 1998): 719–21.

13. Hayden, "Cultural capacities of Neandertals," 123, 133.

14. Klein, *Human Career*, 2nd ed., 422.

15. Gore, "The First Europeans," 110–11.

16. M. B. Roberts, C. B. Stringer, and S. A. Parfitt, "A hominid tibia from Middle Pleistocene sediments at Boxgrove, UK," *Nature* 369 (26 May 1994): 311–13.

17. Peter Aldhous, "England's Oldest Human Bone Steps Out," *Science* 264 (27 May 1994): 1248.

18. F. Mallegni and A. M. Radmilli, "Human Temporal Bone From the Lower Paleolithic Site of Castel di Guido, Near Rome, Italy," *American Journal of Physical Anthropology* 76, no. 2 (June 1988): 177.

19. Ascenzi *et al.*, "A calvarium of late *Homo erectus*," 412.

20. Klein, *Human Career*, 2nd ed., 344, 584.

21. S. Belitzky, Naama Goren-Inbar, and Ella Werker, "A Middle Pleistocene wooden plank with man-made polish," *Journal of Human Evolution* 20 (1991): 349–53.

22. Klein, *Human Career*, 2nd ed., 325–26, 345–46.

23. Paul G. Bahn, "Early teething troubles," *Nature* 337 (23 February 1989): 693.

24. Klein, *Human Career*, 2nd ed., 423.

25. Day, *Guide to Fossil Man*, 4th ed., 85.

26. Jan F. Simek and Fred H. Smith, "Chronological changes in stone tool assemblages from Krapina (Croatia)," *Journal of Human Evolution* 32, no. 6 (June 1997): 561.

27. Hayden, "Cultural capacities of Neandertals," 136.
28. Day, *Guide to Fossil Man*, 4th ed., 39.
29. Hayden, "Cultural capacities of Neandertals," 117, 133.
30. Wesley A. Niewoehner, Aaron Bergstrom, Derrick Eichele, Melissa Zuroff, and Jeffrey T. Clark, "Manual dexterity in Neanderthals," *Nature* 422 (27 March 2003): 395.
31. Hayden, "Cultural capacities of Neandertals," 117.
32. Klein, *Human Career*, 2nd ed., 440.
33. Ibid., 349–50.
34. Hayden, "Cultural capacities of Neandertals," 136.
35. Miller, "New Discoveries," 5.
36. "Germans unearth hoard of Neanderthal remains," 9.
37. Klein, *Human Career*, 2nd ed., 350.
38. Hayden, "Cultural capacities of Neandertals," 133.
39. Day, *Guide to Fossil Man*, 4th ed., 92.
40. Ibid., 121.
41. Hayden, "Cultural capacities of Neandertals," 123.
42. Ibid., 120.
43. Klein, *Human Career*, 2nd ed., 518–19.
44. Folger and Menon, ". . . Or Much Like Us?" 33.
45. Eric Boëda, Jacques Connan, Daniel Dessort, Sultan Muhesen, Norbert Mercier, Hélène Valladas, and Nadine Tisnérat, "Bitumen as a hafting material on Middle Palaeolithic artifacts," *Nature* 380 (28 March 1996): 336–38.
46. Simon Holdaway, "Tool hafting with a mastic," *Nature* 380 (28 March 1996): 288.
47. M. Kretzoi and L. Vértes, "Lower Palaeolithic Hominid and Pebble-Industry in Hungary," *Nature* 208 (9 October 1965): 205.
48. Klein, *Human Career*, 2nd ed., 332, 351.
49. Avis Lang, "French School, 300th Century B.C.," *Natural History* (March 2004): 23.
50. Douglas Palmer, "Neanderthal art alters the face of archaeology," *NewScientist* (6 December 2003): 11.
51. Adrian Cho, "But Is It Art?" *Science* 302 (12 December 2003): 1890.
52. Klein, *Human Career*, 2nd ed., 440.
53. Kate Wong, "Neanderthal Notes," *Scientific American* 277, no. 3 (September 1997): 28–30. "Early Music," 205.
54. Klein, *Human Career*, 2nd ed., 350.
55. Mallegni and Radmilli, "Human Temporal Bone," 177.
56. Klein, *Human Career*, 2nd ed., 344.

Chapter 25: In Life and Death: So Very Human

1. Marcel Otte, Isin Yalcinkaya, Janusz Kozlowski, Ofer Bar-Yosef, Ignacio López Bayón, and Harun Taskiran, "Long-term technical evolution and human remains in the Anatolian Palaeolithic," *Journal of Human Evolution* 34, no. 4 (April 1998): 426.
2. Rick Gore, "Neandertals," *National Geographic* 189, no. 1 (January 1996): 21.
3. Arsuaga, *The Neanderthal's Necklace*, 184, 187.
4. Day, *Guide to Fossil Man*, 4th ed., 49.
5. "Homo erectus in the Pyrenees," *Scientific American* 241, no. 5 (November 1979): 91–92.

6. Arsuaga, *The Neanderthal's Necklace*, 189–90.
7. Gore, "Neandertals," 19.
8. Otte *et al.*, "Long-term technical evolution," 426.
9. Klein, *Human Career*, 2nd ed., 423.
10. "Stone tips on ancient hunting," 13.
11. Oakley, Campbell, and Molleson, *Catalogue of Fossil Hominids* Part II, 162.
12. Arthur Jelinek and André Debénath, "Recent excavations at La Quinta (Charente), France," *Journal of Human Evolution* 34, no. 3 (March 1998): A10.
13. Miller, "New Discoveries," 5.
14. "Germans unearth hoard of Neanderthal remains," 9.
15. Klein, *Human Career*, 2nd ed., 356.
16. Day, *Guide to Fossil Man*, 4th ed., 121.
17. "Stone tips on ancient hunting."
18. Arsuaga, *The Neanderthal's Necklace*, 189.
19. Ibid., 184–87.
20. Ibid., 187–88.
21. Hartmut Thieme, "Lower Palaeolithic hunting spears from Germany," *Nature* 385 (27 February 1987): 807–10.
22. Arsuaga, *The Neanderthal's Necklace*, 192.
23. Ibid., 182.
24. Ibid., 279.
25. Ibid., 272–73.
26. Hayden, "Cultural capacities of Neandertals," 113–46.
27. Ralph S. Solecki, *Shanidar: The First Flower People* (New York: Alfred A. Knopf, 1971), 69.
28. Klein, *Human Career*, 2nd ed., 468–69.
29. Ibid., 467.
30. Ibid., 469.
31. Robert S. Corruccini, "Metrical Reconsideration of the Skhul IV and IX and Border Cave 1 Crania in the Context of Modern Human Origins," *American Journal of Physical Anthropology* 87, no. 4 (April 1992): 433–45.
32. Ibid., 440–42.
33. R. M. Quam and F. H. Smith, "Reconsideration of the Tabun C2 'Neandertal'," *American Journal of Physical Anthropology* Supplement 22 (1996): 192.
34. N. Minugh-Purvis and J. Radovcic, "Krapina A: Neandertal or Not?" *American Journal of Physical Anthropology* Supplement 12 (1991): 132.
35. Minugh-Purvis, Radovcic, and Smith, "Krapina 1," 393–424.
36. Y. Rak, W. H. Kimbel, and E. Hovers, "A Neandertal infant from Amud Cave, Israel," *Journal of Human Evolution* 26, no. 4 (April 1994): 314–15.
37. Mike Parker Pearson, *The Archaeology of Death and Burial* (College Station: Texas A&M University Press, 2000), 149.
38. Arsuaga, *The Neanderthal's Necklace*, 273.
39. Milford H. Wolpoff, "The Krapina Dental Remains," *American Journal of Physical Anthropology* 50, no. 1 (January 1979): 103–4.
40. Klein, *Human Career*, 2nd ed., 468–69.
41. Pearson, *Archaeology of Death and Burial*, 148.
42. Klein, *Human Career*, 2nd ed., 467.

43. Hayden, "Cultural capacities of Neandertals," 117.
44. Arsuaga, *The Neanderthal's Necklace*, 273.
45. Klein, *Human Career*, 2nd ed., 467.
46. Pearson, *Archaeology of Death and Burial*, 149.
47. Arsuaga, *The Neanderthal's Necklace*, 273.
48. Hayden, "Cultural capacities of Neandertals," 120.
49. Arsuaga, *The Neanderthal's Necklace*, 273; Hayden, "Cultural capacities of Neandertals," 120.
50. Arsuaga, *The Neanderthal's Necklace*, 271–72.
51. Ibid., 224.
52. Pearson, *Archaeology of Death and Burial*, 149.
53. Arsuaga, *The Neanderthal's Necklace*, 273.
54. Oakley, *Frameworks for Dating Fossil Man*, 312.
55. Oakley, Campbell, and Molleson, *Catalogue of Fossil Hominids* Part III, 162–63.
56. Klein, *Human Career*, 2nd ed., 469.
57. Serge Lebel, Erik Trinkaus, Martine Faure, Philippe Fernandez, Claude Guérin, Daniel Richter, Norbert Mercier, Helène Valladas, and Günther A. Wagner, "Comparative morphology and paleobiology of Middle Pleistocene human remains from the Bau de l'Aubesier, Vaucluse, France," *Proceedings of the National Academy of Science* 98 (25 September 2001): 11102.

Chapter 26: Stunning Answers from the Land Down Under

1. James B. Conant, *Science and Common Sense* (New Haven: Yale University Press, 1951), 259–60.
2. Margaret Schabas, "The Idea of the Normal," review of *The Taming of Change*, by Ian Hocking, *Science* 251 (15 March 1991): 1373.
3. Eugenie C. Scott, "'Creation Science' and Philosophy of Science: Reflections," abstract, *American Journal of Physical Anthropology* 75, no. 2 (February 1988): 269.
4. Michael J. Oard, *An Ice Age Caused by the Genesis Flood* (San Diego: Institute for Creation Research, 1990).
5. Ibid., 116.
6. Ibid., 95.
7. Francis Ivanhoe, "Was Virchow Right about Neandertal?" *Nature* 227 (8 August 1970): 578.
8. D. J. M. Wright, "Syphilis and Neanderthal Man," *Nature* 229 (5 February 1971): 409.
9. Klein, *Human Career*, 1st ed., 281–82.
10. Ibid., 279.
11. Ibid., 283.
12. Jeffrey Laitman, "Australia," in Tattersall, Delson, and Van Couvering, *Encyclopedia*, 67.
13. Klein, *Human Career*, 2nd ed., 570.
14. Klein, *Human Career*, 2nd ed., 571.
15. Thorne and Macumber, "Discoveries of Late Pleistocene Man," 319.
16. L. Freedman and M. Lofgren, "The Cossack skull and a dihybrid origin of the Australian Aborigines," *Nature* 282 (15 November 1979): 299.
17. A correspondent, "Late Pleistocene Man at Kow Swamp," *Nature* 238 (11 August 1972): 308. Editorials in *Nature* often accompany significant articles. Until recently, those editorials,

although written by authorities in their particular fields, were unsigned and listed as being from "a correspondent."

18. Laitman, "Australia," 67.

19. Klein, *Human Career*, 2nd ed., 396.

20. Oakley, Campbell, and Molleson, *Catalogue of Fossil Hominids* Part III, 199 (plate 5).

21. Chris Stringer, "Homo Sapiens," in Tattersall, Delson, and Van Couvering, *Encyclopedia*, 274.

22. Phillip J. Habgood, "The Origin of the Australian Aborigines: An Alternative Approach and View," in *Hominid Evolution: Past, Present and Future*, ed. Phillip V. Tobias (New York: Alan R. Liss, Inc., 1985), 375.

23. Delson, *Ancestors*, 298.

24. Freedman and Lofgren, "Human Skeletal remains," 295.

25. D. R. Brothwell, *Digging Up Bones*, 3rd ed. (Ithaca, NY: Cornell University Press, 1981), 49.

26. Adcock *et al.*, "Mitochondrial DNA sequences," 537–42.

27. Ibid., 541.

28. Ibid.

29. Ibid.

30. Ibid., 540.

31. Ibid.

32. Ibid., 541.

33. Ibid., 538.

Chapter 27: A Twenty-First-Century Scientific Revolution

1. Andrew Snelling, "Radioactive 'Dating' Failure," *Creation* 22, no. 1 (December 1999–February 2000): 18–21.

2. John Baumgardner, "Carbon Dating Undercuts Evolution's Long Ages," *Impact* Article #364 (San Diego: Institute for Creation Research, October 2003): ii.

3. Ibid.

4. Ibid.

5. Ibid., iii.

6. Ibid., iv.

7. D. Russell Humphreys, "Nuclear Decay: Evidence for a Young World," *Impact* Article #352 (San Diego: Institute for Creation Research, October 2002): ii.

8. Larry Vardiman, *The Age of the Earth's Atmosphere: A Study of the Helium Flux through the Atmosphere* (San Diego: Institute for Creation Research, 1990), 28.

9. Melvin A. Cook, "Where is the Earth's Radiogenic Helium?" *Nature* 4552 (26 January 1957): 213.

10. Humphreys, "Nuclear Decay," iii.

11. Ibid., iii–iv.

12. D. Russell Humphreys, John R. Baumgardner, Steven A. Austin, and Andrew A. Snelling, "Helium Diffusion Rates Support Accelerated Nuclear Decay," in *Proceedings of the Fifth International Conference on Creationism, 2003, Technical Symposium Sessions,* ed. Robert L. Ivey Jr. (Pittsburgh: Creation Science Fellowship, Inc., 2003), 189.

13. Chandler Burr, "The geophysics of God," *U.S. News & World Report* (16 June 1997): 55–58.

Chapter 28: The Dating Gap: Laying a Really Large Egg

1. A. S. Brooks, P. E. Hare, J. E. Kokis, G. E. Miller, R. D. Ernst, and F. Wendorf, "Dating Pleistocene Archeological Sites by Protein Diagenesis in Ostrich Eggshell," *Science* 248 (6 April 1990): 60.

2. William W. Howells, "*Homo erectus*—Who, When and Where: A Survey," *Yearbook of Physical Anthropology* 23 (1980; supplement 1 to the *American Journal of Physical Anthropology*): 8.

3. Klein, *Human Career*, 2nd ed., 35, 43.

4. Edouard Bard, "Extending the Calibrated Radiocarbon Record," *Science* 292 (29 June 2001): 2443.

5. Klein, *Human Career*, 2nd ed., 37.

Chapter 29: The Pretend Humans: The Nonhuman Fossil Primates

1. Tattersall and Schwartz, *Extinct Humans*, 106.

2. Louis S. B. Leakey, Phillip V. Tobias, and John R. Napier, "A New Species of the Genus *Homo* from Olduvai Gorge," *Nature* 202 (4 April 1964): 7–9.

3. Donald C. Johanson, Fidelis T. Masao, Gerald G. Eck, Tim D. White, Robert C. Walter, William H. Kimbel, Berhane Asfaw, Paul Manega, Prosper Ndessokia, and Gen Suwa, "New partial skeleton of *Homo habilis* from Olduvai Gorge, Tanzania," *Nature* 327 (21 May 1987): 205–9.

4. Bernard Wood, "The age of australopithecines," *Nature* 372 (3 November 1994): 31–32.

5. Richard Leakey and Roger Lewin, *Origins Reconsidered* (New York: Doubleday, 1992), 112.

6. Milford H. Wolpoff, review of *Olduvai Gorge, Volume 4: The Skulls, Endocasts, and Teeth of Homo Habilis*, by Phillip V. Tobias, *American Journal of Physical Anthropology* 89, no. 3 (November 1992): 402.

7. Charles E. Oxnard, "The place of the australopithecines in human evolution: grounds for doubt?" *Nature* 258 (4 December 1975): 389.

8. Matt Cartmill, David Pilbeam, and Glynn Isaac, "One Hundred Years of Paleoanthropology," *American Scientist* 74 (July–August 1986): 419.

9. R. A. Foley, review of *Evolutionary History of the 'Robust' Australopithecines*, by Frederick E. Grine, *American Journal of Physical Anthropology* 82, no. 1 (May 1990): 113.

10. Oxnard, "The place of the australopithecines," 389–95. See also Charles Oxnard, *The Order of Man* (New Haven: Yale University Press, 1984).

11. Sean B. Carroll, "Genetics and the making of *Homo sapiens*," *Nature* 422 (24 April 2003): 850.

Chapter 30: Never Discuss Theology with a Chimp

1. Pat Shipman, *The Evolution of Racism* (Cambridge: Harvard University Press, 1994), 237.

2. Andy Coghlan, "Not such close cousins after all," *NewScientist* (28 September 2002): 20.

3. Elizabeth Pennisi, "Jumbled DNA Separates Chimps and Humans," *Science* 298 (25 October 2002): 719.

4. Coghlan, "Not such close cousins," 20.

5. Robin Dunbar, review of *A Brief History of the Mind: From Apes to Intellect and Beyond*, by William H. Calvin, *Nature* 427 (26 February 2004): 783.

6. "The mouse within," *NewScientist* (7 December 2002): 5.

7. Alison Abbott, "Sorry, dogs—man's got a new best friend," *Nature* 420 (19/26 December 2002): 729.

8. Arsuaga, *The Neanderthal's Necklace*, 3.

9. Stringer, "Out of Ethiopia," 693.

10. "An Interview with Ian Tattersall," *Natural History* (February 2003): 77.

11. John Calvin, *Institutes of the Christian Religion*, 2 vols., trans. Ford Lewis Battles, ed. John T. McNeill (Philadelphia: The Westminster Press, 1960), 1:188–89.

12. John H. Gerstner, *The Rational Biblical Theology of Jonathan Edwards*, 3 vols. (Orlando, FL: Ligonier Ministries, 1991): 1:592.

13. John H. Gerstner, personal communication.

14. Arsuaga, *The Neanderthal's Necklace*, 17.

15. A. H. Schultz, "The Physical Distinction of Man," in *Readings in Anthropology*, ed. Thomas W. McKern (Englewood Cliffs, NJ: Prentice-Hall, 1966).

16. Marvin Harris, *Culture, Man, and Nature* (New York: Thomas Y. Crowell, 1971).

17. Stephen Molnar, *Races, Types, and Ethnic Groups* (Englewood Cliffs, NJ: Prentice-Hall, Inc., 1975): 57.

18. Ralph L. Holloway, "The Casts of Fossil Hominid Brains," *Scientific American* (July 1974): 111.

19. Matt Cartmill, "The Gift of Gab," *Discover* (November 1998): 56.

20. Cited in Cartmill, "Gift of Gab," 63.

21. Cartmill, "Gift of Gab," 62.

22. James Shreeve, *The Neandertal Enigma* (New York: William Morrow and Company, 1995), 275–76.

23. *San Diego Union-Tribune*, 13 May 2003.

Chapter 31: Genesis: The Footnotes of Moses

1. See John Korgan, "Free Radical: a word (or two) about linguist Noam Chomsky," *Scientific American* (May 1990): 40–44.

2. P. J. Wiseman, *Ancient Records and the Structure of Genesis*, ed. Donald J. Wiseman (Nashville: Thomas Nelson, 1985).

3. R. K. Harrison, *Introduction to the Old Testament* (Grand Rapids: Eerdmans, 1969), 542–53.

4. Derek Kidner, *Genesis* (Downers Grove, IL: InterVarsity Press, 1967), 80.

5. Ibid., 24.

6. Some recent evangelical writers question that the *tol^e^dot* phrases are colophons. However, their objections are not weighty and can be adequately explained. See Kidner, *Genesis*, 23–24; Allen P. Ross, *Creation and Blessing* (Grand Rapids: Wm. B. Eerdmans Publishing Company, 1988), 69–74; and Duane Garrett, *Rethinking Genesis* (Grand Rapids: Baker, 1991), 94–96.

7. There is some question as to just when Hebrew came into common use as a language. There is abundant reason to believe that it was in common use in Moses's day.

8. For examples of colophons in *Enuma Elish*, a Babylonian creation account, see Alexander Heidel, *The Babylonian Genesis*, 2nd ed. (Chicago: University of Chicago Press, 1951), 25–45.

Chapter 32: Reality in the Human Fossil Record

1. Rick Gore, "The First Steps," *National Geographic* 191, no. 2 (February 1997): 88.
2. "AnthroQuest," *The Leakey Foundation News* no. 43 (Spring 1991): 13.
3. Lewin, *Bones of Contention*, 160.
4. Walter *et al.*, "Laser-fusion 40Ar/39Ar dating," 145–49.
5. Mary D. Leakey, "Primitive Artifacts from Kanapoi Valley," *Nature* 212 (5 November 1966): 581.
6. Mary D. Leakey, "Footprints in the Ashes of Time," *National Geographic* (April 1979): 446.
7. Russell H. Tuttle, "The Pitted Pattern of Laetoli Feet," *Natural History* (March 1990): 64.
8. R. H. Tuttle and D. M. Webb, "The Pattern of Little Feet," abstract, *American Journal of Physical Anthropology* 78, no. 2 (February 1989): 316.
9. R. H. Tuttle and D. M. Webb, "Did Australopithecus afarensis make the Laetoli G footprint trails?" abstract, *American Journal of Physical Anthropology* (1991 Supplement): 175.
10. Tuttle, "Pitted Pattern," 64.
11. Russell H. Tuttle, review of *Primate Origins and Evolution*, by R. D. Martin, *American Journal of Physical Anthropology* 85, no. 2 (June 1991): 244.
12. Tuttle, "Pitted Pattern," 64.
13. Howells, "*Homo erectus* in human descent," 79–80.
14. Mark Ridley, "Who Doubts Evolution?" *NewScientist* 25 (June 1981): 831.
15. Ibid., 830.
16. Ibid., 831.
17. Charles Darwin, *The Origin of Species*, Everyman's Library (London: J. M. Dent and Sons, Ltd., 1967): 292–93.

Anatomically Modern *Homo Sapiens* and *Sapiens*-like Fossils

1. Susannah Patton, "Cave footprints oldest in Europe, scientists say," *Coloradoan* (Fort Collins, CO), 11 June 1999.
2. Tattersall, Delson, and Van Couvering, *Encyclopedia*, 271.
3. N. S. Minugh, "The Mladec 3 child: Aspects of cranial ontogeny in early anatomically modern Europeans," *American Journal of Physical Anthropology* 60, no. 2 (February 1983): 228.
4. Oakley, Campbell, and Molleson, *Catalogue of Fossil Hominids* Part II, 274–75.
5. Stringer and Grün, "Time for the last Neanderthals," 701.
6. Ibid.; Fred H. Smith, "A Fossil Hominid Frontal from Velika Pecina (Croatia) and a Consideration of Upper Pleistocene Hominids from Yugoslavia," *American Journal of Physical Anthropology* 44 (January 1976): 127–34.
7. Tattersall, Delson, and Van Couvering, *Encyclopedia*, 55.
8. Ibid., 56.
9. Kenneth P. Oakley, Bernard G. Campbell, and Theya I. Molleson, *Catalogue of Fossil Hominids* Part I, 2nd ed. (London: Trustees of The British Museum—Natural History, 1977): 101.
10. Erik Trinkaus and Christopher B. Ruff, "Early modern human remains from eastern Asia: the Yamashita-cho 1 immature postcrania," *Journal of Human Evolution* 30, no. 4 (April 1996): 299–314.

11. Wu and Poirier, *Human Evolution in China*, 194–99.

12. James M. Bowler, Harvey Johnston, Jon M. Olley, John R. Prescott, Richard G. Roberts, Wilfred Shawcross, and Nigel A. Spooner, "New ages for human occupation and climate change at Lake Mungo, Australia," *Nature* 421 (20 February 2003): 837–40.

13. Tattersall, Delson, and Van Couvering, *Encyclopedia*, 383.

14. Oakley, Campbell, and Molleson, *Catalogue of Fossil Hominids* Part II, 48–49.

15. Ibid., 101–2.

16. Klein, *Human Career*, 2nd ed., 398.

17. Wu and Poirier, *Human Evolution in China*, 171–74.

18. Smith, "A Fossil Hominid Frontal," 127–34.

19. Miller, "New Discoveries," 5; "Germans unearth hoard of Neanderthal remains," 9.

20. Klein, *Human Career*, 2nd ed., 398.

21. Gunter Bräuer, Yuji Yokoyama, Christophe Falguères, and Emma Mbua, "Modern human origins backdated," *Nature* 386 (27 March 1997): 337–38.

22. Tattersall, Delson, and Van Couvering, *Encyclopedia*, 87.

23. Adcock *et al.*, "Mitochondrial DNA sequences," 537–42.

24. Wu and Poirier, *Human Evolution in China*, 186–93.

25. Bräuer *et al.*, "Modern human origins backdated," 337–38.

26. Frederick E. Grine, Richard G. Klein, and Thomas P. Volman, "Dating, archaeology and human fossils from the Middle Stone Age levels of Die Kelders, South Africa," *Journal of Human Evolution* 21, no. 5 (November 1991): 363–95.

27. Klein, *Human Career*, 1st ed., 226–27.

28. Klein, *Human Career*, 2nd ed., 398.

29. "Human Origins," *Discover* (November 1998): 28.

30. Klein, *Human Career*, 2nd ed., 398.

31. C. B. Stringer, R. Grün, H. P. Schwarcz, and P. Goldberg, "ESR dates for the hominid burial site of Es Skhul in Israel," *Nature* 338 (27 April 1989): 756–58.

32. F. McDermott, R. Grün, C. B. Stringer, and C. J. Hawkesworth, "Mass-spectrometric U-series dates for Israeli Neanderthal/early modern hominid sites," *Nature* 363 (20 May 199): 252–55.

33. A. J. Nelson, "Newly described material from Wadjak, Java," *American Journal of Physical Anthropology* 78, no. 2 (February 1989): 279; Dirk Albert Hooijer, "The Geological Age of *Pithecanthropus*, *Meganthropus*, and *Gigantopithecus*," *American Journal of Physical Anthropology* 9, no. 3 (September 1951): 274–75.

34. Klein, *Human Career*, 1st ed., 289–90.

35. Tattersall, *Last Neanderthal*, rev. ed., 175.

36. Zhang Yinyun, "Chouhu and the question of *Homo erectus Homo sapiens* contemporaneity in China," *American Journal of Physical Anthropology* Supplement 14 (1992): 179.

37. Rick Gore, "Tracking the First of Our Kind," *National Geographic* 192, no. 3 (September 1997): 92–99.

38. R. M. Quam and F. H. Smith, "Reconsideration of the Tabun C2 'Neandertal'," *American Journal of Physical Anthropology* Supplement 22 (1996): 192.

39. Gunter Bräuer and Michael J. Mehlman, "Hominid Molars From a Middle Stone Age Level at the Mumba Rock Shelter, Tanzania," *American Journal of Physical Anthropology* 75, no. 1 (January 1988): 69–76.

40. Tattersall, Delson, and Van Couvering, *Encyclopedia*, 55.

41. Klein, *Human Career*, 2nd ed., 398.

42. Klein, *Human Career*, 2nd ed., 275–76.

43. "Humanity's Baby Steps," *Science* 282 (27 November 1998): 1635.

44. Stringer, "Archaic Character," 115.

45. Tattersall, *Last Neanderthal*, rev. ed., 67.

46. Ibid., 60.

47. Ibid.

48. Anna K. Behrensmeyer and Leo F. Laporte, "Footprints of a Pleistocene hominid in northern Kenya," *Nature* 289 (15 January 1981): 167–69.

49. Brigitte Senut, "Humeral Outlines in Some Hominoid Primates and in Plio-Pleistocene Hominids," *American Journal of Physical Anthropology* 56, no. 3 (November 1981): 281.

50. Day, *Guide to Fossil Man*, 4th ed., 178; Klein, *Human Career*, 1st ed., 120.

51. Craig S. Feibel, Francis H. Brown, and Ian McDougall, "Stratigraphic Context of Fossil Hominids From the Omo Group Deposits: Northern Turkana Basin, Kenya and Ethiopia," *American Journal of Physical Anthropology* 78 (April 1989): 613.

52. Klein, *Human Career*, 1st ed., 120–21.

53. Feibel, Brown, and McDougall, "Stratigraphic Context," 595–622.

54. Ibid.

55. Ibid.

56. Ibid.

57. Howells, "*Homo erectus* in human descent," 70–71.

58. Meave G. Leakey, Craig S. Feibel, Ian McDougall, and Alan Walker, "New four-million-year-old hominid species from Kanapoi and Allia Bay, Kenya," *Nature* 376 (17 August 1995): 565–71. Also see Meave G. Leakey, Craig S. Feibel, Ian McDougall, Carol Ward, and Alan Walker, "New specimens and confirmation of an early age for *Australopithecus anamensis*," *Nature* 393 (7 May 1998): 62–66.

59. Day, *Guide to Fossil Man*, 4th ed., 186.

Neandertal and Pre-Neandertal Fossils

1. Oakley, Campbell, and Molleson, *Catalogue of Fossil Hominids* Part II, 288; Rainer Berger and W. F. Libby, *Radiocarbon Journal* 8 (1966): 480.

2. Cidalia Duarte, Joao Mauricio, Paul B. Pettitt, Pedro Souto, Erik Trinkaus, Hans van der Plicht, and Joao Zilhao, "The early Upper Paleolithic human skeleton from the Abrigo do Lagar Velho (Portugal) and modern human emergence in Iberia," *Proceedings of the National Academy of Sciences* 96 (June 1999): 7604–9.

3. Constance Holden, "Modern Humans Had Neandertal Neighbors," *Science* 286 (5 November 1999): 1079.

4. Igor V. Ovchinnikov, Anders Götherström, Galina P. Romanova, Vitally M. Kharitonov, Kerstin Lidén, and William Goodwin, "Molecular analysis of Neanderthal DNA from the northern Caucasus," *Nature* 404 (30 March 2000): 490–93.

5. James Shreeve, *The Neandertal Enigma* (New York: William Morrow and Company, 1995), 1–8.

6. Tattersall, *Last Neanderthal*, rev. ed., 145.

7. Stringer and Grün, "Time for the last Neanderthals," 701.

8. Stringer and Gamble, *In Search of the Neanderthals*, 179–80.

9. Jean-Jacques Hublin, Fred Spoor, Marc Braun, Frans Zonneveld, and Silvana Condemi, "A late Neanderthal associated with Upper Palaeolithic artifacts," *Nature* 381 (16 May 1996): 224–26; Oakley, Campbell, and Molleson, *Catalogue of Fossil Hominids* Part II, 74–78.

10. Oakley, Campbell, and Molleson, *Catalogue of Fossil Hominids* Part III, 151–52; Kennedy, *Neanderthal Man*, 47.

11. Tattersall, Delson, and Van Couvering, *Encyclopedia*, 56.

12. Mercier *et al.*, "Thermoluminescence dating," 737–39.

13. Oakley, Campbell, and Molleson, *Catalogue of Fossil Hominids* Part III, 163–64; Stringer and Grün, "Time for the last Neanderthals," 701.

14. Oakley, Campbell, and Molleson, *Catalogue of Fossil Hominids* Part II, 102–3; Stringer and Grün, "Time for the last Neanderthals," 701.

15. Erksin Gulec, "A Fossil Cranium from Southern Anatolia," *American Journal of Physical Anthropology* Supplement 24 (1997): 121.

16. "Germans unearth hoard of Neanderthal remains," 9.

17. Miller, "New Discoveries," 5.

18. Y. Rak and W. H. Kimbel, "Diagnostic Neandertal characters in the Amud 7 infant," *American Journal of Physical Anthropology* Supplement 20 (1995): 177–78.

19. Tattersall, *Last Neanderthal*, rev. ed., 117.

20. Oakley, Campbell, and Molleson, *Catalogue of Fossil Hominids* Part II, 237–39.

21. Tattersall, *Last Neanderthal*, rev. ed., 31.

22. Milford H. Wolpoff, *Paleoanthropology* (New York: Alfred A Knopf, 1980), 258–59.

23. Oakley, Campbell, and Molleson, *Catalogue of Fossil Hominids* Part II, 8.

24. Ibid., 306.

25. Alexander Dorozynski and Alun Anderson, "Collagen: A New Probe Into Prehistoric Diet," *Science* 254 (25 October 1991): 520–21.

26. Oakley, Campbell, and Molleson, *Catalogue of Fossil Hominids* Part III, 162–63; Kenneth Oakley, *Frameworks for Dating Fossil Man* (Chicago: Aldine Publishing Company, 1966), 21, 312.

27. J. L. Thompson and A. Bilsborough, "Le Moustier 1: Characteristics of a late western European Neanderthal," *American Journal of Physical Anthropology* Supplement 22 (1996): 229.

28. Tattersall, *Last Neanderthal*, rev. ed., 90.

29. Folger and Menon, ". . . Or Much Like Us?" 33; Boëda *et al.*, "Bitumen as a hafting material," 336–38; Holdaway, "Tool hafting with a mastic," 288–89.

30. Klein, *Human Career*, 2nd ed., 558, 728.

31. Ibid.

32. Oakley, Campbell, and Molleson, *Catalogue of Fossil Hominids* Part II, 51–52.

33. Francesco Mallegni and Erik Trinkaus, "A reconsideration of the Archi 1 Neandertal mandible," *Journal of Human Evolution* 33, no. 6 (December 1997): 651–68.

34. L. Bonfiglio, P. F. Cassoli, F. Mallegni, M. Piperno, and A. Solano, "Neanderthal Parietal, Vertebrate Fauna, and Stone Artifacts From the Upper Pleistocene Deposits of Contrada Ianni de San Calogero (Catanzaro, Calabria, Italy)," *American Journal of Physical Anthropology* 70, no. 2 (June 1986): 241–50.

35. Oakley, Campbell, and Molleson, *Catalogue of Fossil Hominids* Part II, 154–55.

36. Ibid., 73.

37. Ibid., 266.

38. Ibid., 290.

39. Ibid., 278.
40. Ibid., 155–56.
41. Ibid., 160.
42. Ibid., 278–79.
43. Ibid., 209.
44. Ibid., 186.
45. Day, *Guide to Fossil Man*, 4th ed., 134.
46. Oakley, Campbell, and Molleson, *Catalogue of Fossil Hominids* Part II, 37–38.
47. Susan C. Antón, "Endocranial Hyperostosis in Sangiran 2, Gibraltar 1, and Shanidar 5," *American Journal of Physical Anthropology* 102, no. 1 (January 1997): 117.
48. Tattersall, *Last Neanderthal*, rev. ed., 80.
49. Stringer and Grün, "Time for the last Neanderthals," 701.
50. Ian Tattersall, *The Last Neanderthal* (New York: Macmillan, 1995), 204.
51. Oakley, Campbell, and Molleson, *Catalogue of Fossil Hominids* Part II, 61–62.
52. Ibid., 99–100.
53. Ibid., 167–68.
54. Ofer Bar-Yosef and Bernard Vandermeersch, "Modern Humans in the Levant," *Scientific American* (April 1993): 98–99.
55. Klein, *Human Career*, 2nd ed., 427, 440, 467.
56. Oakley, Campbell, and Molleson, *Catalogue of Fossil Hominids* Part II, 161–63.
57. Otte *et al.*, "Long-term technical evolution," 413–31.
58. Oakley, Campbell, and Molleson, *Catalogue of Fossil Hominids* Part II, 90.
59. Ibid., 6–7.
60. Tattersall, *Last Neanderthal*, rev. ed., 79.
61. Oakley, Campbell, and Molleson, *Catalogue of Fossil Hominids* Part II, 318–19.
62. F. Mallegni and A. T. Ronchitelli, "Deciduous Teeth of the Neandertal Mandible From Molare Shelter, Near Scario (Salerno, Italy)," *American Journal of Physical Anthropology* 79 (August 1989): 475–82.
63. Oakley, Campbell, and Molleson, *Catalogue of Fossil Hominids* Part II, 12–13.
64. Ibid., 316–17.
65. Ibid., 116–17.
66. Ibid., 10–11.
67. Ibid., 245–46.
68. Ibid., 55–56.
69. Ibid., 285.
70. Ibid., 215.
71. Ibid., 164–65.
72. Tattersall, *Last Neanderthal*, rev. ed., 167.
73. Takeru Akazawa, Sultan Muhesen, Yukio Dodo, Osamu Kondo, and Yuji Mizoguchi, "Neanderthal infant burial," *Nature* 377 (19 October 1995): 585–86.
74. Oakley, Campbell, and Molleson, *Catalogue of Fossil Hominids* Part II, 142–43.
75. Day, *Guide to Fossil Man*, 4th ed., 134.
76. Tattersall, *Last Neanderthal*, rev. ed., 110.
77. Klein, *Human Career*, 1st ed., 337.
78. Oakley, Campbell, and Molleson, *Catalogue of Fossil Hominids* Part II, 180.
79. Ibid., 259.

80. Alan Mann and Bernard Vandermeersch, "An Adolescent Female Neandertal Mandible From Montgaudier Cave, Charente, France," *American Journal of Physical Anthropology* 103, no. 4 (August 1997): 507–27; Oakley, Campbell, and Molleson, *Catalogue of Fossil Hominids* Part II, 145–46.

81. Oakley, Campbell, and Molleson, *Catalogue of Fossil Hominids* Part II, 250–51.

82. F. Mallegni, M. Piperno, and A. Segre, "Human Remains of *Homo sapiens neanderthalensis* From the Pleistocene Deposit of Santa Croce Cave, Bisceglie (Apulia), Italy," *American Journal of Physical Anthropology* 72, no. 4 (April 1987): 421–29, and 75, no. 1 (January 1988): 143–44.

83. Oakley, Campbell, and Molleson, *Catalogue of Fossil Hominids* Part II, 294.

84. Ibid., 137–38.

85. Robert S. Corruccini, "Metrical Reconsideration of the Skhul IV and IX and the Border Cave 1 Crania in the Context of Modern Human Origins," *American Journal of Physical Anthropology* 87, no. 4 (April 1992): 433–45.

86. W. L. Moore, "A test of the two species hypothesis in the Levant using a cluster analysis of mandibular measurements," *American Journal of Physical Anthropology* Supplement 26 (1998): 166.

87. Corruccini, "Metrical Reconsideration," 433–45.

88. Moore, "Test of the two species hypothesis," 166.

89. Oakley, Campbell, and Molleson, *Catalogue of Fossil Hominids* Part II, 60.

90. Klein, *Human Career*, 2nd ed., 269, 629 reference 548.

91. Oakley, Campbell, and Molleson, *Catalogue of Fossil Hominids* Part II, 51–52; Wolpoff, *Paleoanthropology*, 258.

92. Alban Defleur, Tim White, Patricia Valensi, Ludovic Slimak, and Évelyne Crégut-Bonnoure, "Neanderthal Cannibalism at Moula-Guercy, Ardèche, France," *Science* 286 (1 October 1999): 128–31.

93. Quam and Smith, "Reconsideration of the Tabun C2 'Neandertal'," 192.

94. Oakley, Campbell, and Molleson, *Catalogue of Fossil Hominids* Part II, 209–10; Myra Shackley, *Neanderthal Man* (Hamden, CT: Archon Books, 1980), 10.

95. Klein, *Human Career*, 1st ed., 270–71.

96. Oakley, Campbell, and Molleson, *Catalogue of Fossil Hominids* Part II, 292–93; Wolpoff, *Paleoanthropology*, 258–59.

97. Klein, *Human Career*, 2nd ed., 376.

98. Wolpoff, "Krapina Dental Remains," 103–4.

99. Klein, *Human Career*, 2nd ed., 268, 270, 326.

100. Ibid., 268, 270, 349.

101. Ibid., 268, 350.

102. Tattersall, Delson, and Van Couvering, *Encyclopedia*, 370.

103. Klein, *Human Career*, 2nd ed., 268, 270, 277.

104. Serge Lebel, Erik Trinkaus, Martine Faure, Philippe Fernandez, Claude Guérin, Daniel Richter, Norbert Mercier, Helène Valladas, and Günther A. Wagner, "Comparative morphology and paleobiology of Middle Pleistocene human remains from the Bau de l'Aubesier, Vaucluse, France," *Proceedings of the National Academy of Science* 98 (25 September 2001): 11097–102; Constance Holden, "Neandertals Not Caring?" *Science* 301 (5 September 2003): 1319.

105. Klein, *Human Career*, 2nd ed., 268, 270, 277.

106. Tattersall and Schwartz, *Extinct Humans*, 176.

107. Klein, *Human Career*, 2nd ed., 266, 268, 326.

108. Otte *et al.*, "Long-term technical evolution," 413–31.

109. Tattersall, Delson, and Van Couvering, *Encyclopedia*, 179.

110. Tattersall, *Last Neanderthal*, rev. ed., 135.

111. Giorgio Manzi, Loretana Salvadei, and Pietro Passarello, "The Casal de' Pazzi archaic parietal: comparative analysis of new fossil evidence from the late Middle Pleistocene of Rome," *Journal of Human Evolution* 19, no. 8 (December 1990): 751–59.

112. Klein, *Human Career*, 2nd ed., 269.

113. D. Dean, J. J. Hublin, R. Ziegler, and R. Holloway, "The middle Pleistocene preneandertal partial skull from Reilingen (Germany)," *American Journal of Physical Anthropology* Supplement 18 (1994): 77.

114. Klein, *Human Career*, 2nd ed., 269–70, 277, 393.

115. Ibid., 268, 270, 393.

116. Tattersall, *Last Neanderthal*, rev. ed., 71, 97.

117. Songy Sohn and Milford H. Wolpoff, "Zuttiyeh Face: A View From the East," *American Journal of Physical Anthropology* 91, no. 3 (July 1993): 325–47.

118. Otte *et al.*, "Long-term technical evolution," 413–31.

119. Oakley, Campbell, and Molleson, *Catalogue of Fossil Hominids* Part II, 315.

120. Klein, *Human Career*, 2nd ed., 338.

121. Juan Arsuaga, "Requiem for a Heavyweight," *Natural History* (December 2002/January 2003): 44.

122. Klein, *Human Career*, 2nd ed., 269–70.

123. Ibid., 268, 270, 298.

124. D. Q. Bowen and G. A. Sykes, "How old is 'Boxgrove Man'?" *Nature* 371 (27 October 1994): 751.

125. Tattersall, *Last Neanderthal*, rev. ed., 131.

126. Klein, *Human Career*, 2nd ed., 268, 270.

127. Alexander Dorozynski, "Possible Neandertal Ancestor Found," *Science* 262 (12 February 1993): 991.

128. Rick Gore, "The First Europeans," *National Geographic* 192, no. 1 (July 1997): 110–11.

129. Klein, *Human Career*, 2nd ed., 269, 339–41, 349–53, 440–41.

130. Francesco Mallegni, Renato Mariani-Costantini, Gino Fornaciari, Ernesto T. Longo, Giacomo Giacobini, and Antonio M. Radmilli, "New European Fossil Hominid Material from an Acheulean Site Near Rome (Castel de Guido)," *American Journal of Physical Anthropology* 62, no. 3 (November 1983): 263–74.

131. Mallegni and Radmilli, "Human Temporal Bone," 175–82.

132. Klein, *Human Career*, 2nd ed., 269–70.

133. Ibid., 268, 270, 339, 344.

134. Rainer Grün, "A re-analysis of electron spin resonance dating results associated with the Petralona hominid," *Journal of Human Evolution* 30, no. 3 (March 1996): 227–41.

135. Tattersall, *Last Neanderthal*, rev. ed., 67.

136. M. B. Roberts, C. B. Stringer, and S. A. Parfitt, "A hominid tibia from Middle Pleistocene sediments at Boxgrove, UK," *Nature* 369 (26 May 1994): 311–13.

137. Klein, *Human Career*, 2nd ed., 268, 270.

138. Tattersall, Delson, and Van Couvering, *Encyclopedia*, 331.

139. Klein, *Human Career*, 2nd ed., 269–70.

140. Bruce M. Rothschild, Israel Hershkovitz, and Christine Rothschild, "Origin of yaws in the Pleistocene," *Nature* 378 (23 November 1995): 343–44.

141. Klein, *Human Career*, 2nd ed., 268.

142. Ibid., 269, 320.

143. Ibid., 158, 325–26, 338, 345–46.

144. Ascenzi *et al.*, "A calvarium of late *Homo erectus*," 409–23.

145. Klein, *Human Career*, 2nd ed., 270.

146. E. Carbonell, J. M. Bermudez de Castro, J. L. Arsuaga, J. C. Diez, A. Rosas, G. Cuenca-Bescos, R. Sala, M. Mosquera, and X. P. Rodriguez, "Lower Pleistocene Hominids and Artifacts from Atapuerca-TD6 (Spain)," *Science* 269 (11 August 1995): 826–30.

147. M. Faerman, U. Zilberman, P. Smith, V. Kharitonov, V. Batsevitz, "A Neanderthal infant from the Barakai Cave, Western Caucasus," *Journal of Human Evolution* 27 (1994): 405–15.

148. Marcellin Boule and Henri V. Vallois, *Fossil Men* (New York: The Dryden Press, 1957), 198.

149. Oakley, Campbell, and Molleson, *Catalogue of Fossil Hominids* Part III, 136.

150. Ibid., 164.

151. Klein, *Human Career*, 2nd ed., 373.

152. Tattersall, *Last Neanderthal*, rev. ed., 84.

Early African/Asian *Homo Sapiens* Fossils

1. Kennedy, *Neanderthal Man*, 55–56.

2. D. C. Cornell and R. L. Jantz, "The morphometric relationship of Upper Cave 101 to modern *Homo sapiens*," *American Journal of Physical Anthropology* Supplement 24 (1997): 95.

3. Klein, *Human Career*, 1st ed., 287.

4. Klein, *Human Career*, 2nd ed., 398.

5. Susan Pfeiffer and Marie K. Zehr, "A morphological and histological study of the human humerus from Border Cave," *Journal of Human Evolution* 31, no. 1 (July 1996): 49–59.

6. Klein, *Human Career*, 2nd ed., 398.

7. Alison S. Brooks and Bernard Wood, "The Chinese side of the story," *Nature* 344 (22 March 1990): 288–89.

8. Klein, *Human Career*, 1st ed., 289–90.

9. Brooks and Wood, "Chinese side of the story," 288–89.

10. Percy Cohen, "Fitting a face to Ngaloba," *Journal of Human Evolution* 30, no. 4 (April 1996): 373–79.

11. Tattersall, *Last Neanderthal*, rev. ed., 177.

12. Klein, *Human Career*, 1st ed., 288–89.

13. Brooks and Wood, "Chinese side of the story," 288–89.

14. Gunter Bräuer, Yuji Yokoyama, Christophe Falguères, and Emma Mbua, "Modern human origins backdated," *Nature* 386 (27 March 1997): 337–38.

15. Brooks and Wood, "Chinese side of the story," 288–89.

16. White *et al.*, "Pleistocene *Homo sapiens*," 742–47.

17. Wilford, "Fossilized skulls."

18. Wu and Poirier, *Human Evolution in China*, 134–36.

19. Brooks and Wood, "Chinese side of the story," 288–89.

20. Ibid.

21. Ibid.

22. Day, *Guide to Fossil Man*, 4th ed., 149.

23. Bräuer *et al.*, "Modern human origins backdated," 337–38.

24. Brooks and Wood, "Chinese side of the story," 288–89.

25. Klein, *Human Career*, 2nd ed., 266, 270.

26. Brooks and Wood, "Chinese side of the story," 288–89.

27. Rainer Grün, James S. Brink, Nigel A. Spooner, Lois Taylor, Chris B. Stringer, Robert G. Franciscus, and Andrew S. Murray, "Direct dating of Florisbad hominid," *Nature* 382 (8 August 1996): 500–501.

28. Bräuer *et al.*, "Modern human origins backdated," 337–38.

29. Lee R. Berger and John E. Parkington, "300,000 year-old fossil human remains from Hoedjiespunt, South Africa," *American Journal of Physical Anthropology* Supplement 24 (1997): 75.

30. Bräuer *et al.*, "Modern human origins backdated," 337–38.

31. Klein, *Human Career*, 1st ed., 191.

32. Bräuer *et al.*, "Modern human origins backdated," 337–38.

33. Ibid.

34. Klein, *Human Career*, 2nd ed., 270.

35. Tattersall, *Last Neanderthal*, rev. ed., 67.

36. Stringer, "Archaic Character," 115.

37. Wu and Poirier, *Human Evolution in China*, 150–53.

38. Bräuer *et al.*, "Modern human origins backdated," 337–38; Klein, *Human Career*, 2nd ed., 266, 270.

39. Klein, *Human Career*, 2nd ed., 266, 270.

40. A. R. Sankhyan, "Fossil clavicle of a Middle Pleistocene hominid from the Central Narmada Valley, India," *Journal of Human Evolution* 32, no. 1 (January 1997): 3–16.

41. Klein, *Human Career*, 2nd ed., 266, 270.

42. G. Philip Rightmire, "The human cranium from Bodo, Ethiopia: evidence for speciation in the Middle Pleistocene?" *Journal of Human Evolution* 31, no. 1 (July 1996): 21–39.

43. Klein, *Human Career*, 2nd ed., 266.

Homo Erectus Fossils

1. Laitman, "Australia," 67.

2. Ibid.

3. Ibid., 65–67.

4. Ibid., 65.

5. Ibid.

6. Ibid., 67; K. J. Weinstein and S. C. Antón, "Artificial deformation, frontal recession, and sagittal vault contours in early Australians," *American Journal of Physical Anthropology* Supplement 24 (1997): 237.

7. Laitman, "Australia," 67.

8. Swisher *et al.*, "Latest *Homo erectus* of Java," 1870–74.

9. Hisao Baba, Fachroal Aziz, Yousuke Kaifu, Gen Suwa, Reiko Kono, and Teuku Jacob, "*Homo erectus* Calvarium from the Pleistocene of Java," *Science* 299 (28 February 2003): 1384–88.

10. Laitman, "Australia," 67.

11. Gunter Bräuer, "A Craniological Approach to the Origin of Anatomically Modern *Homo sapiens* in Africa and Implications for the Appearance of Modern Europeans," in *The*

Origins of Modern Humans, eds. Fred H. Smith and Frank Spencer (New York: Alan R. Liss, 1984), 347.

12. Swisher *et al.*, "Latest *Homo erectus* of Java," 1870–74.

13. Tattersall, *Last Neanderthal,* rev. ed., 97.

14. Oakley, Campbell, and Molleson, *Catalogue of Fossil Hominids* Part III, 135.

15. Klein, *Human Career,* 1st ed., 123–24.

16. Wu and Poirier, *Human Evolution in China,* 83–91.

17. Rainer Grün, Pei-Hua Huang, Xinzhi Wu, Chris B. Stringer, Alan G. Thorne, and Malcolm McCulloch, "ESR analysis of teeth from the palaeoanthropological site of Zhoukoudian, China," *Journal of Human Evolution* 32, no. 1 (January 1997): 83–91.

18. Klein, *Human Career,* 2nd ed., 261.

19. Klein, *Human Career,* 1st ed., 191; Klein, *Human Career,* 2nd ed., 270.

20. Wu and Poirier, *Human Evolution in China,* 111–13.

21. Ibid., 101–3.

22. Ibid., 97–101.

23. Ibid., 96–97.

24. Ibid., 103–7.

25. Oakley, Campbell, and Molleson, *Catalogue of Fossil Hominids* Part III, 192.

26. Christine Soares, "Talking Heads," *NewScientist* (14 April 2001): 26–29.

27. M. Solan and M. H. Day, "The Baringo (Kapthurin) ulna," *Journal of Human Evolution* 22, nos. 4, 5 (April/May 1992): 307–13.

28. R. L. Ciochon, R. Grün, H. Schwarcz, and R. Larick, "*Homo* and *Gigantopithecus* in Vietnam and China: ESR and paleomag dates show 1.5 million year co-occurrence," *American Journal of Physical Anthropology* Supplement 22 (1996): 85–86.

29. R. L. Ciochon, "Lang Trang Caves: A New Middle Pleistocene Hominid Site from Northern Vietnam," *American Journal of Physical Anthropology* 81, no. 2 (February 1990): 205.

30. Oakley, Campbell, and Molleson, *Catalogue of Fossil Hominids* Part III: 104–5.

31. Guy Nolch, "New Dates Reignite Human Evolution Debate," *Australasian Science* 22, no. 6 (May 2001): 26.

32. D. E. Tyler and S. Sartono, "A New *Homo erectus* Skull from Sangiran, Java," *American Journal of Physical Anthropology* Supplement 18 (1994): 198–99.

33. Klein, *Human Career,* 2nd ed., 270, 275.

34. Ibid., 270.

35. Ibid., 265.

36. Wu and Poirier, *Human Evolution in China,* 6.

37. Sigmon and Cybulski, *Homo erectus,* 228.

38. Oakley, *Frameworks for Dating Fossil Man,* 7.

39. Tattersall, *Last Neanderthal,* rev. ed., 60.

40. Klein, *Human Career,* 2nd ed., 262, 286.

41. James Shreeve, *The Neandertal Enigma* (New York: William Morrow and Company, Inc., 1995), 102.

42. Oakley, Campbell, and Molleson, *Catalogue of Fossil Hominids* Part III, 2nd ed., 8.

43. Klein, *Human Career,* 1st ed., 120; Day, *Guide to Fossil Man,* 4th ed., 177.

44. Klein, *Human Career,* 2nd ed., 270, 275.

45. Klein, *Human Career,* 1st ed., 129; Mary Leakey, *Disclosing The Past* (Garden City, NY: Doubleday and Company, Inc., 1984), 56.

46. Klein, *Human Career,* 2nd ed., 270; Day, *Guide to Fossil Man,* 4th ed., 177.

47. Tattersall, Delson, and Van Couvering, *Encyclopedia,* 392.

48. Wu and Poirier, *Human Evolution in China,* 109–10.

49. Klein, *Human Career,* 2nd ed., 265, 270.

50. Ibid.

51. Ibid., 120; Day, *Guide to Fossil Man,* 4th ed.,177.

52. Ernesto Abbate *et al.*, "A one-million-year-old *Homo* cranium from the Danakil (Afar) Depression of Eritrea," *Nature* 393 (4 June 1998): 458–60.

53. Klein, *Human Career,* 1st ed., 120; Day, *Guide to Fossil Man,* 4th ed., 178.

54. Klein, *Human Career,* 2nd ed., 265, 270, 333.

55. B. Asfaw, T. Assebework, Y. Beyene, J. D. Clark, G. Curtis, J. deHeinzelin, Y. Haile-Selassie, P. Renne, K. Schick, G. Suwa, E. Vrba, T. D. White, G. WoldeGabriel, and Y. Zeleke, "Fossil Hominids, Fauna, and Artifacts from Bouri, Middle Awash, Ethiopia," *American Journal of Physical Anthropology* Supplement 24 (1997): 69.

56. Li Tianyuan and Dennis A. Etler, "New Middle Pleistocene hominid crania from Yunxian in China," *Nature* 357 (4 June 1992): 404–7.

57. Li Tianyuan, Li Wensen, and Wu Xianzhu, "The age of the Yunxian hominid fauna," *Journal of Human Evolution* 34, no. 3 (March 1998): A21–A22.

58. Wu and Poirier, *Human Evolution in China,* 6.

59. Russell L. Ciochon and Adam A. Mizelle, "Human Evolution in China," *American Journal of Physical Anthropology* 105, no. 2 (February 1998): 254.

60. Klein, *Human Career,* 2nd ed., 265, 270.

61. Klein, *Human Career,* 1st ed., 120; Day, *Guide to Fossil Man,* 4th ed., 177.

62. Charles A. Repenning and Oldrich Fejfar, "Evidence for earlier date of 'Ubeidiya, Israel, hominid site," *Nature* 299 (23 September 1982): 344–47.

63. Donald Johanson and Blake Edgar, *From Lucy to Language* (New York: Simon & Schuster Editions, 1996), 46.

64. Klein, *Human Career,* 2nd ed., 265, 270.

65. Berhane Asfaw, Yonas Beyene, Gen Suwa, Robert C. Walter, Tim D. White, Giday WoldeGabriel, and Tesfaye Yemane, "The earliest Acheulean from Konso-Gardula," *Nature* 360 (24/31 December 1992): 732–35.

66. Susan C. Antón, "Endocranial Hyperostosis in Sangiran 2, Gibraltar 1, and Shanidar 5," *American Journal of Physical Anthropology* 102, no. 1 (January 1997): 116.

67. Feibel, Brown, and McDougall, "Stratigraphic Context," 611; Day, *Guide to Fossil Man,* 4th ed., 203.

68. Feibel, Brown, and McDougall, "Stratigraphic Context," 613; Michael H. Day, "Hominid Postcranial Material from Bed I, Olduvai Gorge," in *Human Origins: Louis Leakey and the East African Evidence,* eds. Glynn L. Isaac and Elizabeth R. McCown (Menlo Park, CA: W. A. Benjamin, Inc., 1976), 369.

69. Klein, *Human Career,* 1st ed., 120; Day, *Guide to Fossil Man,* 4th ed., 177.

70. Feibel, Brown, and McDougall, "Stratigraphic Context," 611; Sigmon and Cybulski, *Homo erectus,* 229–30.

71. Feibel, Brown, and McDougall, "Stratigraphic Context," 613; Bernard A. Wood, "Evidence on the locomotor pattern of *Homo* from early Pleistocene of Kenya," *Nature* 251 (13 September1974): 135–36.

72. Feibel, Brown, and McDougall, "Stratigraphic Context," 611; B. Holly Smith, "Dental development in *Australopithecus* and early *Homo,*" *Nature* 323 (25 September 1986): 327–30.

73. Feibel, Brown, and McDougall, "Stratigraphic Context," 611; G. Philip Rightmire, *The Evolution of Homo Erectus* (Cambridge: Cambridge University Press, 1990), 99.

74. Frank Brown, John Harris, Richard Leakey, and Alan Walker, "Early *Homo erectus* skeleton from west Lake Turkana, Kenya," *Nature* 316 (29 August 1985): 788–92.

75. Feibel, Brown, and McDougall, "Stratigraphic Context," 611; Rightmire, *Evolution of Homo Erectus*, 99.

76. Alan Walker, M. R. Zimmerman, and R. E. F. Leakey, "A possible case of hypervitaminosis A in *Homo erectus*," *Nature* 296 (18 March 1982): 248–50.

77. Feibel, Brown, and McDougall, "Stratigraphic Context," 611; Rightmire, *Evolution of Homo Erectus*, 99.

78. Day, *Guide to Fossil Man*, 4th ed., 200–201, 212.

79. Gunter Bräuer and Michael Schultz, "The morphological affinities of the Plio-Pleistocene mandible from Dmanisi, Georgia," *Journal of Human Evolution* 30, no. 5 (May 1996): 445–81.

80. Leo Gabunia, Abesalom Vekua, David Lordkipanidze, Carl C. Swisher III, Reid Ferring, Antje Justus, Medea Nioradze, Merab Tvalchrelidze, Susan C. Antón, Gerhard Bosinski, Olaf Jöris, Marie-A.-de Lumley, Givi Majsuradze, and Aleksander Mouskhelishvili, "Earliest Pleistocene Hominid Cranial Remains from Dmanisi, Republic of Georgia: Taxonomy, Geological Setting, and Age," *Science* 288 (12 May 2000): 1019–25.

81. Kate Wong, "Stranger in a New Land," *Scientific American* (November 2003): 74–83.

82. Day, *Guide to Fossil Man*, 4th ed., 200, 209, 216.

83. Feibel, Brown, and McDougall, "Stratigraphic Context," 613; Rightmire, *Evolution of Homo Erectus*, 108.

84. Feibel, Brown, and McDougall, "Stratigraphic Context," 611; Smith, "Dental Development," 327–30.

85. Wu and Poirier, *Human Evolution in China*, 12–16.

86. Rick Gore, "The Dawn of Humans: Expanding Worlds," *National Geographic* (May 1997): 102.

87. Randall L. Susman, "New Hominid Fossils from the Swartkrans Formation (1979–1986 Excavations): Postcranial Specimens," *American Journal of Physical Anthropology* 79 (August 1989): 451–74; Frederick E. Grine, "New Hominid Fossils From the Swartkrans Formation (1979–1986 Excavations): Craniodental Specimens," *American Journal of Physical Anthropology* 79 (August 1989): 446.

88. William H. Kimbel, Donald C. Johanson, and Yoel Rak, "Systematic Assessment of a Maxilla of *Homo* From Hadar, Ethiopia," *American Journal of Physical Anthropology* 103, no. 2 (June 1997): 254–55; Day, *Guide to Fossil Man*, 4th ed., 303.

89. Susman, "New Hominid Fossils," 451–74.

90. Ibid.; Grine, "New Hominid Fossils," 409–49; Fred Spoor, Bernard Wood, and Frans Zonneveld, "Implications of early hominid labyrinthine morphology for evolution of human bipedal locomotion," *Nature* 369 (23 June 1994): 645–48; Tattersall, Delson, and Van Couvering, *Encyclopedia*, 557.

91. Ciochon *et al.*, "*Homo* and *Gigantopithecus*," 85–86.

92. Susan C. Antón, "Developmental Age and Taxonomic Affinity of the Mojokerto Child, Java, Indonesia," *American Journal of Physical Anthropology* 102, no. 4 (April 1997): 497–514.

93. Constance Holden, "Man, the Toothpick User," *Science* 288 (28 April 2000): 607.

94. Feibel, Brown, and McDougall, "Stratigraphic Context," 611; Rightmire, *Evolution of Homo Erectus*, 100.

95. Feibel, Brown, and McDougall, "Stratigraphic Context," 613; Frank Brown, John Harris, Richard Leakey, and Alan Walker, "Early *Homo erectus* skeleton from west Lake Turkana, Kenya," *Nature* 316 (29 August 1985): 791.

Homo Erectus / Homo Habilis—Contemporaneousness

1. Day, *Guide to Fossil Man*, 4th ed., 177; Klein, *Human Career*, 1st ed., 120.
2. Ibid.
3. Bernard Wood, "Who is the 'real' Homo habilis?" *Nature* 327 (21 May 1987): 187.
4. Gabunia *et al.*, "Earliest Pleistocene Hominid Cranial Romans," 1019–25.
5. Abesalom Vekua, David Lordkipanidae, G. Philip Rightmire, Jordi Agusti, Reid Ferring, Givi Maisuradze, Alexander Mouskhelishvili, Medea Nioradze, Marcia Ponce de Leon, Martha Tappen, Merab Tvalchrelidze, and Christoph Zollikofer, "A new skull of Early *Homo* from Dmanisi, Georgia," *Science* 297 (5 July 2002): 85–89.
6. Wong, "Stranger in a New Land," 74–83.
7. Alun R. Hughes and Phillip V. Tobias, "A fossil skull probably of the genus Homo from Sterkfontein, Transvaal," *Nature* 265 (27 January 1977): 310–12.
8. Day, *Guide to Fossil Man*, 4th ed., 177–78; R. L. Susman and J. T. Stern Jr., "Functional Affinities of the *Homo habilis* postcranial remains from FLK, Olduvai Gorge," abstract, *American Journal of Physical Anthropology* 57, no. 2 (February 1982): 234–35; Walter *et al.*, "Laser-fusion 40Ar/39Ar dating," 145–49.
9. Ibid.
10. Ibid.
11. Ibid.
12. Ibid.
13. Walter *et al.*, "Laser-fusion 40Ar/39Ar dating," 145–49; Johanson *et al.*, "New partial skeleton," 205–9.
14. Noel T. Boaz and F. Clark Howell, "A Gracile Hominid Cranium from Upper Member G of the Shungura Formation, Ethiopia," *American Journal of Physical Anthropology* 46, no. 1 (January 1977): 93–108.
15. Feibel, Brown, and McDougall, "Stratigraphic Context," 611; Bennett Blumenberg, "Population Characteristics of Extinct Hominid Endocranial Volume," *American Journal of Physical Anthropology* 68, no. 2 (October 1985): 270.
16. Ibid.
17. Day, *Guide to Fossil Man*, 4th ed., 177–78; Walter *et al.*, "Laser-fusion 40Ar/39Ar dating," 145–49.
18. Ibid.
19. Ibid.
20. Feibel, Brown, and McDougall, "Stratigraphic Context," 611; Day, *Guide to Fossil Man*, 4th ed., 214; Donald C. Johanson and M. Taieb, "Plio-Pleistocene hominid discoveries in Hadar, Ethiopia," *Nature* 260 (25 March 1976): 297.
21. Susman, "New Hominid Fossils From the Swartkrans Formation," 451; P. D. Gingerich and B. H. Smith, "Early Homo from Member l, Swartkrans," abstract, *American Journal of Physical Anthropology* 72, no. 2 (February 1987): 203–4.
22. Klein, *Human Career*, 1st ed., 134; and Feibel, Brown, and McDougall, "Stratigraphic Context," 611–12.

Index of Persons

INDEX OF FOSSILS AND FOSSIL SITES

INDEX OF CULTURAL SITES WITHOUT FOSSILS

Index of Topics